Renormalization and
Effective Field Theory

Mathematical
Surveys
and
Monographs

Volume 170

Renormalization and Effective Field Theory

Kevin Costello

American Mathematical Society
Providence, Rhode Island

EDITORIAL COMMITTEE

Ralph L. Cohen, Chair Michael A. Singer
Eric M. Friedlander Benjamin Sudakov
Michael I. Weinstein

2010 *Mathematics Subject Classification.* Primary 81T13, 81T15, 81T17, 81T18, 81T20, 81T70.

The author was partially supported by NSF grant 0706954 and an Alfred P. Sloan Fellowship.

For additional information and updates on this book, visit
www.ams.org/bookpages/surv-170

Library of Congress Cataloging-in-Publication Data

Costello, Kevin.
  Renormalization and effective field theory / Kevin Costello.
    p. cm. — (Mathematical surveys and monographs ; v. 170)
  Includes bibliographical references.
  ISBN 978-0-8218-5288-0 (alk. paper)
  1. Renormalization (Physics) 2. Quantum field theory. I. Title.

QC174.17.R46C67 2011
530.14′3—dc22
                                                    2010047463

Copying and reprinting. Individual readers of this publication, and nonprofit libraries acting for them, are permitted to make fair use of the material, such as to copy a chapter for use in teaching or research. Permission is granted to quote brief passages from this publication in reviews, provided the customary acknowledgment of the source is given.

Republication, systematic copying, or multiple reproduction of any material in this publication is permitted only under license from the American Mathematical Society. Requests for such permission should be addressed to the Acquisitions Department, American Mathematical Society, 201 Charles Street, Providence, Rhode Island 02904-2294 USA. Requests can also be made by e-mail to reprint-permission@ams.org.

© 2011 by the American Mathematical Society. All rights reserved.
The American Mathematical Society retains all rights
except those granted to the United States Government.
Printed in the United States of America.

∞ The paper used in this book is acid-free and falls within the guidelines
established to ensure permanence and durability.
Visit the AMS home page at http://www.ams.org/

        10 9 8 7 6 5 4 3 2 1   16 15 14 13 12 11

Contents

Chapter 1.	Introduction	1
1.	Overview	1
2.	Functional integrals in quantum field theory	4
3.	Wilsonian low energy theories	6
4.	A Wilsonian definition of a quantum field theory	13
5.	Locality	13
6.	The main theorem	16
7.	Renormalizability	19
8.	Renormalizable scalar field theories	21
9.	Gauge theories	22
10.	Observables and correlation functions	27
11.	Other approaches to perturbative quantum field theory	27
	Acknowledgements	28
Chapter 2.	Theories, Lagrangians and counterterms	31
1.	Introduction	31
2.	The effective interaction and background field functional integrals	32
3.	Generalities on Feynman graphs	34
4.	Sharp and smooth cut-offs	42
5.	Singularities in Feynman graphs	44
6.	The geometric interpretation of Feynman graphs	47
7.	A definition of a quantum field theory	53
8.	An alternative definition	55
9.	Extracting the singular part of the weights of Feynman graphs	57
10.	Constructing local counterterms	61
11.	Proof of the main theorem	67
12.	Proof of the parametrix formulation of the main theorem	69
13.	Vector-bundle valued field theories	71
14.	Field theories on non-compact manifolds	80
Chapter 3.	Field theories on $\mathbb{R}^n$	91
1.	Some functional analysis	92
2.	The main theorem on $\mathbb{R}^n$	99
3.	Vector-bundle valued field theories on $\mathbb{R}^n$	104
4.	Holomorphic aspects of theories on $\mathbb{R}^n$	107
Chapter 4.	Renormalizability	113

1. The local renormalization group flow   113
2. The Kadanoff-Wilson picture and asymptotic freedom   122
3. Universality   125
4. Calculations in ϕ^4 theory   126
5. Proofs of the main theorems   131
6. Generalizations of the main theorems   135

Chapter 5. Gauge symmetry and the Batalin-Vilkovisky formalism   139
1. Introduction   139
2. A crash course in the Batalin-Vilkovisky formalism   141
3. The classical BV formalism in infinite dimensions   155
4. Example: Chern-Simons theory   160
5. Example : Yang-Mills theory   162
6. D-modules and the classical BV formalism   164
7. BV theories on a compact manifold   170
8. Effective actions   173
9. The quantum master equation   175
10. Homotopies between theories   178
11. Obstruction theory   186
12. BV theories on $\mathbb{R}^n$ 189
13. The sheaf of BV theories on a manifold   196
14. Quantizing Chern-Simons theory   203

Chapter 6. Renormalizability of Yang-Mills theory   207
1. Introduction   207
2. First-order Yang-Mills theory   207
3. Equivalence of first-order and second-order formulations   210
4. Gauge fixing   213
5. Renormalizability   214
6. Universality   217
7. Cohomology calculations   218

Appendix 1: Asymptotics of graph integrals   227
1. Generalized Laplacians   227
2. Polydifferential operators   229
3. Periods   229
4. Integrals attached to graphs   230
5. Proof of Theorem 4.0.2   233

Appendix 2 : Nuclear spaces   243
1. Basic definitions   243
2. Examples   244
3. Subcategories   245
4. Tensor products of nuclear spaces from geometry   247
5. Algebras of formal power series on nuclear Fréchet spaces   247

Bibliography 249

CHAPTER 1

Introduction

1. Overview

Quantum field theory has been wildly successful as a framework for the study of high-energy particle physics. In addition, the ideas and techniques of quantum field theory have had a profound influence on the development of mathematics.

There is no broad consensus in the mathematics community, however, as to what quantum field theory actually *is*.

This book develops another point of view on perturbative quantum field theory, based on a novel axiomatic formulation.

Most axiomatic formulations of quantum field theory in the literature start from the Hamiltonian formulation of field theory. Thus, the Segal (Seg99) axioms for field theory propose that one assigns a Hilbert space of states to a closed Riemannian manifold of dimension $d-1$, and a unitary operator between Hilbert spaces to a d-dimensional manifold with boundary. In the case when the d-dimensional manifold is of the form $M \times [0, t]$, we should view the corresponding operator as time evolution.

The Haag-Kastler (Haa92) axioms also start from the Hamiltonian formulation, but in a slightly different way. They take as the primary object not the Hilbert space, but rather a $C^\star$ algebra, which will act on a vacuum Hilbert space.

I believe that the Lagrangian formulation of quantum field theory, using Feynman's sum over histories, is more fundamental. The axiomatic framework developed in this book is based on the Lagrangian formalism, and on the ideas of low-energy effective field theory developed by Kadanoff (Kad66), Wilson (Wil71), Polchinski (Pol84) and others.

1.1. The idea of the definition of quantum field theory I use is very simple. Let us assume that we are limited, by the power of our detectors, to studying physical phenomena that occur below a certain energy, say Λ. The part of physics that is visible to a detector of resolution Λ we will call the low-energy effective field theory. This low-energy effective field theory is succinctly encoded by the energy Λ version of the Lagrangian, which is called the low-energy effective action $S^{eff}[\Lambda]$.

The notorious infinities of quantum field theory only occur if we consider phenomena of arbitrarily high energy. Thus, if we restrict attention to phenomena occurring at energies less than Λ, we can compute any quantity we would like in terms of the effective action $S^{eff}[\Lambda]$.

If $\Lambda' < \Lambda$, then the energy Λ' effective field theory can be deduced from knowledge of the energy Λ effective field theory. This leads to an equation expressing the scale Λ' effective action $S^{eff}[\Lambda']$ in terms of the scale Λ effective action $S^{eff}[\Lambda]$. This equation is called the *renormalization group equation*.

If we do have a continuum quantum field theory (whatever that is!) we should, in particular, have a low-energy effective field theory for every energy. This leads to our definition : a continuum quantum field theory is a sequence of low-energy effective actions $S^{eff}[\Lambda]$, for all $\Lambda < \infty$, which are related by the renormalization group flow. In addition, we require that the $S^{eff}[\Lambda]$ satisfy a *locality* axiom, which says that the effective actions $S^{eff}[\Lambda]$ become more and more local as $\Lambda \to \infty$.

This definition aims to be as parsimonious as possible. The only assumptions I am making about the nature of quantum field theory are the following:

(1) The action principle: physics at every energy scale is described by a Lagrangian, according to Feynman's sum-over-histories philosophy.
(2) Locality: in the limit as energy scales go to infinity, interactions between fields occur at points.

1.2. In this book, I develop complete foundations for perturbative quantum field theory in Riemannian signature, on any manifold, using this definition.

The first significant theorem I prove is an existence result: there are as many quantum field theories, using this definition, as there are Lagrangians.

Let me state this theorem more precisely. Throughout the book, I will treat $\hbar$ as a formal parameter; all quantities will be formal power series in $\hbar$. Setting $\hbar$ to zero amounts to passing to the classical limit.

Let us fix a classical action functional S^{cl} on some space of fields $\mathscr{E}$, which is assumed to be the space of global sections of a vector bundle on a manifold M^1. Let $\mathscr{T}^{(n)}(\mathscr{E}, S^{cl})$ be the space of quantizations of the classical theory that are defined modulo $\hbar^{n+1}$. Then,

THEOREM 1.2.1.
$$\mathscr{T}^{(n+1)}(\mathscr{E}, S^{cl}) \to \mathscr{T}^{(n)}(\mathscr{E}, S^{cl})$$

is a torsor for the abelian group of Lagrangians under addition (modulo those Lagrangians which are a total derivative).

Thus, any quantization defined to order n in $\hbar$ can be lifted to a quantization defined to order $n+1$ in $\hbar$, but there is no canonical lift; any two lifts differ by the addition of a Lagrangian.

---

[1]The classical action needs to satisfy some non-degeneracy conditions

1. OVERVIEW

If we choose a section of each torsor $\mathscr{T}^{(n+1)}(\mathscr{E}, S^{cl}) \to \mathscr{T}^{(n)}(\mathscr{E}, S^{cl})$ we find an isomorphism

$$\mathscr{T}^{(\infty)}(\mathscr{E}, S^{cl}) \cong \text{ series } S^{cl} + \hbar S^{(1)} + \hbar^2 S^{(2)} + \cdots$$

where each $S^{(i)}$ is a local functional, that is, a functional which can be written as the integral of a Lagrangian. Thus, this theorem allows one to quantize the theory associated to any classical action functional S^{cl}. However, there is an ambiguity to quantization: at each term in $\hbar$, we are free to add an arbitrary local functional to our action.

1.3. The main results of this book are all stated in the context of this theorem.

In Chapter 4, I give a definition of an action of the group $\mathbb{R}_{>0}$ on the space of theories on $\mathbb{R}^n$. This action is called the *local renormalization group flow*, and is a fundamental part of the concept of renormalizability developed by Wilson and others. The action of group $\mathbb{R}_{>0}$ on the space of theories on $\mathbb{R}^n$ simply arises from the action of this group on $\mathbb{R}^n$ by rescaling.

The coefficients of the action of this local renormalization group flow on any particular theory are the β functions of that theory. I include explicit calculations of the β function of some simple theories, including the ϕ^4 theory on $\mathbb{R}^4$.

This local renormalization group flow leads to a concept of renormalizability. Following Wilson and others, I say that a theory is *perturbatively renormalizable* if it has "critical" scaling behaviour under the renormalization group flow. This means that the theory is fixed under the renormalization group flow except for logarithmic corrections. I then classify all possible renormalizable scalar field theories, and find the expected answer. For example, the only renormalizable scalar field theory in four dimensions, invariant under isometries and under the transformation $\phi \to -\phi$, is the ϕ^4 theory.

In Chapter 5, I show how to include gauge theories in my definition of quantum field theory, using a natural synthesis of the Wilsonian effective action picture and the Batalin-Vilkovisky formalism. Gauge symmetry, in our set up, is expressed by the requirement that the effective action $S^{eff}[\Lambda]$ at each energy Λ satisfies a certain scale Λ Batalin-Vilkovisky quantum master equation. The renormalization group flow is compatible with the Batalin-Vilkovisky quantum master equation: the flow from scale Λ to scale Λ' takes a solution of the scale Λ master equation to a solution to the scale Λ' equation.

I develop a cohomological approach to constructing theories which are renormalizable and which satisfy the quantum master equation. Given any classical gauge theory, satisfying the classical analog of renormalizability, I prove a general theorem allowing one to construct a renormalizable quantization, providing a certain cohomology group vanishes. The dimension of the space of possible renormalizable quantizations is given by a different cohomology group.

In Chapter 6, I apply this general theorem to prove renormalizability of pure Yang-Mills theory. To apply the general theorem to this example, one needs to calculate the cohomology groups controlling obstructions and deformations. This turns out to be a lengthy (if straightforward) exercise in Gel'fand-Fuchs Lie algebra cohomology.

Thus, in the approach to quantum field theory presented here, to prove renormalizability of a particular theory, one simply has to calculate the appropriate cohomology groups. No manipulation of Feynman graphs is required.

2. Functional integrals in quantum field theory

Let us now turn to giving a detailed overview of the results of this book.

First I will review, at a basic level, some ideas from the functional integral point of view on quantum field theory.

2.1. Let M be a manifold with a metric of Lorentzian signature. We will think of M as space-time. Let us consider a quantum field theory of a single scalar field $\phi : M \to \mathbb{R}$.

The space of fields of the theory is $C^\infty(M)$. We will assume that we have an action functional of the form

$$S(\phi) = \int_{x \in M} \mathscr{L}(\phi)(x)$$

where $\mathscr{L}(\phi)$ is a Lagrangian. A typical Lagrangian of interest would be

$$\mathscr{L}(\phi) = -\tfrac{1}{2}\phi(\mathrm{D} + m^2)\phi + \tfrac{1}{4!}\phi^4$$

where D is the Lorentzian analog of the Laplacian operator.

A field $\phi \in C^\infty(M, \mathbb{R})$ can describes one possible history of the universe in this simple model.

Feynman's sum-over-histories approach to quantum field theory says that the universe is in a quantum superposition of all states $\phi \in C^\infty(M, \mathbb{R})$, each weighted by $e^{iS(\phi)/\hbar}$.

An observable – a measurement one can make – is a function

$$O : C^\infty(M, \mathbb{R}) \to \mathbb{C}.$$

If $x \in M$, we have an observable O_x defined by evaluating a field at x:

$$O_x(\phi) = \phi(x).$$

More generally, we can consider observables that are polynomial functions of the values of ϕ and its derivatives at some point $x \in M$. Observables of this form can be thought of as the possible observations that an observer at the point x in the space-time manifold M can make.

The fundamental quantities one wants to compute are the correlation functions of a set of observables, defined by the heuristic formula

$$\langle O_1, \ldots, O_n \rangle = \int_{\phi \in C^\infty(M)} e^{iS(\phi)/\hbar} O_1(\phi) \cdots O_n(\phi) \mathscr{D}\phi.$$

Here $\mathscr{D}\phi$ is the (non-existent!) Lebesgue measure on the space $C^\infty(M)$.

The non-existence of a Lebesgue measure (i.e. a non-zero translation invariant measure) on an infinite dimensional vector space is one of the fundamental difficulties of quantum field theory.

We will refer to the picture described here, where one imagines the existence of a Lebesgue measure on the space of fields, as the *naive functional integral picture*. Since this measure does not exist, the naive functional integral picture is purely heuristic.

2.2. Throughout this book, I will work in Riemannian signature, instead of the more physical Lorentzian signature. Quantum field theory in Riemannian signature can be interpreted as statistical field theory, as I will now explain.

Let M be a compact manifold of Riemannian signature. We will take our space of fields, as before, to be the space $C^\infty(M, \mathbb{R})$ of smooth functions on M. Let $S : C^\infty(M, \mathbb{R}) \to \mathbb{R}$ be an action functional, which, as before, we assume is the integral of a Lagrangian. Again, a typical example would be the ϕ^4 action

$$S(\phi) = -\tfrac{1}{2} \int_{x \in M} \phi(\mathrm{D} + m^2)\phi + \tfrac{1}{4!}\phi^4.$$

Here D denotes the non-negative Laplacian. [2]

We should think of this field theory as a statistical system of a random field $\phi \in C^\infty(M, \mathbb{R})$. The energy of a configuration ϕ is $S(\phi)$. The behaviour of the statistical system depends on a temperature parameter T: the system can be in any state with probability

$$e^{-S(\phi)/T}.$$

The temperature T plays the same role in statistical mechanics as the parameter $\hbar$ plays in quantum field theory.

I should emphasize that time evolution does not play a role in this picture: quantum field theory on d-dimensional space-time is related to statistical field theory on d-dimensional *space*. We must assume, however, that the statistical system is in equilibrium.

As before, the quantities one is interested in are the correlation functions between observables, which one can write (heuristically) as

$$\langle O_1, \ldots, O_n \rangle = \int_{\phi \in C^\infty(M)} e^{-S(\phi)/T} O_1(\phi) \cdots O_n(\phi) \mathscr{D}\phi.$$

---

[2]Our conventions are such that the quadratic part of the action is negative-definite.

The only difference between this picture and the quantum field theory formulation is that we have replaced $i\hbar$ by T.

If we consider the limiting case, when the temperature T in our statistical system is zero, then the system is "frozen" in some extremum of the action functional $S(\phi)$. In the dictionary between quantum field theory and statistical mechanics, the zero temperature limit corresponds to classical field theory. In classical field theory, the system is frozen at a solution to the classical equations of motion.

Throughout this book, I will work perturbatively. In the vocabulary of statistical field theory, this means that we will take the temperature parameter T to be infinitesimally small, and treat everything as a formal power series in T. Since T is very small, the system will be given by a small excitation of an extremum of the action functional.

In the language of quantum field theory, working perturbatively means we treat $\hbar$ as a formal parameter. This means we are considering small quantum fluctuations of a given solution to the classical equations of motions.

Throughout the book, I will work in Riemannian signature, but will otherwise use the vocabulary of quantum field theory. Our sign conventions are such that $\hbar$ can be identified with the negative of the temperature.

3. Wilsonian low energy theories

Wilson (Wil71; Wil72), Kadanoff (Kad66), Polchinski (Pol84) and others have studied the part of a quantum field theory which is seen by detectors which can only measure phenomena of energy below some fixed Λ. This part of the theory is called the *low-energy effective theory*.

There are many ways to define "low energy". I will start by giving a definition which is conceptually very simple, but difficult to work with. In this definition, the low energy fields are those functions on our manifold M which are sums of low-energy eigenvectors of the Laplacian.

In the body of the book, I will use a definition of effective field theory based on length rather than energy. The great advantage of this definition is that it relates better to the concept of locality. I will explain the renormalization group flow from the length-scale point of view shortly.

In this introduction, I will only discuss scalar field theories on compact Riemannian manifolds. This is purely for expository purposes. In the body of the book I will work with a general class of theories on a possibly non-compact manifold, although always in Riemannian signature.

3.1. Let M be a compact Riemannian manifold. For any subset $I \subset [0, \infty)$, let $C^\infty(M)_I \subset C^\infty(M)$ denote the space of functions which are sums of eigenfunctions of the Laplacian with eigenvalue in I. Thus, $C^\infty(M)_{\leq \Lambda}$ denotes the space of functions that are sums of eigenfunctions with eigenvalue $\leq \Lambda$. We can think of $C^\infty(M)_{\leq \Lambda}$ as the space of fields with energy at most Λ.

Detectors that can only see phenomena of energy at most Λ can be represented by functions
$$O : C^\infty(M)_{\leq \Lambda} \to \mathbb{R}[[\hbar]],$$
which are extended to $C^\infty(M)$ via the projection $C^\infty(M) \to C^\infty(M)_{\leq \Lambda}$.

Let us denote by $\mathrm{Obs}_{\leq \Lambda}$ the space of all functions on $C^\infty(M)$ that arise in this way. Elements of $\mathrm{Obs}_{\leq \Lambda}$ will be referred to as observables of energy $\leq \Lambda$.

The fundamental quantities of the low-energy effective theory are the correlation functions $\langle O_1, \ldots, O_n \rangle$ between low-energy observables $O_i \in \mathrm{Obs}_{\leq \Lambda}$. It is natural to expect that these correlation functions arise from some kind of statistical system on $C^\infty(M)_{\leq \Lambda}$. Thus, we will assume that there is a measure on $C^\infty(M)_{\leq \Lambda}$, of the form
$$e^{S^{eff}[\Lambda]/\hbar} \mathscr{D}\phi$$
where $\mathscr{D}\phi$ is the Lebesgue measure, and $S^{eff}[\Lambda]$ is a function on $\mathrm{Obs}_{\leq \Lambda}$, such that
$$\langle O_1, \ldots, O_n \rangle = \int_{\phi \in C^\infty(M)_{\leq \Lambda}} e^{S^{eff}[\Lambda](\phi)/\hbar} O_1(\phi) \cdots O_n(\phi) \mathscr{D}\phi$$
for all low-energy observables $O_i \in \mathrm{Obs}_{\leq \Lambda}$.

The function $S^{eff}[\Lambda]$ is called the *low-energy effective action*. This object completely describes all aspects of a quantum field theory that can be seen using observables of energy $\leq \Lambda$.

Note that our sign conventions are unusual, in that $S^{eff}[\Lambda]$ appears in the functional integral via $e^{S^{eff}[\Lambda]/\hbar}$, instead of $e^{-S^{eff}[\Lambda]/\hbar}$ as is more usual. We will assume the quadratic part of $S^{eff}[\Lambda]$ is negative-definite.

3.2. If $\Lambda' \leq \Lambda$, any observable of energy at most Λ' is in particular an observable of energy at most Λ. Thus, there are inclusion maps
$$\mathrm{Obs}_{\leq \Lambda'} \hookrightarrow \mathrm{Obs}_{\leq \Lambda}$$
if $\Lambda' \leq \Lambda$.

Suppose we have a collection $O_1, \ldots, O_n \in \mathrm{Obs}_{\leq \Lambda'}$ of observables of energy at most Λ'. The correlation functions between these observables should be the same whether they are considered to lie in $\mathrm{Obs}_{\leq \Lambda'}$ or $\mathrm{Obs}_{\leq \Lambda}$. That is,
$$\int_{\phi \in C^\infty(M)_{\leq \Lambda'}} e^{S^{eff}[\Lambda'](\phi)/\hbar} O_1(\phi) \cdots O_n(\phi) \mathrm{d}\mu$$
$$= \int_{\phi \in C^\infty(M)_{\leq \Lambda}} e^{S^{eff}[\Lambda](\phi)/\hbar} O_1(\phi) \cdots O_n(\phi) \mathrm{d}\mu.$$
It follows from this that
$$S^{eff}[\Lambda'](\phi_L) = \hbar \log \left(\int_{\phi_H \in C^\infty(M)_{(\Lambda', \Lambda]}} \exp\left(\frac{1}{\hbar} S^{eff}[\Lambda](\phi_L + \phi_H) \right) \right)$$

where the low-energy field ϕ_L is in $C^\infty(M)_{\leq \Lambda'}$. This is a finite dimensional integral, and so (under mild conditions) is well defined as formal power series in $\hbar$.

This equation is called the *renormalization group equation*. It says that if $\Lambda' < \Lambda$, then $S^{eff}[\Lambda']$ is obtained from $S^{eff}[\Lambda]$ by averaging over fluctuations of the low-energy field $\phi_L \in C^\infty(M)_{\leq \Lambda'}$ with energy between Λ' and Λ.

3.3. Recall that in the naive functional-integral point of view, there is supposed to be a measure on the space $C^\infty(M)$ of the form
$$e^{S(\phi)/\hbar} \mathrm{d}\phi,$$
where $\mathrm{d}\phi$ refers to the (non-existent) Lebesgue measure on the vector space $C^\infty(M)$, and $S(\phi)$ is a function of the field ϕ.

It is natural to ask what role the "original" action S plays in the Wilsonian low-energy picture. The answer is that S is supposed to be the "energy infinity effective action". The low energy effective action $S^{eff}[\Lambda]$ is supposed to be obtained from S by integrating out all fields of energy greater than Λ, that is
$$S^{eff}[\Lambda](\phi_L) = \hbar \log \left(\int_{\phi_H \in C^\infty(M)_{(\Lambda, \infty)}} \exp\left(\frac{1}{\hbar} S(\phi_L + \phi_H) \right) \right).$$

This is a functional integral over the infinite dimensional space of fields with energy greater than Λ. This integral doesn't make sense; the terms in its Feynman graph expansion are divergent.

However, one would not expect this expression to be well-defined. The infinite energy effective action should not be defined; one would not expect to have a description of how particles behave at infinite energy. The infinities in the naive functional integral picture arise because the classical action functional S is treated as the infinite energy effective action.

3.4. So far, I have explained how to define a renormalization group equation using the eigenvalues of the Laplacian. This picture is very easy to explain, but it has many disadvantages. The principal disadvantage is that this definition is not local on space-time. Thus, it is difficult to integrate the locality requirements of quantum field theory into this version of the renormalization group flow.

In the body of this book, I will use a version of the renormalization group flow that is based on length rather than on energy. A complete account of this will have to wait until Chapter 2, but I will give a brief description here.

The version of the renormalization group flow based on length is not derived directly from Feynman's functional integral formulation of quantum field theory. Instead, it is derived from a different (though ultimately equivalent) formulation of quantum field theory, again due to Feynman (Fey50).

Let us consider the propagator for a free scalar field ϕ, with action $S_{free}(\phi) = S_k(\phi) = -\frac{1}{2} \int \phi (\mathrm{D} + m^2) \phi$. This propagator P is defined to be the integral kernel for the inverse of the operator $\mathrm{D} + m^2$ appearing in the action. Thus, P is a distribution on M^2. Away from the diagonal in M^2, P is a distribution. The value $P(x,y)$ of P at distinct points x,y in the space-time manifold M can be interpreted as the correlation between the value of the field ϕ at x and the value at y.

Feynman realized that the propagator can be written as an integral

$$P(x,y) = \int_{\tau=0}^{\infty} e^{-\tau m^2} K_\tau(x,y) \mathrm{d}\tau$$

where $K_\tau(x,y)$ is the heat kernel. The fact that the heat kernel can be interpreted as the transition probability for a random path allows us to write the propagator $P(x,y)$ as an integral over the space of paths in M starting at x and ending at y:

$$P(x,y) = \int_{\tau=0}^{\infty} e^{-\tau m^2} \int_{\substack{f:[0,\tau]\to M \\ f(0)=x, f(\tau)=y}} \exp\left(-\int_0^\tau \|\mathrm{d}f\|^2\right).$$

(This expression can be given a rigorous meaning using the Wiener measure).

From this point of view, the propagator $P(x,y)$ represents the probability that a particle starts at x and transitions to y along a random path (the worldline). The parameter τ is interpreted as something like the proper time: it is the time measured by a clock travelling along the worldline. (This expression of the propagator is sometimes known as the Schwinger representation).

Any reasonable action functional for a scalar field theory can be decomposed into kinetic and interacting terms,

$$S(\phi) = S_{free}(\phi) + I(\phi)$$

where $S_{free}(\phi)$ is the action for the free theory discussed above. From the space-time point of view on quantum field theory, the quantity $I(\phi)$ prescribes how particles interact. The local nature of $I(\phi)$ simply says that particles only interact when they are at the same point in space-time. From this point of view, Feynman graphs have a very simple interpretation: they are the "world-graphs" traced by a family of particles in space-time moving in a random fashion, and interacting in a way prescribed by $I(\phi)$.

This point of view on quantum field theory is the one most closely related to string theory (see e.g. the introduction to (GSW88)). In string theory, one replaces points by 1-manifolds, and the world-graph of a collection of interacting particles is replaced by the world-sheet describing interacting strings.

3.5. Let us now briefly describe how to treat effective field theory from the world-line point of view.

In the energy-scale picture, physics at scales less than Λ is described by saying that we are only allowed fields of energy less than Λ, and that the action on such fields is described by the effective action $S^{eff}[\Lambda]$.

In the world-line approach, instead of having an effective action $S^{eff}[\Lambda]$ at each energy-scale Λ, we have an effective *inter*action $I^{eff}[L]$ at each length-scale L. This object encodes all physical phenomena occurring at *lengths* greater than L. (The effective interaction can also be considered in the energy-scale picture also: the relationship between the effective action $S^{eff}[\Lambda]$ and the effective interaction $I^{eff}[\Lambda]$ is simply

$$S^{eff}[\Lambda](\phi) = -\tfrac{1}{2} \int_M \phi \operatorname{D} \phi + I^{eff}[\Lambda](\phi)$$

for fields $\phi \in C^\infty(M)_{[0,\Lambda)}$. The reason for introducing the effective interaction is that the world-line version of the renormalization group flow is better expressed in these terms.

In the world-line picture of physics at lengths greater than L, we can only consider paths which evolve for a proper time greater than L, and then interact via $I^{eff}[L]$. All processes which involve particles moving for a proper time of less than L between interactions are assumed to be subsumed into $I^{eff}[L]$.

The renormalization group equation for these effective interactions can be described by saying that quantities we compute using this prescription are independent of L. That is,

DEFINITION 3.5.1. *A collection of effective interactions $I^{eff}[L]$ satisfies the renormalization group equation if, when we compute correlation functions using $I^{eff}[L]$ as our interaction, and allow particles to travel for a proper time of at least L between any two interactions, the result is independent of L.*

If one works out what this means, one sees that the scale L effective interaction $I^{eff}[L]$ can be constructed in terms of $I^{eff}[\varepsilon]$ by allowing particles to travel along paths with proper-time between ε and L, and then interact using $I^{eff}[\varepsilon]$.

More formally, $I^{eff}[L]$ can be expressed as a sum over Feynman graphs, where the edges are labelled by the propagator

$$P(\varepsilon, L) = \int_\varepsilon^L e^{-\tau m^2} K_\tau$$

and where the vertices are labelled by $I^{eff}[\varepsilon]$.

This effective interaction $I^{eff}[L]$ is an $\hbar$-dependent functional on the space $C^\infty(M)$ of fields. We can expand $I^{eff}[L]$ as a formal power series

$$I^{eff}[L] = \sum_{i,k \geq 0} \hbar^i I^{eff}_{i,k}[L]$$

where

$$I^{eff}_{i,k}[L] : C^\infty(M) \to \mathbb{R}$$

is homogeneous of order k. Thus, we can think of $I_{i,k}[L]$ as being a symmetric linear map

$$I^{eff}_{i,k}[L] : C^\infty(M)^{\otimes k} \to \mathbb{R}.$$

3. WILSONIAN LOW ENERGY THEORIES

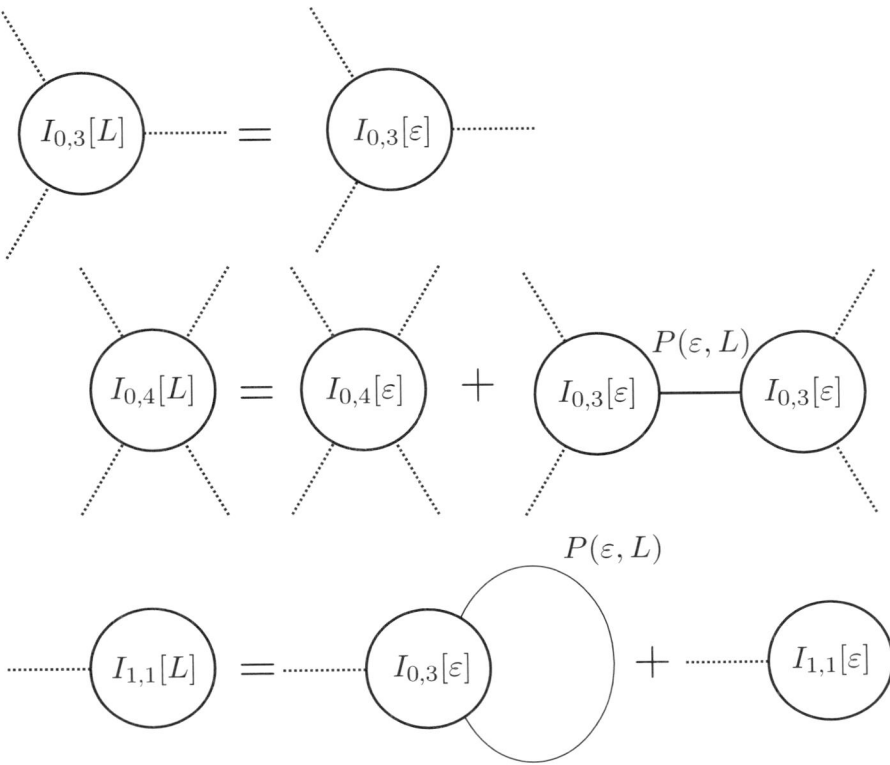

FIGURE 1. The first few expressions in the renormalization group flow from scale ε to scale L. The dotted lines indicate incoming particles. The blobs indicate interactions between these particles. The symbol $I^{eff}_{i,k}[L]$ indicates the $\hbar^i$ term in the contribution to the interaction of k particles at length-scale L. The solid lines indicate the propagation of a particle between two interactions; particles are allowed to propagate with proper time between ε and L.

We should think of $I^{eff}_{i,k}[L]$ as being a contribution to the interaction of k particles which come together in a region of size around L.

Figure 1 shows how to express, graphically, the world-line version of the renormalization group flow.

3.6. So far in this section, we have sketched the definition of two versions of the renormalization group flow: one based on energy, and one based on length. There is a more general definition of the renormalization group flow which includes these two as special cases. This more general version is based on the concept of *parametrix*.

DEFINITION 3.6.1. *A parametrix for the Laplacian* D *on a manifold is a symmetric distribution* P *on* M^2 *such that* $(D \otimes 1)P - \delta_M$ *is a smooth function on* M^2 *(where δ_M refers to the δ-distribution on the diagonal of* M.

This condition implies that the operator $\Phi_P : C^\infty(M) \to C^\infty(M)$ associated to the kernel P is an inverse for D, up to smoothing operators: both $P \circ D - \text{Id}$ and $D \circ P - \text{Id}$ are smoothing operators.

If K_t is the heat kernel for the Laplacian D, then $P_{length}(0, L) = \int_0^L K_t dt$ is a parametrix. This family of parametrices arises when one considers the world-line picture of quantum field theory.

Similarly, we can define an energy-scale parametrix

$$P_{energy}[\Lambda, \infty) = \sum_{\lambda \geq \Lambda} \frac{1}{\lambda} e_\lambda \otimes e_\lambda$$

where the sum is over an orthonormal basis of eigenfunctions e_λ for the Laplacian D, with eigenvalue λ.

Thus, we see that in either the world-line picture or the momentum scale picture one has a family of parametrices ($P_{length}(0, L)$ and $P_{energy}[\Lambda, \infty)$, respectively) which converge (in the $L \to 0$ and $\Lambda \to \infty$ limits, respectively) to the zero distribution. The renormalization group equation in either case is written in terms of the one-parameter family of parametrices.

This suggests a more general version of the renormalization group flow, where an arbitrary parametrix P is viewed as defining a "scale" of the theory. In this picture, one should have an effective action $I^{eff}[P]$ for each parametrix. If P, P' are two different parametrices, then $I^{eff}[P]$ and $I^{eff}[P']$ must be related by a certain renormalization group equation, which expresses $I^{eff}[P]$ in terms of a sum over graphs whose vertices are labelled by $I^{eff}[P']$ and whose edges are labelled by $P - P'$.

If we restrict such a family of effective interactions to parametrices of the form $P_{length}(0, L)$ one finds a solution to the world-line version of the renormalization group equation. If we only consider parametrices of the form $P_{energy}[\Lambda, \infty)$, and then define

$$S^{eff}[\Lambda](\phi) = -\tfrac{1}{2} \int_M \phi \, D \, \phi + I^{eff}[P_{energy}[\Lambda, \infty)](\phi),$$

for $\phi \in C^\infty(M)_{[0,\Lambda)}$, one finds a solution to the energy scale version of the renormalization group flow.

A general definition of a quantum field theory along these lines is explained in detail in Chapter 2, Section 8. This definition is equivalent to one based only on the world-line version of the renormalization group flow, which is the definition used for most of the book.

4. A Wilsonian definition of a quantum field theory

Any detector one could imagine has some finite resolution, and so only probes some low-energy effective theory, described by some $S^{eff}[\Lambda]$. However, one could imagine building detectors of arbitrarily high (but finite) resolution, and so one could imagine probing $S^{eff}[\Lambda]$ for arbitrarily high (but finite) Λ.

As is usual in physics, one should only consider those objects which can in principle be observed. Thus, one should say that *all aspects of a quantum field theory are encoded in its various low-energy effective theories.*

Let us make this into a (rough) definition. A more precise version of this definition is given later in this introduction; a completely precise version is given in the body of the book.

DEFINITION 4.0.1. *A (continuum) quantum field theory is:*

(1) An effective action
$$S^{eff}[\Lambda] : C^\infty(M)_{[0,\Lambda]} \to \mathbb{R}[[\hbar]]$$
for all $\Lambda \in (0,\infty)$. More precisely, $S^{eff}[\Lambda]$ should be a formal power series both in the field $\phi \in C^\infty(M)_{[0,\Lambda]}$ and in the variable $\hbar$.

(2) Modulo $\hbar$, each $S^{eff}[\Lambda]$ must be of the form
$$S^{eff}[\Lambda](\phi) = -\tfrac{1}{2}\int_M \phi\, \mathrm{D}\,\phi +\ \text{cubic and higher terms}.$$
where D is the positive-definite Laplacian. (If we want to consider a massive scalar field theory, we can replace D by $\mathrm{D} + m^2$).

(3) If $\Lambda' < \Lambda$, $S^{eff}[\Lambda']$ is determined from $S^{eff}[\Lambda]$ by the renormalization group equation (which makes sense in the formal power series setting).

(4) The effective actions $S^{eff}[\Lambda]$ satisfy a locality axiom, which we will sketch below.

Earlier I described several different versions of the renormalization group equation; one based on the world-line formulation of quantum field theory, and one defined by considering arbitrary parametrices for the Laplacian. One gets an equivalent definition of quantum field theory using either of these versions of the renormalization group flow.

5. Locality

Locality is one of the fundamental principles of quantum field theory. Roughly, locality says that any interaction between fundamental particles occurs at a point. Two particles at different points of space-time cannot spontaneously affect each other. They can only interact through the medium of other particles. The locality requirement thus excludes any "spooky action at a distance".

Locality is easily understood in the naive functional integral picture. Here, the theory is supposed to be described by a functional measure of the form
$$e^{S(\phi)/\hbar} \mathrm{d}\phi,$$
where $\mathrm{d}\phi$ represents the non-existent Lebesgue measure on $C^\infty(M)$. In this picture, locality becomes the requirement that the action function S is a *local action functional*.

DEFINITION 5.0.1. *A function*
$$S : C^\infty(M) \to \mathbb{R}[[\hbar]]$$
is a local action functional, *if it can be written as a sum*
$$S(\phi) = \sum S_k(\phi)$$
where $S_k(\phi)$ is of the form
$$S_k(\phi) = \int_M (D_1\phi)(D_2\phi) \cdots (D_k\phi) \mathrm{d}Vol_M$$
where D_i are differential operators on M.

Thus, a local action functional S is of the form
$$S(\phi) = \int_{x \in M} \mathscr{L}(\phi)(x)$$
where the Lagrangian $\mathscr{L}(\phi)(x)$ only depends on Taylor expansion of ϕ at x.

5.1. Of course, the naive functional integral picture doesn't make sense. If we want to give a definition of quantum field theory based on Wilson's ideas, we need a way to express the idea of locality in terms of the finite energy effective actions $S^{eff}[\Lambda]$.

As $\Lambda \to \infty$, the effective action $S^{eff}[\Lambda]$ is supposed to encode more and more "fundamental" interactions. Thus, the first tentative definition is the following.

DEFINITION 5.1.1 (Tentative definition of asymptotic locality). *A collection of low-energy effective actions $S^{eff}[\Lambda]$ satisfying the renormalization group equation is* asymptotically local *if there exists a large Λ asymptotic expansion of the form*
$$S^{eff}[\Lambda](\phi) \simeq \sum f_i(\Lambda)\Theta_i(\phi)$$
where the Θ_i are local action functionals. (The $\Lambda \to \infty$ limit of $S^{eff}[\Lambda]$ does not exist, in general).

This asymptotic locality axiom turns out to be a good idea, but with a fundamental problem. If we suppose that $S^{eff}[\Lambda]$ is close to being local for some large Λ, then for all $\Lambda' < \Lambda$, the renormalization group equation implies that $S^{eff}[\Lambda']$ is entirely non-local. In other words, the renormalization group flow is not compatible with the idea of locality.

This problem, however, is an artifact of the particular form of the renormalization group equation we are using. The notion of "energy" is very non-local: high-energy eigenvalues of the Laplacian are spread out all over the manifold. Things work much better if we use the version of the renormalization group flow based on *length* rather than energy.

The length-based version of the renormalization group flow was sketched earlier. It will be described in detail in Chapter 2, and used throughout the rest of the book.

This length scale version of the renormalization group equation is essentially equivalent to the version based on energy, in the following sense:

Any solution to the length-scale RGE can be translated into a solution to the energy-scale RGE and conversely[3].

Under this transformation, large length will correspond (roughly) to low energy, and vice-versa.

The great advantage of working with length scales, however, is that one can make sense of locality. Unlike the energy-scale renormalization group flow, the length-scale renormalization group flow diffuses from local to non-local. We have seen earlier that it is more convenient to describe the length-scale version of an effective field theory by an effective *interaction* $I^{eff}[L]$ rather than by an effective action. If the length-scale L effective interaction $I^{eff}[L]$ is close to being local, then $I^{eff}[L+\varepsilon]$ is slightly less local, and so on.

As $L \to 0$, we approach more "fundamental" interactions. Thus, the locality axiom should say that $I^{eff}[L]$ becomes more and more local as $L \to 0$. Thus, one can correct the tentative definition asymptotic locality to the following:

DEFINITION 5.1.2 (Asymptotic locality). *A collection of low-energy effective actions $I^{eff}[L]$ satisfying the length-scale version of the renormalization group equation is* asymptotically local *if there exists a small L asymptotic expansion of the form*

$$S^{eff}[L](\phi) \simeq \sum f_i(L) \Theta_i(\phi)$$

where the Θ_i are local action functionals. (The actual $L \to 0$ limit will not exist, in general).

Because solutions to the length scale and energy scale RGEs are in bijection, this definition applies to solutions to the energy scale RGE as well.

We can now update our definition of quantum field theory:

DEFINITION 5.1.3. *A (continuum) quantum field theory is:*

(1) An effective action

$$S^{eff}[\Lambda] : C^\infty(M)_{[0,\Lambda]} \to \mathbb{R}[[\hbar]]$$

---

[3] The converse requires some growth conditions on the energy-scale effective actions $S^{eff}[\Lambda]$.

for all $\Lambda \in (0, \infty)$. More precisely, $S^{eff}[\Lambda]$ should be a formal power series both in the field $\phi \in C^\infty(M)_{[0,\Lambda]}$ and in the variable $\hbar$.

(2) Modulo $\hbar$, each $S^{eff}[\Lambda]$ must be of the form

$$S^{eff}[\Lambda](\phi) = -\tfrac{1}{2} \int_M \phi \, \mathrm{D} \, \phi + \text{ cubic and higher terms.}$$

where D is the positive-definite Laplacian. (If we want to consider a massive scalar field theory, we can replace D by $\mathrm{D}+m^2$).

(3) If $\Lambda' < \Lambda$, $S^{eff}[\Lambda']$ is determined from $S^{eff}[\Lambda]$ by the renormalization group equation (which makes sense in the formal power series setting).

(4) The effective actions $S^{eff}[\Lambda]$, when translated into a solution to the length-scale version of the RGE, satisfy the asymptotic locality axiom.

Since solutions to the energy and length-scale versions of the RGE are equivalent, one can base this definition entirely on the length-scale version of the RGE. We will do this in the body of the book.

Earlier we sketched a very general form of the RGE, which uses an arbitrary parametrix to define a "scale" of the theory. In Chapter 2, Section 8, we will give a definition of a quantum field theory based on arbitrary parametrices, and we will show that this definition is equivalent to the one described above.

6. The main theorem

Now we are ready to state the first main result of this book.

THEOREM A. *Let $\mathscr{T}^{(n)}$ denote the set of theories defined modulo $\hbar^{n+1}$. Then, $\mathscr{T}^{(n+1)}$ is a principal bundle over $\mathscr{T}^{(n)}$ for the abelian group of local action functionals $S : C^\infty(M) \to \mathbb{R}$.*

Recall that a functional S is a local action functional if it is of the form

$$S(\phi) = \int_M \mathscr{L}(\phi)$$

where $\mathscr{L}$ is a Lagrangian. The abelian group of local action functionals is the same as that of Lagrangians up to the addition of a Lagrangian which is a total derivative.

Choosing a section of each principal bundle $\mathscr{T}^{(n+1)} \to \mathscr{T}^{(n)}$ yields an isomorphism between the space of theories and the space of series in $\hbar$ whose coefficients are local action functionals.

A variant theorem allows one to get a bijection between theories and local action functionals, once one has made an additional universal (but unnatural) choice, that of a *renormalization scheme*. A renormalization

scheme is a way to extract the singular part of certain functions of one variable. We construct a certain subalgebra
$$\mathscr{P}((0,1)) \subset C^\infty((0,1))$$
consisting of functions $f(\varepsilon)$ of a "motivic" nature. Functions in $\mathscr{P}((0,1))$ arise as the periods of families of algebraic varieties over Zariski open subsets $U \subset \mathbb{A}^1_{\mathbb{Q}}$, such that $U(\mathbb{R})$ contains $(0,1)$. (For more details, see Chapter 2, Section 9).

DEFINITION 6.0.1. *A renormalization scheme is a subspace*
$$\mathscr{P}((0,1))_{<0} \subset \mathscr{P}((0,1))$$
of "purely singular" functions, complementary to the subspace
$$\mathscr{P}((0,1))_{\geq 0} \subset \mathscr{P}((0,1))$$
of functions whose $r \to \infty$ limit exists.

The choice of a renormalization scheme gives us a way to extract the singular part of functions in $\mathscr{P}((0,1))$.

The variant theorem is the following.

THEOREM B. *The choice of a renormalization scheme leads to a bijection between the space of theories and the space of local action functionals*
$$S : C^\infty(M) \to \mathbb{R}[[\hbar]].$$

Equivalently, there is a bijection between the space of theories and the space of Lagrangians up to the addition of a Lagrangian which is a total derivative.

Theorem B implies theorem A, but theorem A is the more natural formulation.

There are certain caveats:
(1) Like the effective actions $S^{eff}[\Lambda]$, the local action functional S is a formal power series both in $\phi \in C^\infty(M)$ and in $\hbar$.
(2) Modulo $\hbar$, we require that S is of the form
$$S(\phi) = -\tfrac{1}{2} \int_M \phi \, \mathrm{D} \, \phi + \text{ cubic and higher terms.}$$

There is a more general formulation of this theorem, where the space of fields is allowed to be the space of sections of a graded vector bundle. In the more general formulation, the action functional S must have a quadratic term which is elliptic in a certain sense.

6.1. Let me sketch how to prove theorem A. Given the action S, we construct the low-energy effective action $S^{eff}[\Lambda]$ by renormalizing a certain functional integral. The formula for the functional integral is
$$S^{eff}[\Lambda](\phi_L) = \hbar \log \left\{ \int_{\phi_H \in C^\infty(M)_{(\Lambda,\infty)}} e^{S(\phi_L + \phi_H)/\hbar} \right\}.$$

This expression is the renormalization group flow from infinite energy to energy Λ. This is an infinite dimensional integral, as the field ϕ_H has unbounded energy.

This functional integral is renormalized using the technique of counterterms. This involves first introducing a *regulating parameter* r into the functional integral, which tames the singularities arising in the Feynman graph expansion. One choice would be to take the regularized functional integral to be an integral only over the finite dimensional space of fields $\phi \in C^\infty(M)_{(\Lambda, r]}$.

Sending $r \to \infty$ recovers the original integral. This limit won't exist, but one renormalizes this limit by introducing *counterterms*. Counterterms are functionals $S^{CT}(r, \phi)$ of both r and the field ϕ, such that the limit

$$\lim_{r \to \infty} \int_{\phi_H \in C^\infty(M)_{(\Lambda, r]}}^{r} \exp\left(\frac{1}{\hbar} S(\phi_L + \phi_H) - \frac{1}{\hbar} S^{CT}(r, \phi_L + \phi_H)\right)$$

exists. These counterterms are local, and are uniquely defined once one chooses a renormalization scheme.

The effective action $S^{eff}[\Lambda]$ is then defined by this limit:

$$S^{eff}[\Lambda](\phi_L) =$$
$$\lim_{r \to \infty} \int_{\phi_H \in C^\infty(M)_{(\Lambda, r]}} \exp\left(\frac{1}{\hbar} S(\phi_L + \phi_H) - \frac{1}{\hbar} S^{CT}(r, \phi_L + \phi_H)\right)$$

6.2. In practise, we don't use the energy-scale regulator r but rather a length-scale regulator ε. The reason is the same as before: it is easier to construct local theories using the length-scale regulator than the energy-scale regulator. In what follows, I will ignore this rather technical point; to make the following discussion completely accurate, the reader should replace the energy-scale regulator r by the length-scale regulator we will use later.

The counterterms S^{CT} are constructed by a simple inductive procedure, and are local action functionals of the field $\phi \in C^\infty(M)$.

Once we have chosen such a renormalization scheme, we find a set of counterterms $S^{CT}(r, \phi)$ for any local action functional S. These counterterms are uniquely determined by the requirements that firstly, the $r \to \infty$ limit above exists, and secondly, they are purely singular as a function of the regulating parameter r.

6.3. What we see from this is that the bijection between theories and local action functionals is not canonical, but depends on the choice of a renormalization scheme. Thus, theorem A is the most natural formulation: there is no natural bijection between theories and local action functionals. Theorem A implies that the space of theories is an infinite dimensional manifold, modelled on the topological vector space of $\mathbb{R}[[\hbar]]$-valued local action functionals on $C^\infty(M)$.

7. Renormalizability

We have seen that the space of theories is an infinite dimensional manifold, modelled on the space of $\mathbb{R}[[\hbar]]$-valued local action functionals on $C^\infty(M)$.

A physicist would find this unsatisfactory. Because the space of theories is infinite dimensional, to specify a particular theory, it would take an infinite number of experiments. Thus, we can't make any predictions.

We need to find a natural finite-dimensional submanifold of the space of all theories, consisting of "well-behaved" theories. These well-behaved theories will be called *renormalizable*.

7.1. An old-fashioned viewpoint is the following:

A local action functional (or Lagrangian) is renormalizable if it has only finitely many counterterms:

$$S^{CT}(r) = \sum_{finite} f_i(r) S_i^{CT}$$

In general, this definition picks out a finite dimensional subspace of the infinite dimensional space of theories. However, it is not natural: the specific counterterms will depend on the choice of renormalization scheme, and therefore this definition may depend on the choice of renormalization scheme.

More fundamentally, any definitions one makes should be directly in terms of the only physical quantities one can measure, namely the low-energy effective actions $S^{eff}[\Lambda]$. Thus, we would like a definition of renormalizability using only the $S^{eff}[\Lambda]$.

The following is the basic idea of the definition we suggest, following Wilson and others.

DEFINITION 7.1.1 (Rough definition). *A theory, defined by effective actions $S^{eff}[\Lambda]$, is renormalizable if the $S^{eff}[\Lambda]$ don't grow too fast as $\Lambda \to \infty$. However, we must measure $S^{eff}[\Lambda]$ in units appropriate to energy scale Λ.*

For instance, if $S^{eff}[1]$ is measured in joules, then $S^{eff}[10^3]$ should be measured in kilo-joules, and so on.

However, this change of units only makes sense on $\mathbb{R}^n$. Since we can identify energy with length^{-2}, changing the units of energy amounts to rescaling $\mathbb{R}^n$. In addition, the field $\phi \in C^\infty(\mathbb{R}^n)$ can have its own energy (which should be thought of as giving the target of the map $\phi : \mathbb{R}^n \to \mathbb{R}$ some weight). Once we incorporate both of these factors, the procedure of changing units (in a scalar field theory) is implemented by the map

$$R_l : C^\infty(\mathbb{R}^n) \to C^\infty(\mathbb{R}^n)$$
$$\phi(x) \mapsto l^{1-n/2} \phi(l^{-1} x).$$

7.2. As our definition of renormalizability only makes sense on $\mathbb{R}^n$, we will now restrict to considering scalar field theories on $\mathbb{R}^n$. We want to measure $S^{eff}[\Lambda]$ as $\Lambda \to \infty$, after we have changed units. Define $\mathcal{RG}_l(S^{eff}[\Lambda])$ by
$$\mathcal{RG}_l(S^{eff}[\Lambda])(\phi) = S^{eff}[l^{-2}\Lambda](R_l(\phi))$$
Thus, $\mathcal{RG}_l(S^{eff}[\Lambda])$ is the effective action $S^{eff}[l^2\Lambda]$, but measured in units that have been rescaled by l.

We can use the map $\mathcal{RG}_l$ to implement precisely the definition of renormalizability suggested above.

DEFINITION 7.2.1. *A theory $\{S^{eff}[\Lambda]\}$ is renormalizable if $\mathcal{RG}_l(S^{eff}[\Lambda])$ grows at most logarithmically as $l \to 0$.*

7.3. It turns out that the map $\mathcal{RG}_l$ defines a flow on the space of theories.

LEMMA 7.3.1. *If $\{S^{eff}[\Lambda]\}$ satisfies the renormalization group equation, then so does $\{\mathcal{RG}_l(S^{eff}[\Lambda])\}$.*

Thus, sending
$$\{S^{eff}[\Lambda]\} \to \{\mathcal{RG}_l(S^{eff}[\Lambda])\}$$
defines a flow on the space of theories: this is the *local renormalization group flow*.

Recall that the choice of a renormalization scheme leads to a bijection between the space of theories and Lagrangians. Under this bijection, the local renormalization group flow acts on the space of Lagrangians. The constants appearing in a Lagrangian (the coupling constants) become functions of l; the dependence of the coupling constants on the parameter l is called the β function. Renormalizability means these coupling constants have at most logarithmic growth in l.

The local renormalization group flow $\mathcal{RG}_l$, as $l \to 0$, can be interpreted geometrically as focusing on smaller and smaller regions of space-time, while always using units appropriate to the size of the region one is considering. In energy terms, applying $\mathcal{RG}_l$ as $l \to 0$ amounts to focusing on phenomena of higher and higher energy.

The logarithmic growth condition thus says the theory doesn't break down completely when we probe high-energy phenomena. If the effective actions displayed polynomial growth, for instance, then one would find that the perturbative description of the theory wouldn't make sense at high energy, because the terms in the perturbative expansion would increase with the energy.

7.4. The definition of renormalizability given above can be viewed as a perturbative approximation to an ideal non-perturbative definition.

DEFINITION 7.4.1 (Ideal definition). *A non-perturbative theory is renormalizable if, as we flow the theory under $\mathcal{RG}_l$ and let $l \to 0$, we converge to a fixed point.*

This fixed point, if it exists, would be a scaling limit of the theory; it would necessarily be a scale-invariant theory. For instance, it is expected that Yang-Mills theory is renormalizable in this sense, and that the scaling limit is a free theory.

This ideal definition is difficult to make sense of perturbatively (when we treat $\hbar$ as a formal parameter). For instance, suppose a coupling constant c changes to
$$c \mapsto l^{\hbar}c = c + \hbar c \log l + \cdots$$
Non-perturbatively, we might think that $\hbar > 0$, so that this flow converges to a fixed point. Perturbatively, however, $\hbar$ is a formal parameter, so it appears to have logarithmic growth.

Our perturbative definition can be interpreted as saying that a perturbative theory is renormalizable if, at first sight, it looks like it might be non-perturbatively renormalizable in this sense. For instance, if it contains coupling constants which are of polynomial growth in l, these will probably persist at the non-perturbative level, implying that the theory does not converge to a fixed point.

One can make a more refined perturbative definition by requiring that the logarithmic growth which does appear is of the correct sign (thus distinguishing between $c \mapsto l^{\hbar}c$ and $c \mapsto l^{-\hbar}c$). This more refined definition leads to *asymptotic freedom*, which is the statement that a theory converges to a free theory as $l \to 0$.

8. Renormalizable scalar field theories

Now that we have a definition of renormalizability, the next question to ask is: which theories are renormalizable?

It turns out to be straightforward to classify all renormalizable scalar field theories.

8.1. Suppose we have a local action functional S of a scalar field on $\mathbb{R}^n$, and suppose that S is translation invariant. We say that S is of *dimension* k if
$$S(R_l(\phi)) = l^k S(\phi).$$
Recall that $R_l(\phi)(x) = l^{1-n/2}\phi(-^1lx)$.

Every translation invariant local action functional S is a finite sum of terms of some dimension. For instance:

$$\int_{\mathbb{R}^4} \phi\, \mathrm{D}\, \phi \ \text{ and } \ \int_{\mathbb{R}^4} \phi^4 \ \ \text{ are of dimension } 0$$

$$\int_{\mathbb{R}^4} \phi^3 \frac{\partial}{\partial x_i}\phi \ \ \text{ is of dimension } -1$$

$$\int_{\mathbb{R}^4} \phi^2 \ \ \text{ is of dimension } 2$$

Now let us state how one classifies scalar field theories, in general.

THEOREM 8.1.1. *Let $\mathscr{R}^{(k)}(\mathbb{R}^n)$ denote the space of renormalizable scalar field theories on $\mathbb{R}^n$, invariant under translation, defined modulo $\hbar^{n+1}$.*
Then,
$$\mathscr{R}^{(k+1)}(\mathbb{R}^n) \to \mathscr{R}^{(k)}(\mathbb{R}^n)$$
is a torsor for the vector space of local action functionals $S(\phi)$ which are a sum of terms of non-negative dimension.

Further, $\mathscr{R}^{(0)}(\mathbb{R}^n)$ is canonically isomorphic to the space of local action functionals of the form
$$S(\phi) = -\tfrac{1}{2} \int_{\mathbb{R}^n} \phi \, \mathrm{D} \, \phi + \text{ cubic and higher terms, of non-negative dimension.}$$

As before, the choice of a renormalization scheme leads to a section of each of the torsors $\mathscr{R}^{(k+1)}(\mathbb{R}^n) \to \mathscr{R}^{(k)}(\mathbb{R}^n)$, and so to a bijection between the space of renormalizable scalar field theories and the space of series
$$-\tfrac{1}{2} \int \phi \, \mathrm{D} \, \phi + \sum \hbar^i S_i$$
where each S_i is a translation invariant local action functional of non-negative dimension, and S_0 is at least cubic.

Applying this to $\mathbb{R}^4$, we find the following.

COROLLARY 8.1.2. *Renormalizable scalar field theories on $\mathbb{R}^4$, invariant under $SO(4) \ltimes \mathbb{R}^4$ and under $\phi \to -\phi$, are in bijection with Lagrangians of the form*
$$\mathscr{L}(\phi) = a\phi \, \mathrm{D} \, \phi + b\phi^4 + c\phi^2$$
for $a, b, c \in \mathbb{R}[[\hbar]]$, where $a = -\tfrac{1}{2}$ modulo $\hbar$ and $b = 0$ modulo $\hbar$.

More generally, there is a finite dimensional space of non-free renormalizable theories in dimensions $n = 3, 4, 5, 6$, an infinite dimensional space in dimensions $n = 1, 2$, and none in dimensions $n > 6$. ("Finite dimensional" means as a formal scheme over $\operatorname{Spec} \mathbb{R}[[\hbar]]$: there are only finitely many $\mathbb{R}[[\hbar]]$-valued parameters).

Thus we find that the scalar field theories traditionally considered to be "renormalizable" are precisely the ones selected by the Wilsonian definition advocated here. However, in this approach, one has a conceptual reason for why these particular scalar field theories, and no others, are renormalizable.

9. Gauge theories

We would like to apply the Wilsonian philosophy to understand gauge theories. In Chapter 5, we will explain how to do this using a synthesis of Wilsonian ideas and the Batalin-Vilkovisky formalism.

9. GAUGE THEORIES

9.1. In mathematical parlance, a gauge theory is a field theory where the space of fields is a stack. A typical example is Yang-Mills theory, where the space of fields is the space of connections on some principal G-bundle on space-time, modulo gauge equivalence.

It is important to emphasize the difference between gauge theories and field theories equipped with some symmetry group. In a gauge theory, the gauge group is *not* a group of symmetries of the theory. The theory does not make any sense before taking the quotient by the gauge group.

One can see this even at the classical level. In classical $U(1)$ Yang-Mills theory on a 4-manifold M, the space of fields (before quotienting by the gauge group) is $\Omega^1(M)$. The action is $S(\alpha) = \int_M \mathrm{d}\alpha * \mathrm{d}\alpha$. The highly degenerate nature of this action means that the classical theory is not predictive: a solution to the equations of motion is not determined by its behaviour on a space-like hypersurface. Thus, classical Yang-Mills theory is not a sensible theory before taking the quotient by the gauge group.

9.2. Let us now discuss gauge theories in effective field theory. Naively, one could imagine that to give a gauge theory would be to give an effective gauge theory at every energy level, in a way related by the renormalization group flow.

One immediate problem with this idea is that the space of low-energy gauge symmetries is *not a group*. The product of low-energy gauge symmetries is no longer low-energy; and if we project this product onto its low-energy part, the resulting multiplication on the set of low-energy gauge symmetries is not associative.

For example, if $\mathfrak{g}$ is a Lie algebra, then the Lie algebra of infinitesimal gauge symmetries on a manifold M is $C^\infty(M) \otimes \mathfrak{g}$. The space of low-energy infinitesimal gauge symmetries is then $C^\infty(M)_{\leq \Lambda} \otimes \mathfrak{g}$. In general, the product of two functions in $C^\infty(M)_{\leq \Lambda}$ can have arbitrary energy; so that $C^\infty(M)_{\leq \Lambda} \otimes \mathfrak{g}$ is not closed under the Lie bracket.

This problem is solved by a very natural union of the Batalin-Vilkovisky formalism and the effective action philosophy.

9.3. The Batalin-Vilkovisky formalism is widely regarded as being the most powerful and general way to quantize gauge theories. The first step in the BV procedure is to introduce extra fields – ghosts, corresponding to infinitesimal gauge symmetries; anti-fields dual to fields; and anti-ghosts dual to ghosts – and then write down an extended classical action functional on this extended space of fields.

This extended space of fields has a very natural interpretation in homological algebra: it describes the *derived* moduli space of solutions to the Euler-Lagrange equations of the theory. The derived moduli space is obtained by first taking a derived quotient of the space of fields by the gauge group, and then imposing the Euler-Lagrange equations of the theory in a derived way. The extended classical action functional on the extended space of fields arises from the differential on this derived moduli space.

In more pedestrian terms, the extended classical action functional encodes the following data:

(1) the original action functional on the original space of fields;
(2) the Lie bracket on the space of infinitesimal gauge symmetries,
(3) the way this Lie algebra acts on the original space of fields.

In order to construct a quantum theory, one asks that the extended action satisfies the *quantum master equation*. This is a succinct way of encoding the following conditions:

(1) The Lie bracket on the space of infinitesimal gauge symmetries satisfies the Jacobi identity.
(2) This Lie algebra acts in a way preserving the action functional on the space of fields.
(3) The Lie algebra of infinitesimal gauge symmetries preserves the "Lebesgue measure" on the original space of fields. That is, the vector field on the original space of fields associated to every infinitesimal gauge symmetry is divergence free.
(4) The adjoint action of the Lie algebra on itself also preserves the "Lebesgue measure". Again, this says that a vector field associated to every infinitesimal gauge symmetry is divergence free.

Unfortunately, the quantum master equation is an ill-defined expression. The 3rd and 4th conditions above are the source of the problem: the divergence of a vector field on the space of fields is a singular expression, involving the same kind of singularities as those appearing in one-loop Feynman diagrams.

9.4. This form of the quantum master equation violates our philosophy: we should always express things in terms of the effective actions. The quantum master equation above is about the original "infinite energy" action, so we should not be surprised that it doesn't make sense.

The solution to this problem is to combine the BV formalism with the effective action philosophy. To give an effective action in the BV formalism is to give a functional $S^{eff}[\Lambda]$ on the energy $\leq \Lambda$ part of the extended space of fields (i.e., the space of ghosts, fields, anti-fields and anti-ghosts). This energy Λ effective action must satisfy a certain *energy Λ quantum master equation*.

The reason that the effective action philosophy and the BV formalism work well together is the following.

LEMMA. *The renormalization group flow from scale Λ to scale Λ' carries solutions of the energy Λ quantum master equation into solutions of the energy Λ' quantum master equation.*

Thus, to give a gauge theory in the effective BV formalism is to give a collection of effective actions $S^{eff}[\Lambda]$ for each Λ, such that $S^{eff}[\Lambda]$ satisfies the scale Λ QME, and such that $S^{eff}[\Lambda']$ is obtained from $S^{eff}[\Lambda]$ by the renormalization group flow. In addition, one requires that the effective actions $S^{eff}[\Lambda]$ satisfy a locality axiom, as before.

This picture also solves the problem that the low energy gauge symmetries are not a group. The energy Λ effective action $S^{eff}[\Lambda]$, satisfying the energy Λ quantum master equation, gives the extended space of low-energy fields a certain homotopical algebraic structure, which has the following interpretation:

(1) The space of low-energy infinitesimal gauge symmetries has a Lie bracket.
(2) This Lie algebra acts on the space of low-energy fields.
(3) The space of low-energy fields has a functional, invariant under the bracket.
(4) The action of the Lie algebra on the space of fields, and on itself, preserves the Lebesgue measure.

However, these axioms don't hold on the nose, but hold *up to a sequence of coherent higher homotopies*.

9.5. Let us now formalize our definition of a gauge theory. As we have seen, whenever we have the data of a classical gauge theory, we get an extended space of fields, that we will denote by $\mathscr{E}$. This is always the space of sections of a graded vector bundle on the manifold M. As before, let $\mathscr{E}_{\leq \Lambda}$ denote the space of low-energy extended fields.

DEFINITION 9.5.1. *A theory in the BV formalism consists of a set of low-energy effective actions*

$$S^{eff}[\Lambda] : \mathscr{E}_{\leq \Lambda} \to \mathbb{R}[[\hbar]],$$

which is a formal series both in $\mathscr{E}_{\leq \Lambda}$ and $\hbar$, and which is such that:

(1) *The renormalization group equation is satisfied.*
(2) *Each $S^{eff}[\Lambda]$ satisfies the energy Λ quantum master equation.*
(3) *The same locality axiom as before holds.*
(4) *There is one more technical restriction : modulo $\hbar$, each $S^{eff}[\Lambda]$ is of the form*

$$S^{eff}[\Lambda](e) = \langle e, Qe \rangle + \text{ cubic and higher terms}$$

where $\langle -, - \rangle$ is a certain canonical pairing on $\mathscr{E}$, and $Q : \mathscr{E} \to \mathscr{E}$ satisfies certain ellipticity conditions.

As before, the locality axiom needs to be expressed in length-scale terms.

The main theorem holds in this context also, but in a slightly modified form. If we remove the requirement that the effective actions satisfy the quantum master equation, we find a bijection between theories and local action functionals, depending on the choice of a renormalization scheme, as before. Requiring that the effective actions satisfy the QME leads to a constraint on the corresponding local action functional, which is called the *renormalized* quantum master equation. This renormalized QME replaces the ill-defined QME appearing in the naive BV formalism.

Physicists often say that a theory satisfying the quantum master equation is free of "gauge anomalies". In general, an anomaly is a symmetry of the classical theory which fails to be a symmetry of the quantum theory. In my opinion, this terminology is misleading: the gauge group action on the space of fields is not a symmetry of the theory, but rather an inextricable part of the theory. The presence of gauge anomalies means that the theory does not exist in a meaningful way.

9.6. Renormalizing gauge theories. It is straightforward to generalize the Wilsonian definition of renormalizability (Definition 7.2.1) to apply to gauge theories in the BV formalism. As before, this definition only works on $\mathbb{R}^n$, because one needs to rescale space-time. This rescaling of space-time leads to a flow on the space of theories, which we call the local renormalization group flow. (This flow respects the quantum master equation). A theory is defined to be renormalizable if it exhibits at most logarithmic growth under the local renormalization group flow.

Now we are ready to state one of the main results of this book.

THEOREM. *Pure Yang-Mills theory on $\mathbb{R}^4$, with coefficients in a simple Lie algebra $\mathfrak{g}$, is perturbatively renormalizable.*

That is, there exists a theory $\{S_{YM}^{eff}[\Lambda]\}$, which is renormalizable, which satisfies the quantum master equation, and which modulo $\hbar$ is given by the classical Yang-Mills action.

The moduli space of such theories is isomorphic to $\hbar\mathbb{R}[[\hbar]]$.

Let me state more precisely what I mean by this. At the classical level (modulo $\hbar$) there are no difficulties with renormalization, and it is straightforward to define pure Yang-Mills theory in the BV formalism[4]. Because the classical Yang-Mills action is conformally invariant in four dimensions, it is a fixed point of the local renormalization group flow.

One is then interested in quantizing this classical theory in a renormalizable way.

The theorem states that one can do this, and that the set of all such renormalizable quantizations is isomorphic (non-canonically) to $\hbar\mathbb{R}[[\hbar]]$.

This theorem is proved by obstruction theory. A lengthy (but straightforward) calculation in Lie algebra cohomology shows that the group of obstructions to finding a renormalizable quantization of Yang-Mills theory vanishes; and that the corresponding deformation group is one-dimensional. Standard obstruction theory arguments then imply that the moduli space of quantizations is $\mathbb{R}[[\hbar]]$, as desired.

---

[4]For technical reasons, we use a first-order formulation of Yang-Mills, which is equivalent to the usual formulation.

This calculation uses the following strange "coincidence" in Lie algebra cohomology: although $H^5(\mathfrak{su}(3))$ is one-dimensional, the outer automorphism group of $\mathfrak{su}(3)$ acts on this space in a non-trivial way. A more direct construction of Yang-Mills theory, not relying on obstruction theory, is desirable.

10. Observables and correlation functions

The key quantities one wants to compute in a quantum field theory are the correlation functions between observables. These are the quantities that can be more-or-less directly related to experiment.

The theory of observables and correlation functions is addressed in the work in progress (CG10), written jointly with Owen Gwilliam. In this sequel, we will show how the observables of a quantum field theory (in the sense of this book) form a rich algebraic structure called a *factorization algebra*. The concept of factorization algebra was introduced by Beilinson and Drinfeld (BD04), as a geometric formulation of the axioms of a vertex algebra. The factorization algebra associated to a quantum field theory is a complete encoding of the theory: from this algebraic object one can reconstruct the correlation functions, the operator product expansion, and so on.

Thus, a proper treatment of correlation functions requires a great deal of preliminary work on the theory of factorization algebras and on the factorization algebra associated to a quantum field theory. This is beyond the scope of the present work.

11. Other approaches to perturbative quantum field theory

Let me finish by comparing briefly the approach to perturbative quantum field theory developed here with others developed in the literature.

11.1. In the last ten years, the perturbative version of algebraic quantum field theory has been developed by Brunetti, Dütsch, Fredenhagen, Hollands, Wald and others: see (BF00; BF09; DF01; HW10). In this work, the authors investigate the problem of constructing a solution to the axioms of algebraic quantum field theory in perturbation theory. These authors prove results which have a very similar form to those proved in this book: term by term in $\hbar$, there is an ambiguity in quantization, described by a certain class of Lagrangians.

The proof of these results relies on a version of the Epstein-Glaser (EG73) construction of counterterms. This construction of counterterms relies, like the approach used in this book, on working directly on real space, as opposed to on momentum space. In the Epstein-Glaser approach to renormalization, as in the approach described here, the proof that the counterterms are local is easy. In contrast, in momentum-space approaches to constructing counterterms – such as that developed by Bogoliubov-Parasiuk (BP57) and Hepp (Hep66) – the problem of constructing local counterterms involves complicated graph combinatorics.

11.2. Another, related, approach to perturbative quantum field theory on Riemannian space-times was developed by Hollands (Hol09) and Hollands-Olbermann (HO09). In this approach the field theory is encoded in a vertex algebra on the space-time manifold.

This seems to be philosophically very closely related to my joint work with Owen Gwilliam (CG10), which uses the renormalization techniques developed in this paper to produce a factorization algebra on the space-time manifold. Thus, the quantization results proved by Hollands-Olbermann should be close analogues of the results presented here.

11.3. D. Tamarkin (Tam03) also develops an approach to renormalization of quantum field theory based on the theory of vertex algebras and on the Batalin-Vilkovisky formalism. Tamarkin's approach, again, seems to be closely related to both the work of Hollands-Olbermann and my joint work with Gwilliam.

11.4. In a lecture at the conference "Renormalization: algebraic, geometric and probabilistic aspects" in Lyon in 2010, Maxim Kontsevich presented an approach to perturbative renormalization which he developed some years before. The output of Kontsevich's construction is (as in (HO09) and (CG10)) a vertex algebra on the space-time manifold. The form of Kontsevich's theorem is very similar to the main theorem of this book: order by order in $\hbar$, the space of possible quantizations is a torsor for an Abelian group constructed from certain Lagrangians. Kontsevich's work relies on a new construction of counterterms which, like the Epstein-Glaser construction and the construction developed here, relies on working in real space rather than on momentum space.

Again, it is natural to speculate that there is a close relationship between Kontsevich's work and the construction of factorization algebras presented in (CG10).

11.5. Let me finally mention an approach to perturbative renormalization developed initially by Connes and Kreimer (CK98; CK99), and further developed by (among others) Connes-Marcolli (CM04; CM08b) and van Suijlekom (vS07). The first result of this approach is that the Bogoliubov-Parasiuk-Hepp-Zimmermann (BP57; Hep66) algorithm has a beautiful interpretation in terms of the Birkhoff decomposition for loops in a certain pro-algebraic group constructed combinatorially from graphs.

In this book, however, counterterms have no intrinsic importance: they are simply a technical tool used to prove the main results. Thus, it is not clear to me if there is any relationship between Connes-Kreimer Hopf algebra and the results of this book.

Acknowledgements

Many people have contributed to the material in this book. Without the constant encouragement and insightful editorial suggestions of Lauren

Weinberg this project would probably never have been finished. Particular thanks are also due to Owen Gwilliam, for many discussions which have greatly clarified the ideas in this book; and to Dennis Sullivan, for his encouragement and many helpful comments. I would also like to thank Alberto Cattaneo, Jacques Distler, Mike Douglas, Giovanni Felder, Arthur Greenspoon, Si Li, Pavel Mnëv, Josh Shadlen, Yuan Shen, Stephan Stolz, Peter Teichner, and A.J. Tolland. Of course, any errors are solely the responsibility of the author.

CHAPTER 2

Theories, Lagrangians and counterterms

1. Introduction

In this chapter, we will make precise the definition of quantum field theory we sketched in Chapter 1. Then, we will show the main theorem:

THEOREM A. *Let $\mathscr{T}^{(n)}(M)$ denote the space of scalar field theories on a manifold M, defined modulo $\hbar^{n+1}$.*

Then $\mathscr{T}^{(n+1)}(M) \to \mathscr{T}^{(n)}(M)$ is (in a canonical way) a principal bundle for the space of local action functionals.

Further, $\mathscr{T}^{(0)}(M)$ is canonically isomorphic to the space of local action functionals which are at least cubic.

This theorem has a less natural formulation, depending on an additional choice, that of a renormalization scheme. A renormalization scheme is an object of a "motivic" nature, defined in Section 9.

THEOREM B. *The choice of a renormalization scheme leads to a section of each principal bundle $\mathscr{T}^{(n+1)}(M) \to \mathscr{T}^{(n)}(M)$, and thus to an isomorphism between the space of theories and the space of local action functionals of the form $\sum \hbar^i S_i$, where S_0 is at least cubic.*

1.1. Let me summarize the contents of this chapter.

The first few sections explain, in a leisurely fashion, the version of the renormalization group flow we use throughout this book. Sections 2 and 4 introduce the heat kernel version of high-energy cut-off we will use throughout the book. Section 3 contains a general discussion of Feynman graphs, and explains how certain finite dimensional integrals can be written as sums over graphs. Section 5 explains why infinities appear in the naive functional integral formulation of quantum field theory. Section 6 shows how the weights attached to Feynman graphs in functional integrals can be interpreted geometrically, as integrals over spaces of maps from graphs to a manifold.

In Section 7, we finally get to the precise definition of a quantum field theory and the statement of the main theorem. Section 8 gives a variant of this definition which doesn't rely on the heat kernel, but instead works with an arbitrary parametrix for the Laplacian. This variant definition is equivalent to the one based on the heat kernel. Section 9 introduces the concept of renormalization scheme, and shows how the choice of renormalization scheme allows one to extract the singular part of the weights attached to

Feynman graphs. Section 10 uses this to construct the local counterterms associated to a Lagrangian, which are needed to render the functional integral finite. Section 11 gives the proof of theorems A and B above.

Finally, we turn to generalizations of the main results. Section 13 shows how everything generalizes, *mutatis mutandis*, to the case when our fields are no longer just functions, but sections of some vector bundle. Section 14 shows how we can further generalize to deal with theories on non-compact manifolds, as long as an appropriate infrared cut-off is introduced.

2. The effective interaction and background field functional integrals

As in the introduction, a quantum field theory in our Wilsonian definition will be given by a collection of effective actions, related by the renormalization group flow. In this section we will write down a version of the renormalization group flow, based on the effective interaction, which we will use throughout the book.

2.1. Let us assume that our energy Λ effective action can be written as

$$S[\Lambda](\phi) = -\tfrac{1}{2}\langle \phi, (\mathrm{D}+m^2)\phi\rangle + I[\Lambda](\phi)$$

where:

(1) The function $I[\Lambda]$ is a formal series in $\hbar$, $I[\Lambda] = I_0[\Lambda] + \hbar I_1[\Lambda] + \cdots$, where the leading term I_0 is at least cubic. Each I_i is a formal power series on the vector space $C^\infty(M)$ of fields (later, I will explain what this means more precisely).

The function $I[\Lambda]$ will be called the *effective interaction*.
(2) $\langle\,,\,\rangle$ denotes the L^2 inner product on $C^\infty(M,\mathbb{R})$ defined by $\langle \phi,\psi\rangle = \int_M \phi\psi$.
(3) D denotes[1] the Laplacian on M, with signs chosen so that the eigenvalues of D are non-negative; and $m \in \mathbb{R}_{>0}$.

Recall that the renormalization group equation relating $S[\Lambda]$ and $S[\Lambda']$ can be written

$$S[\Lambda'](\phi_L) = \hbar \log \left(\int_{\phi_H \in C^\infty(M)_{[\Lambda',\Lambda)}} e^{S[\Lambda](\phi_L+\phi_H)/\hbar} \right).$$

We can rewrite this in terms of the effective interactions, as follows. The spaces $C^\infty(M)_{<\Lambda'}$ and $C^\infty(M)_{[\Lambda',\Lambda)}$ are orthogonal. It follows that

$$S[\Lambda](\phi_L + \phi_H)$$
$$= -\tfrac{1}{2}\langle \phi_L, (\mathrm{D}+m^2)\phi_L\rangle - \tfrac{1}{2}\langle \phi_H, (\mathrm{D}+m^2)\phi_H\rangle + I[\Lambda](\phi_L+\phi_H).$$

---
[1] The symbol Δ will be reserved for the BV Laplacian

Therefore the effective interaction form of the renormalization group equation (RGE) is

$$I[\Lambda'](a) = \hbar \log \left(\int_{\phi \in C^\infty(M)_{[\Lambda', \Lambda]}} \exp\left(-\frac{1}{2\hbar} \langle \phi, (\mathrm{D} + m^2) \phi \rangle + \frac{1}{\hbar} I[\Lambda](\phi + a) \right) \right).$$

Note that in this expression the field a no longer has to be low-energy. We obtain a variant of the renormalization group equation by considering effective interactions $I[\Lambda]$ which are functionals of all fields, not just low-energy fields, and using the equation above. This equation is invertible; it is valid even if $\Lambda' > \Lambda$.

We will always deal with this invertible effective interaction form of the RGE. Henceforth, it will simply be referred to as the RGE.

2.2. We will often deal with integrals of the form

$$\int_{x \in U} \exp\left(\Phi(x)/\hbar + I(x+a)/\hbar \right)$$

over a vector space U, where Φ is a quadratic form (normally negative definite) on U. We will use the convention that the "measure" on U will be the Lebesgue measure normalised so that

$$\int_{x \in U} \exp\left(\Phi(x)/\hbar \right) = 1.$$

Thus, the measure depends on $\hbar$.

2.3. Normally, in quantum field theory textbooks, one starts with an action functional

$$S(\phi) = -\tfrac{1}{2} \langle \phi, (\mathrm{D} + m^2) \phi \rangle + I(\phi),$$

where the interacting term $I(\phi)$ is a *local action functional*. This means that it can be written as a sum

$$I(\phi) = \sum \hbar^i I_{i,k}(\phi)$$

where

$$I_{i,k}(a) = \sum_{j=1}^{s} \int_M D_{1,j}(a) \cdots D_{k,j}(a)$$

and the $D_{i,j}$ are differential operators on M. We also require that $I(\phi)$ is at least cubic modulo $\hbar$.

As I mentioned before, the local interaction I is supposed to be thought of as the scale ∞ effective interaction. Then the effective interaction at scale Λ is obtained by applying the renormalization group flow from energy ∞ down to energy Λ. This is expressed in the functional integral

$$I[\Lambda](a) = \hbar \log \left(\int_{\phi \in C^\infty(M)_{[\Lambda, \infty)}} \exp\left(-\frac{1}{2\hbar} \langle \phi, (\mathrm{D} + m^2) \phi \rangle + \frac{1}{\hbar} I(\phi + a) \right) \right).$$

This functional integral is ill-defined.

3. Generalities on Feynman graphs

In this section, we will describe the Feynman graph expansion for functional integrals of the form appearing in the renormalization group equation. This section will only deal with finite dimensional vector spaces, as a toy model for the infinite dimensional functional integrals we will be concerned with for most of this book. For another mathematical description of the Feynman diagram expansion in finite dimensions, one can consult, for example, (Man99).

3.1.

DEFINITION 3.1.1. *A stable[2] graph is a graph γ, possibly with external edges (or tails); and for each vertex v of γ an element $g(v) \in \mathbb{Z}_{\geq 0}$, called the genus of the vertex v; with the property that every vertex of genus 0 is at least trivalent, and every vertex of genus 1 is at least 1-valent (0-valent vertices are allowed, provided they are of genus > 1).*

If γ is a stable graph, the genus $g(\gamma)$ of γ is defined by

$$g(\gamma) = b_1(\gamma) + \sum_{v \in V(\gamma)} g(v)$$

where $b_1(\gamma)$ is the first Betti number of γ.

More formally, a stable graph γ is determined by the following data.
(1) A finite set $H(\gamma)$ of half-edges of γ.
(2) A finite set $V(\gamma)$ of vertices of γ.
(3) An involution $\sigma : H(\gamma) \to H(\gamma)$. The set of fixed points of this involution is denoted $T(\gamma)$, and is called the set of tails of γ. The set of two-element orbits is denoted $E(\gamma)$, and is called the set of edges.
(4) A map $\pi : H(\gamma) \to V(\gamma)$, which sends a half-edge to the vertex to which it is attached.
(5) A map $g : V(\gamma \to \mathbb{Z}_{\geq 0}$.

From this data we construct a topological space $|\gamma|$ which is the quotient of

$$V(\gamma) \amalg \left(H(\gamma) \times [0, \tfrac{1}{2}] \right)$$

by the relation which identifies $(h, 0) \in H(\gamma) \times [0, \tfrac{1}{2}]$ with $\pi(h) \in V(\gamma)$; and identifies $(h, \tfrac{1}{2})$ with $(\sigma(h), \tfrac{1}{2})$. We say γ is connected if $|\gamma|$ is. A graph γ is stable, as above, if every vertex v of genus 0 is at least trivalent, and every vertex of genus 1 is at least univalent.

We are also interested in automorphisms of stable graphs. It is helpful to give a formal definition. An element of $g \in \text{Aut}(\gamma)$ of the group $\text{Aut}(\gamma)$ is

---

[2]The term "stable" comes from algebraic geometry, where such graphs are used to label the strata of the Deligne-Mumford moduli space of stable curves.

a pair of maps $H(g) : H(\gamma) \to H(\gamma)$, $V(g) : V(\gamma) \to V(\gamma)$, such that $H(g)$ commutes with σ, and such that the diagram

$$\begin{array}{ccc} H(\gamma) & \xrightarrow{H(g)} & H(\gamma) \\ \downarrow & & \downarrow \\ V(\gamma) & \xrightarrow{V(g)} & V(\gamma) \end{array}$$

commutes.

3.2. Let U be a finite-dimensional super vector space, over a ground field $\mathbb{K}$. Let $\mathscr{O}(U)$ denote the completed symmetric algebra on the dual vector space $U^\vee$. Thus, $\mathscr{O}(U)$ is the ring of formal power series in a variable $u \in U$.

Let

$$\mathscr{O}^+(U)[[\hbar]] \subset \mathscr{O}(U)[[\hbar]]$$

be the subspace of those functionals which are at least cubic modulo $\hbar$.

For an element $I \in \mathscr{O}(U)[[\hbar]]$, let us write

$$I = \sum_{i,k \geq 0} \hbar^i I_{i,k}$$

where $I_{i,k} \in \mathscr{O}(U)$ is homogeneous of degree k as a function of $u \in U$.

If $f \in \mathscr{O}(U)$ is homogeneous of degree k, then it defines an S_k-invariant linear map

$$D^k f : U^{\otimes k} \to \mathbb{K}$$

$$u_1 \otimes \cdots \otimes u_k \to \left(\frac{\partial}{\partial u_1} \cdots \frac{\partial}{\partial u_k} f \right)(0).$$

Thus, if we expand $I \in \mathscr{O}(U)[[\hbar]]$ as a sum $I = \sum \hbar^i I_{i,k}$ as above, then we have collection of S_k invariant elements $D^k I_{i,k} \in (U^\vee)^{\otimes k}$.

Let γ be a stable graph, with n tails. Let $\phi : \{1, \ldots, n\} \cong T(\gamma)$ be an ordering of the set of tails of ϕ. Let $P \in \operatorname{Sym}^2 U \subset U^{\otimes 2}$, let $I \in \mathscr{O}^+(U)[[\hbar]]$, and let $a_1, \ldots, a_n \in U$ By contracting the tensors P and a_i with the dual tensor I according to a rule given by γ, we will define

$$w_{\gamma,\phi}(P, I)(a_1, \ldots, a_{T(\gamma)}) \in \mathbb{K}.$$

The rule is as follows. Let $H(\gamma)$, $T(\gamma)$, $E(\gamma)$, and $V(\gamma)$ refer to the sets of half-edges, tails, internal edges, and vertices of γ, respectively. Recall that we have chosen an isomorphism $\phi : T(\gamma) \cong \{1, \ldots, n\}$. Putting a propagator P at each internal edge of γ, and putting a_i at the ith tail of γ, gives an element of

$$U^{\otimes E(\gamma)} \otimes U^{\otimes E(\gamma)} \otimes U^{\otimes T(\gamma)} \cong U^{\otimes H(\gamma)}.$$

Putting $D^k I_{i,k}$ at each vertex of valency k and genus i gives us an element of

$$\operatorname{Hom}(U^{\otimes H(\gamma)}, \mathbb{K}).$$

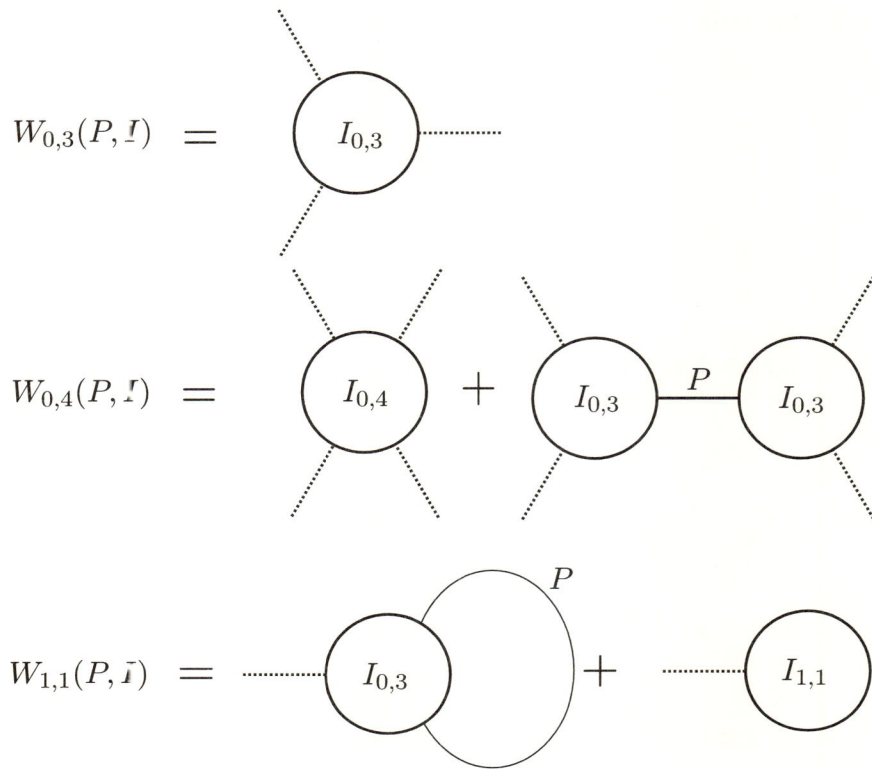

FIGURE 1. The first few graphs in the expansion of $W(P, I)$. The variable $a \in U$ is placed at each external edge.

Contracting these two elements yields the weight $w_{\gamma,\phi}(P, I)(a_1, \ldots, a_n)$.

Define a function
$$w_\gamma(P, I) \in \mathscr{O}(U)$$
by
$$w_\gamma(P, I)(a) = w_{\gamma,\phi}(P, I)(a, \ldots, a)$$
where ϕ is any ordering of the set of tails of γ. Note that $w_\gamma(P, I)$ is homogeneous of degree n, and has the property that, for all $a_1, \ldots, a_n \in U$,
$$\frac{\partial}{\partial a_1} \cdots \frac{\partial}{\partial a_n} w_\gamma(P, I) = \sum_{\phi: \{1,\ldots,n\} \cong T(\gamma)} w_{\gamma,\phi}(P, I)(a_1, \ldots, a_n)$$
where the sum is over ways of ordering the set of tails of γ.

Let $v_{i,k}$ denote the graph with one vertex of genus i and valency k, and with no internal edges. Then our definition implies that
$$w_{v_{i,k}}(P, I) = k! I_{i,k}.$$

3.3. Now that we have defined the functions $w_\gamma(P, I) \in \mathcal{O}(U)$, we will arrange them into a formal power series.

Let us define
$$W(P, I) = \sum_\gamma \frac{1}{|\mathrm{Aut}(\gamma)|} \hbar^{g(\gamma)} w_\gamma(P, I) \in \mathcal{O}^+(U)[[\hbar]]$$
where the sum is over connected stable graphs γ, and $g(\gamma)$ is the genus of the graph γ. The condition that all genus 0 vertices are at least trivalent implies that this sum converges. Figure 1 illustrates the first few terms of the graphical expansion of $W(P, I)$.

Our combinatorial conventions are such that, for all $a_1, \ldots, a_k \in U$,
$$\left(\frac{\partial}{\partial a_1} \cdots \frac{\partial}{\partial a_k} W(P, I) \right)(0) = \sum_{\gamma, \phi} \frac{\hbar^{g(\gamma)}}{|\mathrm{Aut}|(\gamma, \phi)} w_{\gamma, \phi}(P, I)(a_1, \ldots, a_k)$$
where the sum is over graphs γ with k tails and an isomorphism $\phi: \{1, \ldots, k\} \cong T(\gamma)$, and the automorphism group $\mathrm{Aut}(\gamma, \phi)$ preserves the ordering ϕ of the set of tails.

LEMMA 3.3.1.
$$W(0, I) = I.$$

PROOF. Indeed, when $P = 0$ only graphs with no edges can contribute, so that
$$W(0, I) = \sum_{i,k} \frac{\hbar^i}{|\mathrm{Aut}(v_{i,k})|} w_{v_{i,k}}(0, I).$$
Here, as before, $v_{i,k}$ is the graph with a single vertex of genus i and valency k, and with no internal edges. The automorphism group of $v_{i,k}$ is the symmetric group S_k, and we have seen that $w_{v_{i,k}}(0, I) = k! I_{i,k}$. Thus, $W(0, I) = \sum_{i,k} \hbar^i I_{i,k}$, which (by definition) is equal to I. □

3.4. As before, let $P \in \mathrm{Sym}^2 U$, and let us write $P = \sum P' \otimes P''$. Define an order two differential operator $\partial_P : \mathcal{O}(U) \to \mathcal{O}(U)$ by
$$\partial_P = \frac{1}{2} \sum \frac{\partial}{\partial P''} \frac{\partial}{\partial P'}.$$
A convenient way to summarize the Feynman graph expansion $W(P, I)$ is the following.

LEMMA 3.4.1.
$$W(P, I)(a) = \hbar \log \{\exp(\hbar \partial_P) \exp(I/\hbar)\}(a) \in \mathcal{O}^+(U)[[\hbar]].$$

The expression $\exp(\hbar \partial_P) \exp(I/\hbar)$ is the exponential of a differential operator on U applied to a function on U; thus, it is a function on U.

PROOF. We will prove this by first verifying the result for $P = 0$, and then checking that both sides satisfy the same differential equation as a function of P. When $P = 0$, we have seen that

$$W(0, I) = I$$

which of course is the same as $\hbar \log \exp(\hbar \partial_P) \exp(I/\hbar)$.

Now let us turn to proving the general case. It is easier to consider the exponentiated version: so we will actually verify that

$$\exp\left(\hbar^{-1} W(P, I)\right) = \exp(\hbar \partial_{P'}) \exp(I/\hbar).$$

We will do this by verifying that, if ε is a parameter of square zero, and $P' \in \operatorname{Sym}^2 U$,

$$\exp\left(\hbar^{-1} W(P + \varepsilon P', I)\right) = (1 + \hbar \varepsilon \partial_P) \exp\left(\hbar^{-1} W(P, I)\right).$$

Let $a_1, \ldots, a_k \in U$, and let us consider

$$\left(\frac{\partial}{\partial a_1} \cdots \frac{\partial}{\partial a_k} \exp\left(\hbar^{-1} W(P, I)\right)\right)(0).$$

It will suffice to prove a similar differential equation for this expression.

It follows immediately from the definition of the weight function $w_\gamma(P, I)$ of a graph that

$$\left(\frac{\partial}{\partial a_1} \cdots \frac{\partial}{\partial a_k} e^{W(P,I)/\hbar}\right)(0) = \sum_{\gamma, \phi} \frac{\hbar^{g(\gamma)}}{|\operatorname{Aut}(\gamma, \phi)|} w_{\gamma, \phi}(P, I)(a_1, \ldots, a_k)$$

where the sum is over all possibly disconnected stable graphs γ with an isomorphism $\phi : \{1, \ldots, k\} \cong T(\gamma)$. The automorphism group $\operatorname{Aut}(\gamma, \phi)$ consists of those automorphisms preserving the ordering ϕ of the set of tails of γ.

Let ε be a parameter of square zero, and let $P' \in \operatorname{Sym}^2 U$. Let us consider varying P to $P + \varepsilon P'$. We find that

$$\frac{\mathrm{d}}{\mathrm{d}\varepsilon} \left(\frac{\partial}{\partial a_1} \cdots \frac{\partial}{\partial a_k} \exp\left(\hbar^{-1} W(P + \varepsilon P', I)\right)\right)(0)$$

$$= \sum_{\gamma, e, \phi} \frac{\hbar^{g(\gamma)}}{|\operatorname{Aut}(\gamma, e, \phi)|} w_{\gamma, e, \phi}(P, I)(a_1, \ldots, a_k).$$

Here, the sum is over possibly disconnected stable graphs γ with a distinguished edge $e \in E(\gamma)$. The weight $w_{\gamma, e, \phi}$ is defined in the same way as $w_{\gamma, \phi}$ except that the distinguished edge e is labelled by P', whereas all other edges are labelled by P. The automorphism group considered must preserve the edge e as well as ϕ.

Given any graph γ with a distinguished edge e, we can cut along this edge to get another graph γ' with two more tails. These tails can be ordered in two different ways. If we write $P' = \sum u' \otimes u''$ where $u', u'' \in U$, these extra tails are labelled by u' and u''. Thus we find that

$$\frac{d}{d\varepsilon}\left(\frac{\partial}{\partial a_1}\cdots\frac{\partial}{\partial a_k}\exp\left(\hbar^{-1}W\left(P+\varepsilon P', I\right)\right)\right)(0)$$
$$= \frac{1}{2}\sum_{\gamma,\phi}\frac{\hbar^{g(\gamma)}}{|\operatorname{Aut}(\gamma,\phi)|}w_{\gamma,\phi}(P,I)(a_1,\ldots,a_k,u',u'').$$

Here the sum is over graphs γ with $k+2$ tails, and an ordering ϕ of these $k+2$ tails. The factor of $\frac{1}{2}$ arises because of the two different ways to order the new tails. Comparing this to the previous expression, we find that

$$\frac{d}{d\varepsilon}\left(\frac{\partial}{\partial a_1}\cdots\frac{\partial}{\partial a_k}\exp\left(\hbar^{-1}W\left(P+\varepsilon P', I\right)\right)\right)(0)$$
$$= \frac{1}{2}\frac{\partial}{\partial a_1}\cdots\frac{\partial}{\partial a_k}\frac{\partial}{\partial u'}\frac{\partial}{\partial u''}\exp\left(\hbar^{-1}W(P,I)\right)(0)$$

Since
$$\partial_{P'} = \frac{1}{2}\sum\frac{\partial}{\partial u'}\frac{\partial}{\partial u''}$$
this completes the proof. □

This expression makes it clear that, for all $P_1, P_2 \in \operatorname{Sym}^2 U$,
$$W(P_1, W(P_2, I)) = W(P_1 + P_2, I).$$

3.5. Now suppose that U is a finite dimensional vector space over $\mathbb{R}$, equipped with a non-degenerate negative definite quadratic form Φ. Let $P \in \operatorname{Sym}^2 U$ be the inverse to $-\Phi$. Thus, if e_i is an orthonormal basis for $-\Phi$, $P = \sum e_i \otimes e_i$. (When we return to considering scalar field theories, U will be replaced by the space $C^\infty(M)$, the quadratic form Φ will be replaced by the quadratic form $-\langle \phi, (D+m^2)\phi\rangle$, and P will be the propagator for the theory).

The Feynman diagram expansion $W(P,I)$ described above can also be interpreted as an asymptotic expansion for an integral on U.

LEMMA 3.5.1.
$$W(P,I)(a) = \hbar \log \int_{x \in U} \exp\left(\frac{1}{2\hbar}\Phi(x,x) + \frac{1}{\hbar}I(x+a)\right).$$

The integral is understood as an asymptotic series in $\hbar$, and so makes sense whatever the signature of Φ. As I mentioned before, we use the convention that the measure on U is normalized so that

$$\int_{x \in U} \exp\left(\frac{1}{2\hbar}\Phi(x,x)\right) = 1.$$

This normalization accounts for the lack of a graph with one loop and zero external edges in the expansion.

If U is a complex vector space and Φ is a non-degenerate complex linear inner product on U, then the same formula holds, where we integrate over any real slice $U_{\mathbb{R}}$ of U.

PROOF. It suffices to show that, for all functions $f \in \mathscr{O}(U)$,
$$\int_{x \in U} e^{(2\hbar)^{-1}\Phi(x,x)} f(x+a) = e^{\hbar \partial_P} f$$
(where both sides are regarded as functions of $a \in U$).

The result is clear when $f = 1$. Let $l \in U^{\vee}$. Note that
$$e^{\hbar \partial_P}(lf) - l e^{\hbar \partial_P}(lf) = \hbar [\partial_P, l] e^{\hbar \partial_P}(lf).$$
In this expression, $\hbar[\partial_P, l]$ is viewed as an order 1 differential operator on $\mathscr{O}(U)$.

The quadratic form Φ on U provides an isomorphism $U \to U^{\vee}$. If $u \in U$, let $u^{\vee} \in U^{\vee}$ be the corresponding element; and, dually, if $l \in U^{\vee}$, let $l^{\vee} \in U$ be the corresponding element.

Note that
$$[\partial_P, l] = -\frac{\partial}{\partial l^{\vee}}.$$
It suffices to verify a similar formula for the integral. Thus, we need to check that
$$\int_{x \in U} e^{(2\hbar)^{-1}\Phi(x,x)} l(x) f(x+a) = -\hbar \frac{\partial}{\partial l_a^{\vee}} \int_{x \in U} e^{(2\hbar)^{-1}\Phi(x,x)} f(x+a).$$
(The subscript in $l_a^{\vee}$ indicates we are applying this differential operator to the a variable).

Note that
$$\frac{\partial}{\partial l_x^{\vee}} e^{(2\hbar)^{-1}\Phi(x,x)} = \hbar^{-1} l(x) e^{\Phi(x,x)/\hbar}.$$
Thus,
$$\int_{x \in U} e^{(2\hbar)^{-1}\Phi(x,x)} l(x) f(x+a) = \hbar \int_{x \in U} \left(\frac{\partial}{\partial l_x^{\vee}} e^{(2\hbar)^{-1}\Phi(x,x)} \right) f(x+a)$$
$$= -\hbar \int_{x \in U} e^{(2\hbar)^{-1}\Phi(x,x)} \frac{\partial}{\partial l_x^{\vee}} f(x+a)$$
$$= -\hbar \int_{x \in U} e^{(2\hbar)^{-1}\Phi(x,x)} l(x) \frac{\partial}{\partial l_a^{\vee}} f(x+a)$$
$$= -\hbar \frac{\partial}{\partial l_a^{\vee}} \int_{x \in U} e^{(2\hbar)^{-1}\Phi(x,x)} f(x+a)$$
as desired.

□

3.6. One can ask, how much of this picture holds if U is replaced by an infinite dimensional vector space? We can't define the Lebesgue measure in this situation, thus we can't define the integral directly. However, one can still contract tensors using Feynman graphs, and one can still define the expression $W(P, I)$, as long as one is careful with tensor products and dual spaces. (As we will see later, the singularities in Feynman graphs arise because the inverse to the quadratic forms we will consider on infinite dimensional vector spaces do not lie in the correct completed tensor product.)

Let us work over a ground field $\mathbb{K} = \mathbb{R}$ or $\mathbb{C}$. Let M be a manifold and E be a super vector bundle on M over $\mathbb{K}$. Let $\mathscr{E} = \Gamma(M, E)$ be the super nuclear Fréchet space of global sections of E. Let $\otimes$ denote the completed projective tensor product, so that $\mathscr{E} \otimes \mathscr{E} = \Gamma(M \times M, E \boxtimes E)$. (Some details of the symmetric monoidal category of nuclear spaces, equipped with the completed projective tensor product, are presented in Appendix 2).

Let $\mathscr{O}(\mathscr{E})$ denote the algebra of formal power series on $\mathscr{E}$,

$$\mathscr{O}(\mathscr{E}) = \prod_{n \geq 0} \operatorname{Hom}(\mathscr{E}^{\otimes n}, \mathbb{K})_{S_n}$$

where Hom denotes continuous linear maps and the subscript S_n denotes coinvariants. Note that $\mathscr{O}(\mathscr{E})$ is an algebra: direct product of distributions defines a map

$$\operatorname{Hom}(\mathscr{E}^{\otimes n}, \mathbb{K}) \times \operatorname{Hom}(\mathscr{E}^{\otimes m}, \mathbb{K}) \to \operatorname{Hom}(\mathscr{E}^{\otimes n+m}, \mathbb{K}).$$

These maps induce an algebra structure on $\mathscr{O}(\mathscr{E})$.

We can also regard $\mathscr{O}(\mathscr{E})$ as simply the completed symmetric algebra of the dual space $\mathscr{E}^\vee$, that is,

$$\mathscr{O}(\mathscr{E}) = \widehat{\operatorname{Sym}}^*(\mathscr{E}^\vee).$$

Here, $\mathscr{E}^\vee$ is the strong dual of $\mathscr{E}$, and is again a nuclear space. The completed symmetric algebra is taken in the symmetric monoidal category of nuclear spaces, as detailed in Appendix 2.

As before, let

$$\mathscr{O}^+(\mathscr{E})[[\hbar]] \subset \mathscr{O}(\mathscr{E})[[\hbar]]$$

be the subspace of those functionals I which are at least cubic modulo $\hbar$.

Let $\operatorname{Sym}^n \mathscr{E}$ denote the S_n-invariants in $\mathscr{E}^{\otimes n}$. If $P \in \operatorname{Sym}^2 \mathscr{E}$ and $I \in \mathscr{O}^+(\mathscr{E})[[\hbar]]$ then, for any stable graph γ, one can define

$$w_\gamma(P, I) \in \mathscr{O}(\mathscr{E}).$$

The definition is exactly the same as in the finite dimensional situation. Let $T(\gamma)$ be the set of tails of γ, $H(\gamma)$ the set of half-edges of γ, $V(\gamma)$ the set of vertices of γ, and $E(\gamma)$ the set of internal edges of γ. The tensor products of interactions at the vertices of γ define an element of

$$\operatorname{Hom}(\mathscr{E}^{\otimes H(\gamma)}, \mathbb{R}).$$

We can contract this tensor with the element of $\mathscr{E}^{\otimes 2E(\gamma)}$ given by the tensor product of the propagators; the result of this contraction is

$$w_\gamma(P, I) \in Hom(\mathscr{E}^{\otimes T(\gamma)}, \mathbb{R}).$$

Thus, one can define

$$W(P, I) = \sum_\gamma \frac{1}{|\mathrm{Aut}(\gamma)|} w_\gamma(P, I) \in \mathscr{O}^+(\mathscr{E})[[\hbar]]$$

exactly as before.

The interpretation in terms of differential operators works in this situation too. As in the finite dimensional situation, we can define an order two differential operator

$$\partial_P : \mathscr{O}(\mathscr{E}) \to \mathscr{O}(\mathscr{E}).$$

On the direct factor

$$\mathrm{Hom}(\mathscr{E}^{\otimes n}, \mathbb{K})_{S_n} = \mathrm{Sym}^n \mathscr{E}^\vee$$

of $\mathscr{O}(\mathscr{E})$, the operator ∂_P comes from the map

$$\mathrm{Hom}(\mathscr{E}^{\otimes n}, \mathbb{K}) \to \mathrm{Hom}(\mathscr{E}^{\otimes n-2}, \mathbb{K})$$

given by contracting with the tensor $P \in \mathscr{E}^{\otimes 2}$.

Then,

$$W(P, I) = \hbar \log\{\exp(\hbar \partial_P) \exp(I/\hbar)\}$$

as before.

4. Sharp and smooth cut-offs

4.1. Let us return to our scalar field theory, whose action is of the form

$$S(\phi) = -\tfrac{1}{2} \langle \phi, (\mathrm{D} + m^2)\phi \rangle + I(\phi).$$

The propagator P is the kernel for the operator $(\mathrm{D} + m^2)^{-1}$. There are several natural ways to write this propagator. Let us pick a basis $\{e_i\}$ of $C^\infty(M)$ consisting of orthonormal eigenvectors of D, with eigenvalues $\lambda_i \in \mathbb{R}_{\geq 0}$. Then,

$$P = \sum_i \frac{1}{\lambda_i^2 + m^2} e_i \otimes e_i.$$

There are natural cut-off propagators, where we only sum over some of the eigenvalues. For a subset $U \subset \mathbb{R}_{\geq 0}$, let

$$P_U = \sum_{i \text{ such that } \lambda_i \in U} \frac{1}{\lambda_i^2 + m^2} e_i \otimes e_i.$$

Note that, unlike the full propagator P, the cut-off propagator P_U is a smooth function on $M \times M$ as long as U is a bounded subset of $\mathbb{R}_{\geq 0}$.

By analogy with the case of finite-dimensional integrals, we have the formal identity

$$W(P,I)(a) = \hbar \log \int_{\phi \in C^\infty(M)} \exp\left(-\frac{1}{2\hbar}\langle \phi, (D+m^2)\phi\rangle + \frac{1}{\hbar}I(\phi+a)\right)$$

Both sides of this equation are ill-defined. The propagator P is not a smooth function on $C^\infty(M \times M)$, but has singularities along the diagonal; this means that $W(P, I)$ is not well defined. And, of course, the integral on the right hand side is infinite dimensional.

In a similar way, we have the following (actual) identity, for any functional $I \in \mathscr{O}^+(C^\infty(M))[[\hbar]]$:

(†) $\quad W\left(P_{[\Lambda',\Lambda)}, I\right)(a)$
$$= \hbar \log \int_{\phi \in C^\infty(M)_{[\Lambda',\Lambda)}} \exp\left(-\frac{1}{2\hbar}\langle \phi, (D+m^2)\phi\rangle + \frac{1}{\hbar}I(\phi+a)\right).$$

Both sides of this identity are well-defined; the propagator $P_{[\Lambda',\Lambda)}$ is a smooth function on $M \times M$, so that $W\left(P_{[\Lambda',\Lambda)}, I\right)$ is well-defined. The right hand side is a finite dimensional integral.

The equation (†) says that the map

$$\mathscr{O}^+(C^\infty(M))[[\hbar]] \to \mathscr{O}^+(C^\infty(M))[[\hbar]]$$
$$I \mapsto W\left(P_{[\Lambda',\Lambda)}, I\right)$$

is the renormalization group flow from energy Λ to energy Λ'.

4.2. In this book we will use a cut-off based on the heat kernel, rather than the cut-off based on eigenvalues of the Laplacian described above.

For $l \in \mathbb{R}_{>0}$, let $K_l^0 \in C^\infty(M \times M)$ denote the heat kernel for D; thus,

$$\int_{y \in M} K_l^0(x,y)\phi(y) = \left(e^{-lD}\phi\right)(x)$$

for all $\phi \in C^\infty(M)$.

We can write K_l^0 in terms of a basis of eigenvalues for D as

$$K_l^0 = \sum e^{-l\lambda_i} e_i \otimes e_i.$$

Let

$$K_l = e^{-lm^2} K_l^0$$

be the kernel for the operator $e^{-l(D+m^2)}$. Then, the propagator P can be written as

$$P = \int_{l=0}^{\infty} K_l \, dl.$$

For $\varepsilon, L \in [0, \infty]$, let

$$P(\varepsilon, L) = \int_{l=\varepsilon}^{L} K_l \, dl.$$

This is the propagator with an infrared cut-off L and an ultraviolet cut-off ε. Here ε and L are *length* scales rather than energy scales; length behaves as the inverse to energy. Thus, ε is the high-energy cut-off and L is the low-energy cut-off.

The propagator $P(\varepsilon, L)$ damps down the high energy modes in the propagator P. Indeed,

$$P(\varepsilon, L) = \sum_i \frac{e^{-\varepsilon \lambda_i} - e^{-L\lambda_i}}{\lambda_i^2 + m^2} e_i \otimes e_i$$

so that the coefficient of $e_i \otimes e_i$ decays as $\lambda_i^{-2} e^{-\varepsilon \lambda_i}$ for λ_i large.

Because $P(\varepsilon, L)$ is a smooth function on $M \times M$, as long as $\varepsilon > 0$, the expression $W(P(\varepsilon, L), I)$ is well-defined for all $I \in \mathscr{O}^+(C^\infty(M))[[\hbar]]$. (Recall the superscript $+$ means that I must be at least cubic modulo $\hbar$.)

DEFINITION 4.2.1. *The map*

$$\mathscr{O}^+(C^\infty(M))[[\hbar]] \to \mathscr{O}^+(C^\infty(M))[[\hbar]]$$
$$I \mapsto W(P(\varepsilon, L), I)$$

is defined to be the renormalization group flow from length scale ε to length scale L.

From now on, we will be using this length-scale version of the renormalization group flow.

Figure 2 illustrates the first few terms of the renormalization group flow from scale ε to scale L.

As the effective interaction $I[L]$ varies smoothly with L, there is an infinitesimal form of the renormalization group equation, which is a differential equation in $I[L]$. This is illustrated in figure 3.

The expression for the propagator in terms of the heat kernel has a very natural geometric/physical interpretation, which will be explained in Section 6.

5. Singularities in Feynman graphs

5.1. In this section, we will consider explicitly some of the simple Feynman graphs appearing in $W(P(\varepsilon, L), I)$ where $I(\phi) = \frac{1}{3!} \int_M \phi^3$, and try to take the limit as $\varepsilon \to 0$. We will see that, for graphs which are not trees, the limit in general won't exist. The $\frac{1}{3!}$ present in the interaction term simplifies the combinatorics of the Feynman diagram expansion. The Feynman diagrams we will consider are all trivalent, and as explained in Section 3, each vertex is labelled by the linear map

$$C^\infty(M)^{\otimes 3} \to \mathbb{R}$$
$$\phi_1 \otimes \phi_2 \otimes \phi_3 \mapsto \frac{\partial}{\partial \phi_1} \frac{\partial}{\partial \phi_2} \frac{\partial}{\partial \phi_3} I$$
$$= \int_M \phi_1(x) \phi_2(x) \phi_3(x)$$

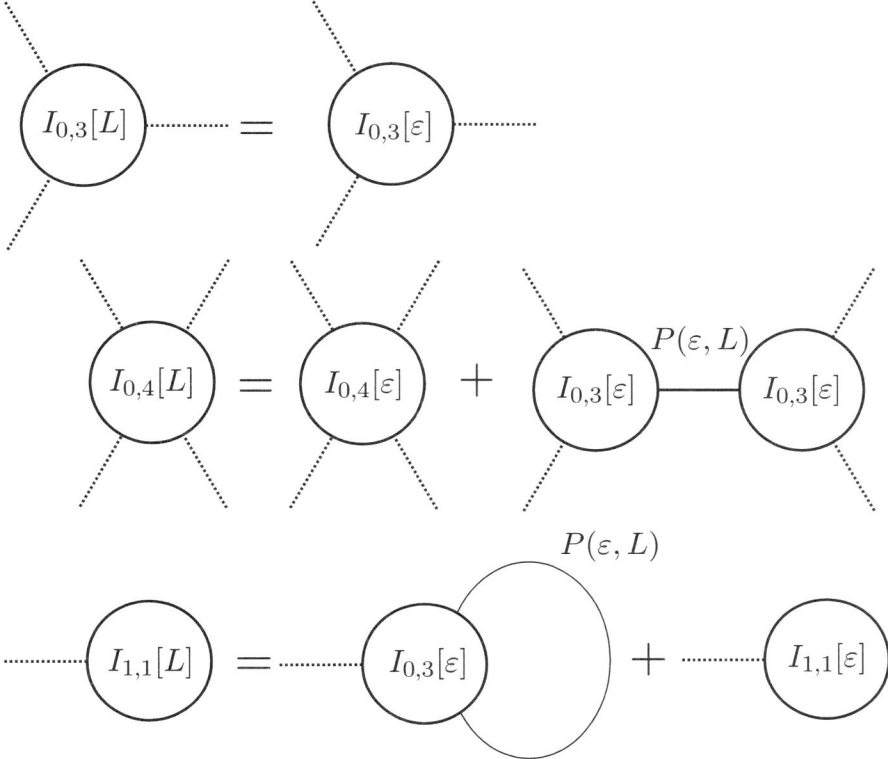

FIGURE 2. The first few expressions in the renormalization group flow from scale ε to scale L.

Consider the graph γ_1, given by

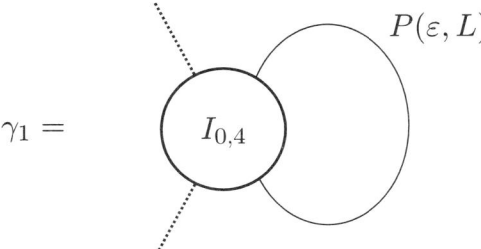

We see that
$$w_{\gamma_1}(P(\varepsilon, L), I)(a) = \int_{l \in [\varepsilon, L]} \int_{x \in M} a(x) K_l(x, x) \, \mathrm{d}\,\mathrm{Vol}_M \, \mathrm{d}l$$
where $\mathrm{d}\,\mathrm{Vol}_M$ is the volume form associated to the metric on M, and $a \in C^\infty(M)$ is the field we place at the external edge of the graph. Since
$$K_l(x, x) \simeq l^{-\dim M/2} + \text{ higher order terms}$$
for l small, the limit $\lim_{\varepsilon \to 0} w_{\gamma_1}(P(\varepsilon, L), I)(a)$ doesn't exist.

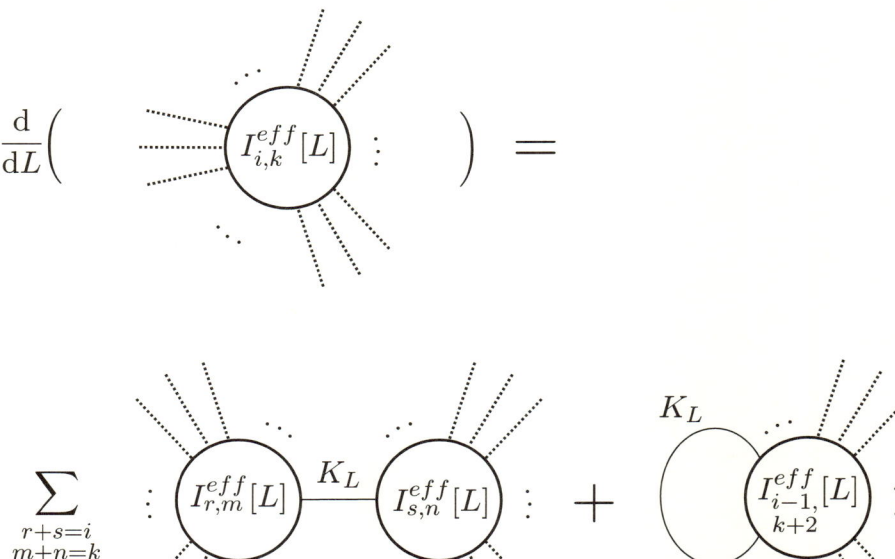

FIGURE 3. The renormalization group differential equation.

Next consider γ_2, given by

$$\gamma_2 = \begin{array}{c} I_{0,3} \end{array} \begin{array}{c} P(\varepsilon, L) \\ P(\varepsilon, L) \\ P(\varepsilon, L) \end{array} \begin{array}{c} I_{0,3} \end{array}$$

Since there are no external edges there is no dependence on $a \in C^\infty(M)$, and we find

$$w_{\gamma_2}(P(\varepsilon, L), I) = \int_{l_1, l_2, l_3 \in [\varepsilon, L]} \int_{x, y \in M} K_{l_1}(x, y) K_{l_2}(x, y) K_{l_3}(x, y) \mathrm{d}\,\mathrm{Vol}_{M \times M}.$$

Using the fact that

$$K_l(x, y) \simeq l^{-\dim M/2} e^{-d(x-y)^2/l} + \text{ higher order terms}$$

for l small, we can see that the limit of $w_{\gamma_2}(P(\varepsilon, L), I)$ as $\varepsilon \to 0$ is singular.

Let γ_3 be the graph

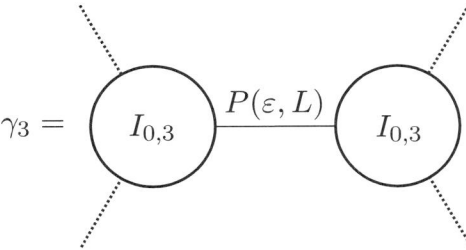

This graph is a tree; the integrals associated to graphs which are trees always admit $\varepsilon \to 0$ limits. Indeed,

$$w_{\gamma_3}(P(\varepsilon, L), I)(a) = \int_{l \in [\varepsilon, L]} \int_{x, y \in M} a(x)^2 K_l(x, y) a(y)^2$$

$$= \left\langle a^2, \int_{\varepsilon}^{L} \left(e^{-l\mathrm{D}} a^2 \right) \mathrm{d}l \right\rangle$$

and the second expression clearly admits an $\varepsilon \to 0$ limit.

All of the calculations above are for the interaction $I(\phi) = \frac{1}{3!} \int \phi^3$. For more general interactions I, one would have to apply a differential operator to both a and $K_l(x, y)$ in the integrands.

6. The geometric interpretation of Feynman graphs

From the functional integral point of view, Feynman graphs are just graphical tools which help in the perturbative calculation of certain functional integrals. In the introduction, we gave a brief account of the world-line interpretation of quantum field theory. In this picture, Feynman graphs describe the trajectories taken by some interacting particles. The length scale version of the renormalization group flow becomes very natural from this point of view.

As before, we will work in Euclidean signature; Lorentzian signature presents significant additional analytical difficulties. We will occasionally comment on the formal picture in Lorentzian signature.

6.1. Let us consider a massless scalar field theory on a compact Riemannian manifold M. Thus, the fields are $C^\infty(M)$ and the action is

$$S(\phi) = -\frac{1}{2} \int_M \phi \, \mathrm{D} \, \phi.$$

The propagator $P(x, y)$ is a distribution on M^2, which can be expressed as a functional integral

$$P(x, y) = \int_{\phi \in C^\infty(M)} e^{S(\phi)} \phi(x) \phi(y).$$

Thus, the propagator encodes the correlation between the values of the fields ϕ at the points x and y.

We will derive an alternative expression of the propagator as a *one-dimensional* functional integral.

Recall that we can write the propagator of a massless scalar field theory on a Riemannian manifold M as an integral of the heat kernel. If $x, y \in M$ are distinct points, then

$$P(x,y) = \int_0^\infty K_t(x,y) \mathrm{d}t$$

where $K_t \in C^\infty(M^2)$ is the heat kernel. This expression of the propagator is sometimes known as the Schwinger representation, and the parameter t as the Schwinger parameter. One can also interpret the parameter t as proper time, as we will see shortly.

The heat kernel $K_t(x,y)$ is the probability density that a particle in Brownian motion on M, which starts at x at time zero, lands at y at time t. Thus, we can rewrite the heat kernel as

$$K_t(x,y) = \int_{\substack{f:[0,t]\to M \\ f(0)=x, f(t)=y}} \mathcal{D}_{\mathrm{Wiener}} f$$

where $\mathcal{D}_{\mathrm{Wiener}} f$ is the Wiener measure on the path space.

We can think of the Wiener measure as the measure for a quantum field theory of maps

$$f : [0,t] \to M$$

with action given by

$$E(f) = \int_0^t \langle \mathrm{d}f, \mathrm{d}f \rangle.$$

Thus, we will somewhat loosely write

$$K_t(x,y) = \int_{\substack{f:[0,t]\to M \\ f(0)=x, f(t)=y}} e^{-E(f)}$$

where we understand that the integral can be given rigorous meaning using the Wiener measure.

Combining these expressions, we find the desired expression for the propagator as a one-dimensional functional integral:

$$P(x,y) = \int_{\phi \in C^\infty(M)} e^{S(\phi)} \phi(x)\phi(y) = \int_{t=0}^\infty \int_{\substack{f:[0,t]\to M \\ f(0)=x, f(t)=y}} e^{-E(f)}.$$

This expression is the core of the world-line formulation of quantum field theory. This expression tells us that the correlation between the values of the fields at points x and y can be expressed in terms of an integral over paths in M which start at x and end at y.

If we work in Lorentzian signature, we find the (formal) identity

$$\int_{\phi \in C^\infty(M)} e^{S(\phi)i} \phi(x)\phi(y) = \int_{t=0}^\infty \int_{\substack{f:[0,t]\to M \\ f(0)=x, f(t)=y}} e^{iE(f)}.$$

This expression is difficult to make rigorous sense of; I don't know of a rigorous treatment of the Wiener measure when the target manifold has Lorentzian signature.

6. THE GEOMETRIC INTERPRETATION OF FEYNMAN GRAPHS

6.2. We should interpret these identities as follows. We should think of particles moving through space-time as equipped with an "internal clock"; as the particle moves, this clock ticks at a rate independent of the time parameter on space-time. The world-line of such a particle is a parameterized path in space-time, that is, a map $f : \mathbb{R} \to M$. This path is completely arbitrary: it can go backwards or forwards in time. Two world-lines which differ by a translation on the source $\mathbb{R}$ should be regarded as the same. In other words, the internal clock of a particle doesn't have an absolute starting point.

If $I = [0, \tau]$ is a closed interval, and if $f : I \to M$ is a path describing part of the world-line of a particle, then the energy of f is, as before,
$$E(f) = \int_{[0,\tau]} \langle \mathrm{d}f, \mathrm{d}f \rangle.$$
In quantum field theory, everything that can happen will happen, but with a probability amplitude of e^{iE} where E is the energy. Thus, to calculate the probability that a particle starts at the point x in space-time and ends at the point y, we must integrate over all paths $f : [0, \tau] \to M$, starting at x and ending at y. We must also integrate over the parameter τ, which is interpreted as the time taken on the internal clock of the particle as it moves from x to y. This leads to the expression (in Lorentzian signature) we discussed earlier,
$$P(x,y) = \int_{t=0}^{\infty} \int_{\substack{f:[0,t]\to M \\ f(0)=x, f(t)=y}} e^{iE(f)}.$$

6.3. We would like to have a similar picture for interacting theories. We will consider, for simplicity, the ϕ^3 theory, where the fields are $\phi \in C^{\infty}(M)$, and the action is
$$S(\phi) = \int_M -\frac{1}{2}\phi \operatorname{D} \phi + \frac{1}{6}\phi^3.$$
We will discuss the heuristic picture first, ignoring the difficulties of renormalization. At the end, we will explain how the renormalization group flow and the idea of effective interactions can be explained in the world-line point of view.

The fundamental quantities one is interested in are the correlation functions, defined by the heuristic functional integral formula
$$\mathbb{E}(x_1, \ldots, x_n) = \int_{\phi \in C^{\infty}(M)} e^{S(\phi)/\hbar} \phi(x_1) \cdots \phi(x_n).$$
We would like to express these correlation functions in the world-line point of view.

6.4. The ϕ^3 theory corresponds, in the world-line point of view, to a theory where three particles can fuse at a point in M. Thus, world-lines in the ϕ^3 theory become world-*graphs*; further, just as the world-lines for the free theory are parameterized, the world-graphs arising in the ϕ^3 theory have a metric, that is, a length along each edge. This length on the edge of the graph corresponds to time traversed by the particle on this edge in its internal clock.

Measuring the value of a field ϕ at a point $x \in M$ corresponds, in the world-line point of view, to observing a particle at a point x. We would like to find an expression for the correlation function $\mathbb{E}(x_1, \ldots, x_n)$ in the world-line point of view. As always in quantum field theory, one should calculate this expectation value by summing over all events that could possibly happen. Such an event is described by a world-graph with end points at $x_1, \ldots, x_n$. Since only three particles can interact at a given point in space-time, such world-graphs are trivalent. Thus, the relevant world-graphs are trivalent, have n external edges, and the end points of these external edges maps to the points $x_1, \ldots, x_n$.

Thus, we find that

$$\mathbb{E}(x_1, \ldots, x_n) = \sum_\gamma \frac{1}{|\operatorname{Aut} \gamma|} \hbar^{-\chi(\gamma)} \int_{g \in \operatorname{Met} \gamma} \int_{f : \gamma \to M} e^{-E(f)}.$$

Here, the sum runs over all world-graphs γ, and the integral is over those maps $f : \gamma \to M$ which take the endpoints of the n external edges of γ to the points $x_1, \ldots, x_n$.

The symbol $\operatorname{Met}(\gamma)$ refers to the space of metrics on γ, in other words, to the space $\mathbb{R}_{>0}^{E(\gamma) \amalg T(\gamma)}$ where $E(\gamma)$ is the set of internal edges of γ, and $T(\gamma)$ is the set of tails.

If γ is a metrized graph, and $f : \gamma \to M$ is a map, then $E(f)$ is the sum of the energies of f restricted to the edges of γ, that is,

$$E(f) = \sum_{e \in E(\gamma)} \int_0^{l(e)} \langle \mathrm{d}f, \mathrm{d}f \rangle.$$

The space of maps $f : \gamma \to M$ is given a Wiener measure, constructed from the usual Wiener measure on path space.

This graphical expansion for the correlation functions is only a formal expression: if γ has a non-zero first Betti number, then the integral over $\operatorname{Met}(\gamma)$ will diverge, as we will see shortly. However, this graphical expansion is precisely the expansion one finds when formally applying Wick's lemma to the functional integral expression for $\mathbb{E}(x_1, \ldots, x_n)$. The point is that we recover the propagator when we consider the integral over all possible maps from a given edge.

In Lorentzian signature, of course, one should use $e^{iE(f)}$ instead of $e^{-E(f)}$.

6.5. As an example, we will consider the path integral $f : \gamma \to M$ where γ is the metrized graph

6. THE GEOMETRIC INTERPRETATION OF FEYNMAN GRAPHS

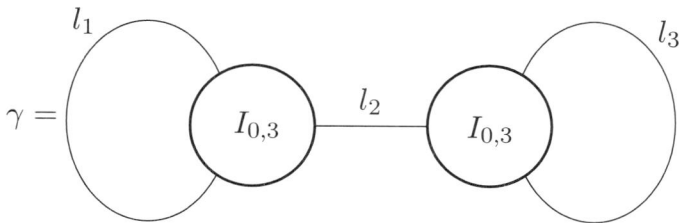

Then γ has two vertices, labelled by interactions $I_{0,3}$.

The integral
$$\int_{f:\gamma \to M} e^{-E(f)}$$
is obtained by putting the heat kernel K_l on each edge of γ of length l, and integrating over the position of the two vertices. Thus, we find
$$\int_{f:\gamma \to M} e^{-E(f)} = \int_{x,y,\in M} K_{l_1}(x,x) K_{l_2}(x,y) K_{l_3}(y,y).$$
However, the second integral, over the space of metrized graphs, does not make sense. Indeed, the heat kernel $K_l(x, y)$ has a small l asymptotic expansion of the form
$$K_l(x,y) = l^{-n/2} e^{-\|x-y\|^2/4l} \sum l^i f_i(x,y).$$
This implies that the integral
$$\int_{l_1,l_2,l_3} \int_{f:\gamma \to M} e^{-E(f)} = \int_{x,y,\in M} K_{l_1}(x,x) K_{l_2}(x,y) K_{l_3}(y,y)$$
does not converge.

This second integral is the weight attached to the graph γ in the Feynman diagram expansion of the functional integral for the $\frac{1}{3!}\phi^3$ interaction.

6.6. Next, we will explain (briefly and informally) how to construct the correlation functions from a general scale L effective interaction $I[L]$. We will not need this construction of the correlation functions elsewhere in this book. A full treatment of observables and correlation functions will appear in (CG10).

The correlation functions will allow us to give a world-line formulation for the renormalization group equation on a collection $\{I[L]\}$ of effective interactions: the renormalization group equation is equivalent to the statement that the correlation functions computed using $I[L]$ are independent of L.

The correlations $\mathbb{E}^n_{I[L]}(f_1, \ldots, f_n)$ we will define will take, as their input, functions $f_1, \ldots, f_n \in C^\infty(M)$. Thus, the correlation functions will give a collection of distributions on M^n:
$$\mathbb{E}^n_{I[L]} : C^\infty(M^n) \to \mathbb{R}[[\hbar]].$$

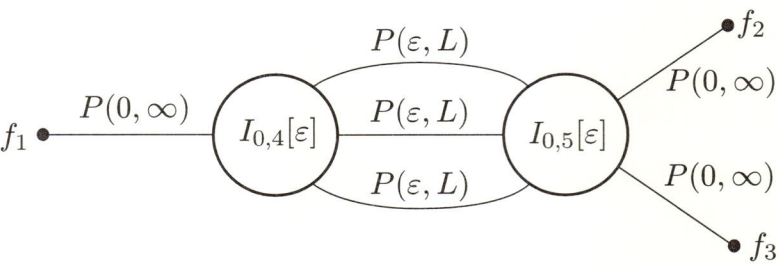

FIGURE 4. The weight $C_\gamma(I[\varepsilon], f_1, f_2, f_3)$ attached to a graph with 3 external vertices.

These correlation functions will be defined as a sum over graphs.

Let Γ_n denote the set of graphs γ with n univalent vertices, which are labelled as $v_1, \ldots, v_n$. These vertices will be referred to as the *external vertices*. The remaining vertices will be called the *internal vertices*. The internal vertices of a graph $\gamma \in \Gamma_n$ can be of any valency, and are labelled by a genus $g(v) \in \mathbb{Z}_{\geq 0}$. The internal vertices of genus 0 must be at least trivalent.

For a graph $\gamma \in \Gamma_n$, and smooth functions $f_1, \ldots, f_n \in C^\infty(M)$, we will define

$$C_\gamma(I[L])(f_1, \ldots, f_n) \in \mathbb{R}$$

by contracting certain tensors attached to the edges and the vertices.

We will label each internal vertex v of genus i and valency k by

$$I_{i,k}[L] : C^\infty(M)^{\otimes H(v)} \to \mathbb{R}.$$

Let $f_1, \ldots, f_n \in C^\infty(M)$ be smooth functions on M. The external vertex v_i of γ will be labelled by the distribution

$$C^\infty(M) \to \mathbb{R}$$

$$\phi \mapsto \int_M f\phi.$$

Any edge e joining two internal vertices will be labelled by

$$P(L, \infty) \in C^\infty(M^2).$$

The remaining edges, which join two external vertices or join an external and an internal vertex, will be labelled by $P(0, \infty)$, which is a distribution on M^2.

As usual, we can contract all these tensors to define an element

$$C_\gamma(I[L])(f_1, \ldots, f_n) \in \mathbb{R}.$$

One may worry that because some of the edge are labelled by the distribution $P(0, \infty)$, this expression is not well defined. However, because the external edges are labelled by smooth functions f_i, there are no problems. Figure 4 describes $C_\gamma(I[\varepsilon], f_1, f_2, f_3)$ for a particular graph γ.

Then, the correlation function for the effective interaction $I[L]$ is defined by
$$\mathbb{E}_{I[L]}(f_1,\ldots,f_n) = \sum_{\gamma \in \Gamma_n} \frac{1}{\operatorname{Aut}(\gamma)} \hbar^{n-g(\gamma)-1} C_\gamma(I[L])(f_1,\ldots,f_n).$$
Unlike the heuristic graphical expansion we gave for the correlation functions of the ϕ^3 theory, this expression is well-defined.

We should interpret this expansion as saying that we can compute the correlation functions from the effective interaction $I[L]$ by allowing particles to propagate in the usual way, and to interact by $I[L]$; except that in between any two interactions, particles must travel for a proper time of at least L. This accounts for the fact that edges which join to internal vertices are labelled by $P(L,\infty)$.

If we have a collection $\{I[L] \mid L \in (0,\infty)\}$ of effective interactions, then the renormalization group equation is equivalent to the statement that all the correlation functions constructed from $I[L]$ using the prescription given above are independent of L.

7. A definition of a quantum field theory

7.1. Now we have some preliminary definitions and an understanding of why the terms in the graphical expansion of a functional integral diverge. This book will describe a method for renormalizing these functional integrals to yield a finite answer.

This section will give a formal definition of a quantum field theory, based on Wilson's philosophy of the effective action; and a precise statement of the main theorem, which says roughly that there's a bijection between theories and Lagrangians.

DEFINITION 7.1.1. *A local action functional $I \in \mathscr{O}(C^\infty(M))$ is a functional which arises as an integral of some Lagrangian. More precisely, if we Taylor expand I as $I = \sum_k I_k$ where*
$$I_k(\lambda a) = \lambda^k I_k(a)$$
(so that I_k is homogeneous of degree k of the variable $a \in C^\infty(M)$), then I_k must be of the form
$$I_k(a) = \sum_{j=1}^s \int_M D_{1,j}(a) \cdots D_{k,j}(a)$$
where $D_{i,j}$ are arbitrary differential operators on M.

Let $\mathscr{O}_{loc}(C^\infty(M)) \subset \mathscr{O}(C^\infty(M))$ be the subspace of local action functionals.

As before, let
$$\mathscr{O}_{loc}^+(C^\infty(M))[[\hbar]] \subset \mathscr{O}_{loc}(C^\infty(M))[[\hbar]]$$
be the subspace of those local action functionals which are at least cubic modulo $\hbar$.

Thus, local action functionals are the same as Lagrangians modulo those Lagrangians which are a total derivative.

DEFINITION 7.1.2. *A perturbative quantum field theory, with space of fields $C^\infty(M)$ and kinetic action $-\frac{1}{2}\langle \phi, (\mathrm{D}+m^2)\phi\rangle$, is given by a set of effective interactions $I[L] \in \mathscr{O}^+(C^\infty(M))[[\hbar]]$ for all $L \in (0,\infty]$, such that*

(1) The renormalization group equation
$$I[L] = W\left(P(\varepsilon, L), I[\varepsilon]\right)$$
is satisfied, for all $\varepsilon, L \in (0,\infty]$.

(2) For each i, k, there is a small L asymptotic expansion
$$I_{i,k}[L] \simeq \sum_{r \in \mathbb{Z}_{\geq 0}} g_r(L) \Phi_r$$
where $g_r(L) \in C^\infty((0,\infty)_L)$ and $\Phi_r \in \mathscr{O}_{loc}(C^\infty(M))$.

Let $\mathscr{T}^{(\infty)}$ denote the set of perturbative quantum field theories, and let $\mathscr{T}^{(n)}$ denote the set of theories defined modulo $\hbar^{n+1}$. Thus, $\mathscr{T}^{(\infty)} = \varprojlim \mathscr{T}^{(n)}$.

Let me explain more precisely what I mean by saying there is a small L asymptotic expansion
$$I_{i,k}[L] \simeq \sum_{j \in \mathbb{Z}_{\geq 0}} g_r(L) \Phi_r.$$

Without loss of generality, we can require that the local action functionals Φ_r appearing here are homogeneous of degree k in the field a.

Then, the statement that there is such an asymptotic expansion means that there is a non-decreasing sequence $d_R \in \mathbb{Z}$, tending to infinity, such that for all R, and for all fields $a \in C^\infty(M)$,
$$\lim_{L \to 0} L^{-d_R} \left(I_{i,k}[L](a) - \sum_{r=0}^{R} g_r(L) \Phi_r(a)\right) = 0.$$

In other words, we are asking that the asymptotic expansion exists in the weak topology on $\mathrm{Hom}(C^\infty(M)^{\otimes k}, \mathbb{R})_{S_k}$.

The main theorem of this chapter (in the case of scalar field theories) is the following.

THEOREM A. *Let $\mathscr{T}^{(n)}$ denote the set of perturbative quantum field theories defined modulo $\hbar^{n+1}$. Then $\mathscr{T}^{(n+1)}$ is, in a canonical way, a principal bundle over $\mathscr{T}^{(n)}$ for the abelian group $\mathscr{O}_{loc}(C^\infty(M))$ of local action functionals on M.*

Further, $\mathscr{T}^{(0)}$ is canonically isomorphic to the space $\mathscr{O}_{loc}^+(C^\infty(M))$ of local action functionals which are at least cubic.

There is a variant of this theorem, which states that there is a bijection between theories and Lagrangians once we choose a *renormalization scheme*, which is a way to extract the singular part of certain functions of one variable. The concept of renormalization scheme will be discussed in Section 9;

this is a choice that only has to be made once, and then it applies to all theories on all manifolds.

THEOREM B. *Let us fix a renormalization scheme.*
Then, we find a section of each torsor $\mathscr{T}^{(n+1)} \to \mathscr{T}^{(n)}$, and so a bijection between the set of perturbative quantum field theories and the set of local action functionals $I \in \mathscr{O}_{loc}^{+}(C^{\infty}(M))[[\hbar]]$. (Recall the superscript $+$ means that I must be at least cubic modulo $\hbar$.)

7.2. We will first prove theorem B, and deduce theorem A (which is the more canonical formulation) as a corollary.

In one direction, the bijection in theorem B is constructed as follows. If $I \in \mathscr{O}_{loc}^{+}(C^{\infty}(M))[[\hbar]]$ is a local action functional, then we will construct a canonical series of counterterms $I^{CT}(\varepsilon)$. These are local action functionals, depending on a parameter $\varepsilon \in (0, \infty)$ as well as on $\hbar$. The counterterms are zero modulo $\hbar$, as the tree-level Feynman graphs all converge. Thus, $I^{CT}(\varepsilon) \in \hbar \mathscr{O}_{loc}(C^{\infty}(M))[[\hbar]] \otimes C^{\infty}((0, \infty))$ where $\otimes$ denotes the completed projective tensor product.

These counterterms are constructed so that the limit
$$\lim_{\varepsilon \to 0} W\left(P(\varepsilon, L), I - I^{CT}(\varepsilon)\right)$$
exists. This limit defines the scale L effective interaction $I[L]$.

Conversely, if we have a perturbative QFT given by a collection of effective interactions $I[l]$, the local action functional I is obtained as a certain renormalized limit of $I[l]$ as $l \to 0$. The actual limit doesn't exist; to construct the renormalized limit we again need to subtract off certain counterterms.

A detailed proof of the theorem, and in particular of the construction of the local counterterms, is given in Section 10.

8. An alternative definition

In the previous section I presented a definition of quantum field theory based on the heat-kernel cut-off. In this section, I will describe an alternative, but equivalent, definition, which allows a much more general class of cut-offs. This alternative definition is a little more complicated, but is conceptually more satisfying. One advantage of this alternative definition is that it does not rely on the heat kernel.

As before, we will consider a scalar field theory where the quadratic term of the action is $\frac{1}{2} \int \phi (\mathrm{D} + m^2) \phi$.

DEFINITION 8.0.1. *A* parametrix *for the operator $\mathrm{D} + m^2$ is a distribution P on $M \times M$, which is symmetric, smooth away from the diagonal, and is such that*
$$((\mathrm{D} + m^2) \otimes 1)P - \delta_M \in C^{\infty}(M \times M)$$
is smooth; where δ_M refers to the delta distribution along the diagonal in $M \times M$.

For any $L > 0$, the propagator $P(0, L)$ is a parametrix. In the alternative definition of a quantum field theory presented in this section, we can use *any* parametrix as the propagator.

Note that if P, P' are two parametrices, the difference $P - P'$ between them is a smooth function. We will give the set of parametrices a partial order, by saying that
$$P \leq P'$$
if $\text{Supp}(P) \subset \text{Supp}(P')$. For any two parametrices P, P', we can find some P'' with $P'' < P$ and $P'' < P'$.

8.1. Before we introduce the alternative definition of quantum field theory, we need to introduce a technical notation. Given any functional $J \in \mathscr{O}(C^\infty(M))$, we get a continuous linear map
$$C^\infty(M) \to \mathscr{O}(C^\infty(M))$$
$$\phi \mapsto \frac{dJ}{d\phi}.$$

DEFINITION 8.1.1. *A function J has* smooth first derivative *if this map extends to a continuous linear map*
$$\mathcal{D}(M) \to \mathscr{O}(C^\infty(M)),$$
where $\mathcal{D}(M)$ is the space of distributions on M.

LEMMA 8.1.2. *Let $\Phi \in C^\infty(M)^{\otimes 2}$ and suppose that $J \in \mathscr{O}^+(C^\infty(M))[[\hbar]]$ has smooth first derivative. Then so does $W(\Phi, J) \in \mathscr{O}^+(C^\infty(M))[[\hbar]]$.*

PROOF. Recall that
$$W(\Phi, J) = \hbar \log \left(e^{\hbar \partial_P} e^{J/\hbar} \right).$$

Thus, it suffices to verify two things. Firstly, that the subspace $\mathscr{O}(C^\infty(M))$ consisting of functionals with smooth first derivative is a subalgebra; this is clear. Secondly, we need to check that ∂_Φ preserves this subalgebra. This is also clear, because ∂_Φ commutes with $\frac{d}{d\phi}$ for any $\phi \in C^\infty(M)$. □

8.2. The alternative definition of quantum field theory is as follows.

DEFINITION 8.2.1. *A quantum field theory is a collection of functionals*
$$I[P] \in \mathscr{O}^+(C^\infty(M))[[\hbar]],$$
one for each parametrix P, such that the following properties hold.

(1) If P, P' are parametrices, then
$$W(P - P', I[P']) = I[P].$$

This expression makes sense, because $P - P'$ is a smooth function on $M \times M$.

(2) The functionals $I[P]$ satisfy the following locality axiom. For any (i,k), the support of
$$\operatorname{Supp} I_{i,k}[P] \subset M^k$$
can be made as close as we like to the diagonal by making the parametrix P small. More precisely, we require that, for any open neighbourhood U of the small diagonal $M \subset M^k$, we can find some P such that
$$\operatorname{Supp} I_{i,k}[P'] \subset U$$
for all $P' \leq P$.

(3) Finally, the functionals $I[P]$ all have smooth first derivative.

THEOREM 8.2.2. *This definition of a quantum field theory is equivalent to the previous one, presented in definition 7.1.2.*

More precisely, if $I[L]$ is a set of effective interactions satisfying the heat-kernel definition of a QFT, then if we set
$$I[P] = W\left(P - P(0,L), I[L]\right)$$
for any parametrix P, the functionals $I[P]$ satisfy the parametrix definition of quantum field theory presented in this section. Further, every collection $I[P]$ of functionals satisfying the parametrix definition arises uniquely in this way.

The proof of this theorem will be presented in Section 12, after the proof of Theorem A.

9. Extracting the singular part of the weights of Feynman graphs

9.1. In order to construct the local counterterms needed for theorem A, we need a method for extracting the singular part of the finite-dimensional integrals $w_\gamma(P(\varepsilon, L), I)$ attached to Feynman graphs. This section will describe such a method, which relies on an understanding of the behaviour of of the functions $w_\gamma(P(\varepsilon, L), I)$ as $\varepsilon \to 0$. We will see that $w_\gamma(P(\varepsilon, L), I)$ has a small ε asymptotic expansion
$$w_\gamma(P(\varepsilon, L), I)(a) \simeq \sum g_i(\varepsilon) \Phi_i(L, a),$$
where the $\Phi_i(L, a)$ are well-behaved functions of the field a and of L. Further, the $\Phi_i(L, a)$ have a small L asymptotic expansion in terms of local action functionals.

The functionals $g_i(\varepsilon)$ appearing in this expansion are of a very special form: they are periods of algebraic varieties. For the purposes of this book, the fact that these functions are periods is not essential. Thus, the reader may skip the definition of periods without any loss. However, given the interest in the relationship between periods and quantum field theory in the mathematics literature (see (KZ01), for example) I felt that this point is worth mentioning.

Before we state the theorem precisely, we need to explain what makes a function of ε a period.

9.2. According to Kontsevich and Zagier (KZ01), most or all constants appearing in mathematics should be periods.

DEFINITION 9.2.1. *A number $\alpha \in \mathbb{C}$ is a* period *if there exists an algebraic variety X of dimension d, a normal crossings divisor $D \subset X$, and a form $\omega \in \Omega^d(X)$ vanishing on D, all defined over $\mathbb{Q}$; and a homology class*

$$\gamma \in H_d(X(\mathbb{C}), D(\mathbb{C})) \otimes \mathbb{Q}$$

such that

$$\alpha = \int_\gamma \omega.$$

We are interested in periods which depend on a variable $\varepsilon \in (0, 1)$. Such families of periods arise from families of algebraic varieties over the affine line.

Suppose we have the following data.
(1) an algebraic variety X over $\mathbb{Q}$;
(2) a normal crossings divisor $D \subset X$;
(3) a Zariski open subset $U \subset \mathbb{A}^1_{\mathbb{Q}}$, defined over $\mathbb{Q}$, such that $U(\mathbb{R})$ containts $(0, 1)$.
(4) a smooth map $X \to U$, of relative dimension d, also defined over $\mathbb{Q}$, whose restriction to D is flat.
(5) a relative d-form $\omega \in \Omega^d(X/U)$, defined over $\mathbb{Q}$, and vanishing along D.
(6) a homology class $\gamma \in H_d((X_{1/2}(\mathbb{C}), D_{1/2}(\mathbb{C})), \mathbb{Q})$, where $X_{1/2}$ and $D_{1/2}$ are the fibres of X and D over $1/2 \in U(\mathbb{R})$. We assume that γ is invariant under the complex conjugation map on the pair $(X_{1/2}(\mathbb{C}), D_{1/2}(\mathbb{C}))$.

Let us assume that the maps

$$X(\mathbb{C}) \to U(\mathbb{C})$$
$$D(\mathbb{C}) \to U(\mathbb{C})$$

are locally trivial fibrations. For $t \in (0, 1) \subset U(\mathbb{R})$, we will let $X_t(\mathbb{C})$ and $D_t(\mathbb{C})$ denote the fibre over $s(t) \in U(\mathbb{R})$.

We can transfer the homology class $\gamma \in H_*(X_{1/2}(\mathbb{C}), D_{1/2}(\mathbb{C}))$ to any fibre $(X_t(\mathbb{C}), D_t(\mathbb{C}))$ for $t \in (0, 1)$. This allows us to define a function f on $(0, 1)$ by

$$f(t) = \int_{\gamma_t} \omega_t.$$

The function f is real analytic. Further, f is real valued, because the cycle

$$\gamma_t \in H_d(X_{1/2}(\mathbb{C}), D_{1/2}(\mathbb{C}))$$

is invariant under complex conjugation.

DEFINITION 9.2.2. *Let $\mathscr{P}_{\mathbb{Q}}((0,1)) \subset C^\infty((0,1))$ be the subalgebra of functions of this form. Elements of this subalgebra will be called rational periods.*

Note that $\mathscr{P}_{\mathbb{Q}}((0,1))$ is, indeed, a subalgebra; the sum of functions corresponds to the disjoint union of the pairs (X, D) of algebraic varieties, and the product of functions corresponds to fibre product of the algebraic varieties (X, D) over U. Further, $\mathscr{P}_{\mathbb{Q}}((0,1))$ is of countable dimension over $\mathbb{Q}$.

Note that, if f is a rational period, then, for every rational number $t \in \mathbb{Q} \cap (0,1)$, $f(t)$ is a period in the sense of Kontsevich and Zagier.

DEFINITION 9.2.3. *Let*
$$\mathscr{P}((0,1)) = \mathscr{P}_{\mathbb{Q}}((0,1)) \otimes \mathbb{R} \subset C^\infty((0,1))$$
be the real vector space spanned by the space of rational periods. Elements of $\mathscr{P}((0,1))$ will be called periods.

9.3. Now we are ready to state the theorem on the small ε asymptotic expansions of the functions $w_\gamma(P(\varepsilon, L), I)(a)$.

We will regard the functional $w_\gamma(P(\varepsilon, L), I)(a)$ as a function of the three variables ε, L and $a \in C^\infty(M)$. It is an element of the space of functionals
$$\mathscr{O}(C^\infty(M), C^\infty((0,1)_\varepsilon) \otimes C^\infty((0,\infty)_L))).$$

The subscripts ε and L indicate the coordinates on the intervales $(0,1)$ and $(0,\infty)$. If we fix ε but allow L and a to vary, we get a functional
$$w_\gamma(P(\varepsilon, L), I) \in \mathscr{O}(C^\infty(M), C^\infty((0,\infty)_L)).$$

This is a topological vector space; we are interested in the behaviour of $w_\gamma(P(\varepsilon, L), I)$ as $\varepsilon \to 0$.

The following theorem describes the small ε behaviour of $w_\gamma(P(\varepsilon, L), I)$.

THEOREM 9.3.1. *Let $I \in \mathscr{O}_{loc}(C^\infty(M))[[\hbar]]$ be a local functional, and let γ be a connected stable graph.*

(1) There exists a small ε asymptotic expansion
$$w_\gamma(P(\varepsilon, L), I) \simeq \sum_{i=0}^{\infty} g_i(\varepsilon) \Psi_i$$
where the
$$g_i \in \mathscr{P}((0,1)_\varepsilon)$$
are periods, and $\Psi_i \in \mathscr{O}(C^\infty(M), C^\infty((0,\infty)_L))$.

The precise meaning of "asymptotic expansion" is as follows: there is a non-decreasing sequence $d_R \in \mathbb{Z}$, indexed by $R \in \mathbb{Z}_{>0}$, such that $d_R \to \infty$ as $R \to \infty$, and such that for all R,
$$\lim_{\varepsilon \to 0} \varepsilon^{-d_R} \left(w_\gamma(P(\varepsilon, L), I) - \sum_{i=0}^{R} g_i(\varepsilon) \Psi_i \right) = 0.$$

where the limit is taken in the topological vector space
$$\mathscr{O}(C^\infty(M), C^\infty((0,\infty)_L)).$$

(2) The $g_i(\varepsilon)$ appearing in this asymptotic expansion have a finite order pole at zero: for each i there is a k such that $\lim_{\varepsilon \to 0} \varepsilon^k g_i(\varepsilon) = 0$.

(3) Each Ψ_i appearing in the asymptotic expansion above has a small L asymptotic expansion of the form
$$\Psi_i \simeq \sum_{j=0}^{\infty} f_{i,j}(L) \Phi_{i,j}$$
where the $\Phi_{i,j}$ are local action functionals, that is, elements of $\mathscr{O}_l(C^\infty(M))$; and each $f_{i,k}(L)$ is a smooth function of $L \in (0,\infty)$.

This theorem is proved in Appendix 1. All of the hard work required to construct counterterms is encoded in this theorem. The theorem is proved by using the small t asymptotic expansion for the heat kernel to approximate each $w_\gamma(P(\varepsilon, L), I)$ for small ε.

9.4. These results allow us to extract the singular part of the finite-dimensional integral $w_\gamma(P(\varepsilon, L), I)$. Of course, the notion of singular part is not canonical, but depends on a choice.

DEFINITION 9.4.1. *Let $\mathscr{P}((0,1))_{\geq 0} \subset \mathscr{P}((0,1))$ be the subspace of those functions f of ε which are periods and which admit a limit as $\varepsilon \to 0$.*

A renormalization scheme *is a complementary subspace*
$$\mathscr{P}((0,1))_{<0} \subset \mathscr{P}((0,1))$$
to $\mathscr{P}((0,1))_{\geq 0}$.

Thus, once we have chosen a renormalization scheme we have a direct sum decomposition
$$\mathscr{P}((0,1)) = \mathscr{P}((0,1))_{\geq 0} \oplus \mathscr{P}((0,1))_{<0}.$$

A renormalization scheme is the data one needs to define the singular part of a function in $\mathscr{P}((0,1))$.

DEFINITION 9.4.2. *If $f \in \mathscr{P}((0,1))$, define the singular $\mathrm{Sing}(f)$ of f to be the projection of f onto $\mathscr{P}((0,1))_{<0}$.*

9.5. We can now use this definition to extract the singular part of the functions $w_\gamma(P(\varepsilon, L), I)$. As before, let us think of $w_\gamma(P(\varepsilon, L), I)$ as a distribution on M^k. Then, Theorem 9.3.1 shows that $w_\gamma(P(\varepsilon, L), I)$ has a small ε asymptotic expansion of the form
$$w_\gamma(P(\varepsilon, L), I) \simeq \sum_{i=0}^{\infty} g_i(\varepsilon) \Phi_i$$
where the $g_i(\varepsilon)$ are periods and the $\Phi_i \in \mathrm{Hom}(C^\infty(M^k), C^\infty((0,\infty)_L))$. Theorem 9.3.1 also implies that there exists an $N \in \mathbb{Z}_{\geq 0}$ such that, for all $n > N$, $g_n(\varepsilon)$ admits an $\varepsilon \to 0$ limit.

Denote the Nth partial sum of the asymptotic expansion by
$$\Psi_N(\varepsilon) = \sum_{i=0}^{N} g_i(\varepsilon)\Phi_i.$$
Then, we can define the singular part of $w_\gamma(P(\varepsilon,L),I)$ simply by
$$\operatorname{Sing}_\varepsilon w_\gamma(P(\varepsilon,L),I) = \operatorname{Sing}_\varepsilon \Psi_N(\varepsilon) = \sum_{i=0}^{N} \left(\operatorname{Sing}_\varepsilon g_i(\varepsilon)\right) \Phi_i.$$
This singular part is independent of N, because if N is increased the function $\Psi_N(\varepsilon)$ is modified only by the addition of functions of ε which are periods and which tend to zero as $\varepsilon \to 0$.

Theorem 9.3.1 implies that $\operatorname{Sing}_\varepsilon w_\gamma(P(\varepsilon,L),I)$ has the following properties.

THEOREM 9.5.1. *Let $I \in \mathscr{O}_{loc}(C^\infty(M))[[\hbar]]$ be a local functional, and let γ be a connected stable graph.*
 (1) $\operatorname{Sing}_\varepsilon w_\gamma(P(\varepsilon,L),I)$ *is a finite sum of the form*
$$\operatorname{Sing}_\varepsilon w_\gamma(P(\varepsilon,L),I) = \sum f_i(\varepsilon)\Phi_i$$
where
$$\Phi_i \in \mathscr{O}_{loc}(C^\infty(M), C^\infty((0,\infty)_L)),$$
and
$$f_i \in \mathscr{P}((0,1))_{<0}$$
are purely singular periods.
 (2) *The limit*
$$\lim_{\varepsilon \to 0} \left(w_\gamma(P(\varepsilon,L),I) - \operatorname{Sing}_\varepsilon w_\gamma(P(\varepsilon,L),I)\right)$$
exists in the topological vector space $\mathscr{O}_{loc}(C^\infty(M), C^\infty((0,\infty)_L))$.
 (3) *Each Φ_i appearing in the finite sum above has a small L asymptotic expansion*
$$\Phi_i \simeq \sum_{j=0}^{\infty} f_{i,j}(L)\Psi_{i,j}$$
where $\Psi_{i,j} \in \mathscr{O}_{loc}(C^\infty(M))$ is local, and $f_{i,j}(L)$ is a smooth function of $L \in (0,\infty)$.

10. Constructing local counterterms

10.1. The heart of the proof of theorem A is the construction of local counterterms for a local interaction $I \in \mathscr{O}_{loc}(C^\infty(M))$. This construction is simple and inductive, without the complicated graph combinatorics of the BPHZ algorithm.

The theorem on the existence of local counterterms is the following.

62   2. THEORIES, LAGRANGIANS AND COUNTERTERMS

THEOREM 10.1.1. *There exists a unique series of local counterterms*
$$I_{i,k}^{CT}(\varepsilon) \in \mathcal{O}_{loc}(C^\infty(M)) \otimes_{alg} \mathscr{P}((0,\infty))_{<0},$$
for all $i > 0, k \geq 0$, with $I_{i,k}^{CT}$ homogeneous of degree k as a function of $a \in C^\infty(M)$, such that, for all $L \in (0, \infty]$, the limit
$$\lim_{\varepsilon \to 0} W\left(P(\varepsilon, L), I - \sum_{i,k} \hbar^i I_{i,k}^{CT}(\varepsilon)\right)$$
exists.

Here the symbol $\otimes_{alg}$ denotes the algebraic tensor product, so only finite sums are allowed.

10.2. We will construct our counterterms using induction on the genus and number of external edges of the Feynman graphs. Later, we will see a very short (though unilluminating) construction of the counterterms, which does not use Feynman graphs. For reasons of exposition, we will introduce the Feynman graph picture first.

Let $\Gamma_{i,k}$ denote the set of all stable graphs of genus i with k external edges. Let
$$W_{i,k}(P, I) = \sum_{\gamma \in \Gamma_{i,k}} w_\gamma(P(\varepsilon, L), I).$$
Thus,
$$W(P, I) = \sum \hbar^i W_{i,k}(P, I).$$

If the graph γ is of genus zero, and so is a tree, then $\lim_{\varepsilon \to 0} w_\gamma(P(\varepsilon, L), I)$ converges. Thus, the first counterterms we need to construct are those from graphs with one loop and one external edge. Let us define
$$I_{1,1}^{CT}(\varepsilon, L) = \operatorname{Sing}_\varepsilon W_{1,1}(P(\varepsilon, L), I).$$
Section 9 explains the meaning of the singular part $\operatorname{Sing}_\varepsilon$ of $W_{1,1}(P(\varepsilon, L), I)$.

We need to check that this has the desired properties. It is immediate from the definition that
$$W_{1,1}\left(P(\varepsilon, L), I - \hbar I_{1,1}^{CT}(\varepsilon, L)\right) = W_{1,1}(P(\varepsilon, L), I) - I_{1,1}^{CT}(\varepsilon, L)$$
and so the limit $\lim_{\varepsilon \to 0} W_{1,1}\left(P(\varepsilon, L), I - \hbar I_{1,1}^{CT}(\varepsilon, L)\right)$ exists.

Next, we need to check that

LEMMA. *$I_{1,1}^{CT}(\varepsilon, L)$ is local.*

First we will show that

LEMMA. *$I_{1,1}^{CT}(\varepsilon, L)$ is independent of L.*

Figure 5 illustrates $\frac{d}{dL} W_{1,1}(P(\varepsilon, L), I)$. This expression is non-singular, as it is obtained by contracting the distribution $I_{0,3}$ on $C^\infty(M^3)$ with the smooth function $K_L \in C^\infty(M^2)$. Therefore $I_{1,1}^{CT}(\varepsilon, L)$, which we defined to be the singular part of $W_{1,1}(P(\varepsilon, L), I)$, is independent of L.

$$\frac{d}{dL} W_{1,1}(P(\varepsilon,L), I)$$

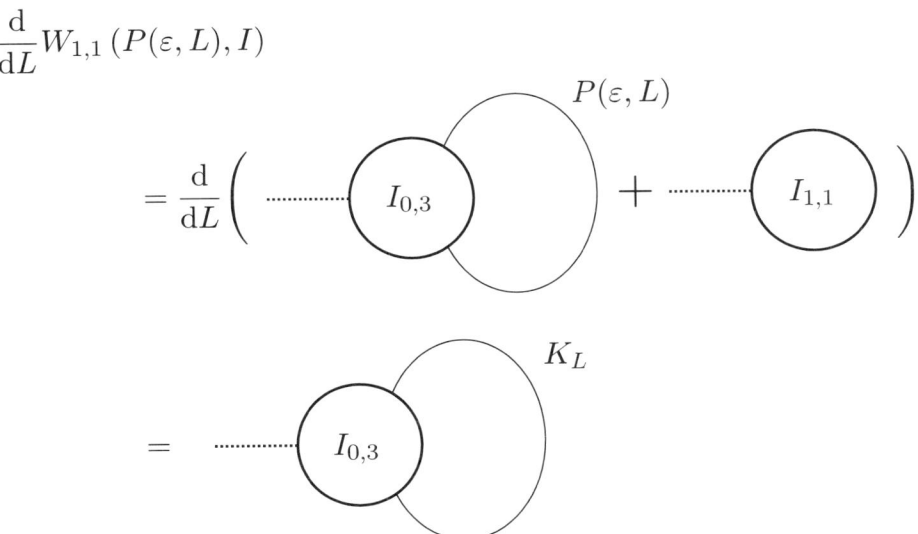

FIGURE 5. Explanation of why $I_{1,1}^{CT}(\varepsilon, L)$ is independent of L.

Since $I_{1,1}^{CT}(\varepsilon, L)$ is independent of L, to verify that it is local we only need to examine the behaviour of $W_{1,1}(P(\varepsilon, L), I)$ at small L. Theorem 9.5.1 implies that $\text{Sing}_{\varepsilon} W_{1,1}(P(\varepsilon, L), I)$ has a small L asymptotic expansion in terms of local action functionals. Therefore, since we know $I_{1,1}^{CT}(\varepsilon, L)$ is independent of L, it follows that it is local.

Now that we know $I_{1,1}^{CT}(\varepsilon, L)$ is independent of L, we will normally drop L from the notation.

10.3. The next step is to construct $I_{1,2}^{CT}(\varepsilon, L)$. However, it is just as simple to construct directly the general counterterm $I_{i,k}^{CT}(\varepsilon, L)$. Let us lexicographically order the set $\mathbb{Z}_{\geq 0} \times \mathbb{Z}_{\geq 0}$, so that $(i,k) < (j,l)$ if $i < j$ or if $i = j$ and $k < l$. Let us write

$$W_{<(i,k)}(P, I) = \sum_{(j,l)<(i,k)} \hbar^j W_{j,l}(P, I).$$

We can write this expression in terms of stable graphs, as follows. Let $\Gamma_{<(i,k)}$ denote the set of stable graphs with genus smaller than i, or with genus equal to i and fewer than k external edges. Then,

$$W_{<(i,k)}(P, I) = \sum_{\gamma \in \Gamma_{<(i,k)}} \frac{\hbar^{g(\gamma)}}{|\text{Aut}\,\gamma|} w_\gamma(P, I).$$

Let us assume, by induction, we have constructed counterterms $I_{j,l}^{CT}(\varepsilon)$ for all $(j,l) < (i,k)$, with the following properties.

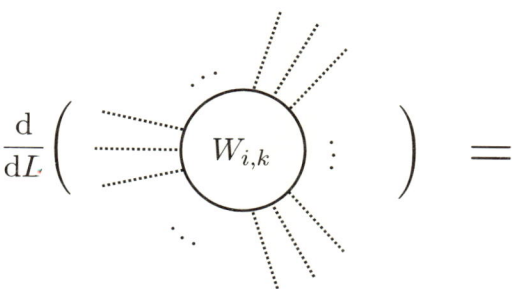

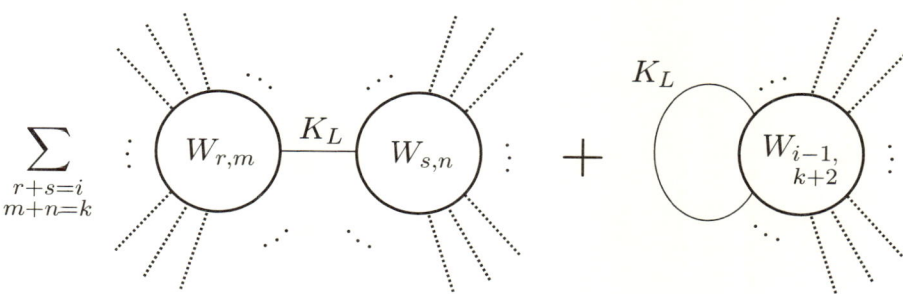

FIGURE 6. This diagram, which is just the infinitesimal version of the renormalization group equation, explains why $I_{i,k}^{CT}(\varepsilon, L)$ is independent of L. In this diagram, $W_{*,*}$ is shorthand for
$$W_{*,*}\left(P(\varepsilon, L), I - \sum_{(a,b)<(i,k)} I_{a,b}^{CT}(\varepsilon)\right).$$

(1) For all L, the limit
$$\lim_{\varepsilon \to 0} W_{<(i,k)}\left(P(\varepsilon, L), I - \sum_{(j,l)<(i,k)} \hbar^j I_{j,l}^{CT}(\varepsilon)\right)$$
exists.
(2) The counterterms $I_{j,l}^{CT}(\varepsilon)$ are local for $(j,k) < (i,k)$.

Then, we define the counterterm
$$I_{i,k}^{CT}(\varepsilon, L) = \mathrm{Sing}_\varepsilon W_{i,k}\left(P(\varepsilon, L), I - \sum_{(j,l)<(i,k)} \hbar^j I_{j,l}^{CT}(\varepsilon)\right).$$

10. CONSTRUCTING LOCAL COUNTERTERMS

As before, it is immediate that

$$W_{i,k}\left(P(\varepsilon,L), I - \sum_{(j,l)<(i,k)} \hbar^j I_{j,l}^{CT}(\varepsilon) - \hbar^i I_{i,k}^{CT}(\varepsilon,L)\right)$$

$$= W_{i,k}\left(P(\varepsilon,L), I - \sum_{(j,l)<(i,k)} \hbar^j I_{j,l}^{CT}(\varepsilon)\right) - I_{i,k}^{CT}(\varepsilon,L)$$

is non-singular.

What we need to show is that
(1) $I_{i,k}^{CT}(\varepsilon,L)$ is independent of L.
(2) $I_{i,k}^{CT}(\varepsilon,L)$ is local.

As before, the second statement follows from the first one. To show independence of L it suffices to show that

(†) $$\frac{\mathrm{d}}{\mathrm{d}L} W_{i,k}\left(P(\varepsilon,L), I - \sum_{(j,l)<(i,k)} \hbar^j I_{j,l}^{CT}(\varepsilon)\right)$$

is non-singular, that is, the limit of this expression as $\varepsilon \to 0$ exists. A proof of this is illustrated in figure 6. In this diagram, an expression for (†) is given as a sum over graphs whose vertices are labeled by

$$W_{j,l}\left(P(\varepsilon,L), I - \sum_{(r,s)\leq(j,l)} \hbar^r I_{r,s}^{CT}(\varepsilon)\right)$$

for various $(j,l) < (i,k)$. We know by induction that these are non-singular. On the unique edge of these graphs we put the heat kernel K_L, which is smooth. Thus, the expression resulting from each graph is non-singular.

10.4. In fact, the use of Feynman graphs is not at all necessary in this proof; I first came up with the argument by thinking of $W(P,I)$ as

$$W(P,I) = \hbar \log\left(\exp(\hbar \partial_P) \exp(I/\hbar)\right).$$

From this expression, one can see that

$$W(P(L,L'), W(P(\varepsilon,L), I)) = W(P(\varepsilon,L'), I)$$
$$W_{i,k}(P(\varepsilon,L), I) = W_{i,k}\left(P(\varepsilon,L), I_{<(i,k)}\right) + I_{i,k}.$$

The first identity is obvious, and the second identity can be seen (for instance) using the expression of $W(P,I)$ in terms of Feynman graphs.

These two identities are all that is really needed for the argument. Indeed, suppose we have constructed counterterms $I_{r,s}^{CT}(\varepsilon)$ for all $(r,s) < (i,k)$, such that, for all L, the limit

$$\lim_{\varepsilon \to 0} W_{<(i,k)}\left(P(\varepsilon,L), I - \sum_{(r,s)<(i,k)} \hbar^r I_{r,s}^{CT}(\varepsilon)\right)$$

66   2. THEORIES, LAGRANGIANS AND COUNTERTERMS

exists. Let us suppose, by induction, that these counterterms are local and independent of L.

Then, we define the next counterterm by

$$I_{i,k}^{CT}(L,\varepsilon) = \mathrm{Sing}_\varepsilon W_{i,k}\left(P(\varepsilon,L), I - \sum_{(r,s)<(i,k)} \hbar^r I_{r,s}^{CT}(\varepsilon)\right).$$

The identity

$$W_{i,k}\left(P(\varepsilon,L), I - \sum_{(r,s)<(i,k)} \hbar^r I_{r,s}^{CT}(\varepsilon) - \hbar^i I_{i,k}^{CT}(L,\varepsilon)\right) =$$

$$W_{i,k}\left(P(\varepsilon,L), I - \sum_{(r,s)<(i,k)} \hbar^r I_{r,s}^{CT}(\varepsilon)\right) - I_{i,k}^{CT}(L,\varepsilon)$$

shows that the limit

$$\lim_{\varepsilon \to 0} W_{\leq(i,k)}\left(P(\varepsilon,L), I - \sum_{(r,s)<(i,k)} \hbar^r I_{r,s}^{CT}(\varepsilon) - \hbar^i I_{i,k}^{CT}(L,\varepsilon)\right)$$

exists.

To show locality of the counterterm $I_{i,k}^{CT}(L,\varepsilon)$, it suffices, as before, to show that it is independent of L. If $L' > L$, we have

$$I_{i,k}^{CT}(L',\varepsilon) = \mathrm{Sing}_\varepsilon W_{i,k}\left(P(\varepsilon,L'), I - \sum_{(r,s)<(i,k)} \hbar^r I_{r,s}^{CT}(\varepsilon)\right)$$

$$= \mathrm{Sing}_\varepsilon W_{i,k}\left(P(L,L'), W\left(P(\varepsilon,L), I - \sum_{(r,s)<(i,k)} \hbar^r I_{r,s}^{CT}(\varepsilon)\right)\right)$$

$$= \mathrm{Sing}_\varepsilon W_{i,k}\left(P(L,L'), W_{<(i,k)}\left(P(\varepsilon,L), I - \sum_{(r,s)<(i,k)} \hbar^r I_{r,s}^{CT}(\varepsilon)\right)\right.$$

$$\left. + \hbar^i W_{i,k}\left(P(\varepsilon,L), I - \sum_{(r,s)<(i,k)} \hbar^r I_{r,s}^{CT}(\varepsilon)\right)\right)$$

$$= \mathrm{Sing}_\varepsilon W_{i,k}\left(P(L,L'), W_{<(i,k)}\left(P(\varepsilon,L), I - \sum_{(r,s)<(i,k)} \hbar^r I_{r,s}^{CT}(\varepsilon)\right)\right)$$

$$+ \mathrm{Sing}_\varepsilon W_{i,k}\left(P(\varepsilon,L), I - \sum_{(r,s)<(i,k)} \hbar^r I_{r,s}^{CT}(\varepsilon)\right)$$

Since
$$W_{i,k}\left(P(L,L'), W_{<(i,k)}\left(P(\varepsilon,L), I - \sum_{(r,s)<(i,k)} \hbar^r I_{r,s}^{CT}(\varepsilon)\right)\right)$$
is non-singular, the last equation reduces to
$$I_{i,k}^{CT}(L',\varepsilon) = \mathrm{Sing}_\varepsilon W_{i,k}\left(P(\varepsilon,L), I - \sum_{(r,s)<(i,k)} \hbar^r I_{r,s}^{CT}(\varepsilon)\right)$$
$$= I_{i,k}^{CT}(L,\varepsilon)$$
as desired.

Thus, $I_{i,k}^{CT}(L,\varepsilon)$ is independent of L, and can be written as $I_{i,k}^{CT}(\varepsilon)$; and we can continue the induction.

11. Proof of the main theorem

Let $I \in \mathscr{O}_{loc}^+(C^\infty(M))[[\hbar]]$ be a local action functional. What we have shown so far allows us to construct the corresponding theory. Let
$$W^R(P(0,L), I) = \lim_{\varepsilon \to 0} W\left(P(\varepsilon, L), I - I^{CT}(\varepsilon)\right)$$
be the renormalized version of the renormalization group flow from scale 0 to scale L applied to I. The theory associated to I has scale L effective interaction $W^R(P(0,L), I)$. It is easy to see that this satisfies all the axioms of a perturbative quantum field theory, as given in Definition 7. The locality axiom of the definition of perturbative quantum field theory follows from Theorem 9.5.1.

We need to show the converse: that to each theory there is a corresponding local action functional. This is a simple induction. Let $\{I[L]\}$ denote a collection of effective interactions defining a theory. Let us assume, by induction on the lexicographic ordering as before, that we have constructed local action functionals $I_{r,s}$ for $(r,s) < (i,k)$ such that
$$W_{a,b}^R\left(P(0,L), \sum_{(r,s)<(i,k)} \hbar^r I_{r,s}\right) = I_{a,b}[L]$$
for all $(a,b) < (i,k)$.

Then, the infinitesimal renormalization group equation implies that
$$W_{i,k}^R\left(P(0,L), \sum_{(r,s)<(i,k)} \hbar^r I_{r,s}\right) - I_{i,k}[L]$$
is independent of L. The locality axiom for the theory $\{I[L]\}$ – which says that each $I_{i,k}[L]$ has a small L asymptotic expansion in terms of local action

functionals – implies that this quantity is local. Thus, let

$$I_{i,k} = I_{i,k}[L] - W_{i,k}^R \left(P(0,L), \sum_{(r,s)<(i,k)} \hbar^r I_{r,s} \right).$$

Then,

$$W_{a,b}^R \left(P(0,L), \sum_{(r,s)\leq(i,k)} \hbar^r I_{r,s} \right) = I_{a,b}[L]$$

for all $(a,b) \leq (i,k)$, and we can continue the induction.

11.1. Recall that we defined $\mathscr{T}^{(n)}$ to be the space of theories defined modulo $\hbar^{n+1}$, and $\mathscr{T}^{(\infty)}$ to be the space of theories. Thus, we have shown that the choice of renormalization scheme sets up a bijection between $\mathscr{T}^{(\infty)}$ and functionals $I \in \mathscr{O}_{loc}^+(C^\infty(M))[[\hbar]]$, and between $\mathscr{T}^{(n)}$ and functionals $I \in \mathscr{O}_{loc}^+(C^\infty(M))[\hbar]/\hbar^{n+1}$.

The more fundamental statement of the theorem on the bijection between theories and local action functionals is that the map $\mathscr{T}^{(n+1)} \to \mathscr{T}^{(n)}$ makes $\mathscr{T}^{(n+1)}$ into a principal bundle for the group $\mathscr{O}_{loc}(C^\infty(M))$, in a canonical way (independent of any arbitrary choices, such as that of a renormalization scheme). In this subsection we will use the bijection constructed above to prove this statement.

The bijection between theories and Lagrangians shows that the map $\mathscr{T}^{(n+1)} \to \mathscr{T}^{(n)}$ is surjective; this is the only place the bijection is used.

To show that $\mathscr{T}^{(n+1)} \to \mathscr{T}^{(n)}$ is a principal bundle, suppose that $\{I[L]\}$, $\{J[L]\}$ are two theories which are defined modulo $\hbar^{n+2}$ and which agree modulo $\hbar^{n+1}$.

Let $I_0[L] \in \mathscr{T}^{(0)}$ be the classical theory corresponding to both $I[L]$ and $J[L]$. Let us consider the tangent space to $\mathscr{T}^{(0)}$ at $I[L]$, which includes infinitesimal deformations of classical theories which do not have to be at least cubic. More precisely, let $T_{I_0[L]}\mathscr{T}^{(0)}$ be the set of $H[L] \in \mathscr{O}(C^\infty(M))$ such that

$$I_0[L] + \delta H[L]$$

satisfies the classical renormalization group equation modulo δ^2,

$$I_{0,i}[L] + \delta H_i[L] = W_{0,i}\left(P(\varepsilon,L), \sum I_{0,j}[\varepsilon] + \delta H_j[\varepsilon]\right) \text{ modulo } \delta^2,$$

and which satisfy the usual locality axiom, that $H[L]$ has a small L asymptotic expansion in terms of local action functionals.

The bijection between classical theories and local action functionals is canonical, independent of the choice of renormalization scheme. This is true even if we include non-cubic terms in our effective interaction, as long as these non-cubic terms are accompanied by nilpotent parameters.

Thus, we have a canonical isomorphism of vector spaces

$$T_{I_0[L]}\mathscr{T}^{(0)} \cong \mathscr{O}_{loc}(C^\infty(M)).$$

The following lemma now shows that $\mathscr{T}^{(n+1)} \to \mathscr{T}^{(n)}$ is a torsor for $\mathscr{O}_{loc}(C^\infty(M))$.

LEMMA 11.1.1. *Let $I, J \in \mathscr{T}^{(n+1)}$ be theories which agree in $\mathscr{T}^{(n)}$. Then, the functional*
$$I_0[L] + \delta \hbar^{-(n+1)}(I[L] - J[L]) \in \mathscr{O}(C^\infty(M))$$
satisfies the classical renormalization group equation modulo δ^2, and so defines an element of
$$T_{I_0[L]}(\mathscr{T}^{(0)}) \cong \mathscr{O}_{loc}(C^\infty(M)).$$

Note that $\hbar^{-(n+1)}(I[L] - J[L])$ is well-defined as $I[L]$ and $J[L]$ agree modulo $\hbar^{n+1}$.

PROOF. This is a simple calculation. □

12. Proof of the parametrix formulation of the main theorem

In this page we will prove the equivalence of the definition of theory based on arbitrary parametrices, explained in Section 8 with the definition based on the heat kernel. Since this result is not used elsewhere in the book, I will not give all the details.

Thus, suppose we have a theory in the heat kernel sense, given by a family $I[L]$ of effective interactions satisfying the renormalization group equation and the locality axiom. If P is a parametrix, let us define a functional $I[P]$ by
$$I[P] = W(P - P(0,L), I[L]) \in \mathscr{O}(C^\infty(M))[[\hbar]].$$
Since $P(0, L)$ and P are both parametrices for the operator $D + m^2$, the difference between them is smooth. Thus, $W(P - P(0, L), I[L])$ is well-defined.

LEMMA 12.0.1. *The collection of effective interactions $\{I[P]\}$, defined for each parametrix P, defines a theory using the parametrix definition of theory.*

PROOF. To prove this, we need to verify the following.

(1) If P' is another parametrix, then
$$I[P'] = W(P' - P, I[P])$$
(this is the version of the renormalization group equation for the definition of theory based on parametrices.

(2) By choosing a parametrix P with support close to the diagonal, we can make the distribution
$$I_{i,k}[P] \in \mathcal{D}(M^k)_{S_k}$$
on M^k supported as close as we like to the small diagonal.

(3) The functional $I[P]$ has smooth first derivative. Recall, as explained in Section 8, that this means the following. There is a continuous linear map
$$C^\infty(M) \to \mathscr{O}(C^\infty(M))[[\hbar]]$$
$$\phi \mapsto \frac{\mathrm{d}I[P]}{\mathrm{d}\phi}.$$

Saying that $I[\Phi]$ has smooth first derivative means that this map extends to a continuous linear map
$$\mathcal{D}(M) \to \mathscr{O}(C^\infty(M))[[\hbar]]$$
where $\mathcal{D}(M)$ is the space of distributions on M.

In order to verify these properties, it is convenient to choose a renormalization scheme, so that we can write
$$I[L] = \lim_{\varepsilon \to 0} W\left(P(\varepsilon, L), I - I^{ct}(\varepsilon)\right).$$

Now let us choose a cut-off function $\Psi \in C^\infty(M \times M)$ which is 1 in a neighbourhood of the diagonal, and 0 outside a small neighbourhood of the diagonal. Then, $\Psi P(0, L)$ is a parametrix, which agrees with $P(0, L)$ near the diagonal. Thus, we have
$$I[\Psi P(0, L)] = \lim_{\varepsilon \to 0} W\left(\Psi P(\varepsilon, L), I - I^{ct}(\varepsilon)\right).$$

Since I and $I^{ct}(\varepsilon)$ are local, if we choose the function Ψ to be supported in a very small neighbourhood of the diagonal then we can ensure that $I_{i,k}[\Psi P(0, L)]$ is supported within an arbitrarily small neighbourhood of the small diagonal in M^k.

From this and from the identity
$$I[P] = W\left(P - \Psi P(0, L), I[\Psi P(0, L)]\right)$$
we can check that, by choosing the parametrix P to have support very close to the diagonal, we can ensure that $I_{i,k}[P]$ has support arbitrarily close to the small diagonal in M^k. The point is that the combinatorial formulae for $W(\Phi, J)$ (where $\Phi \in C^\infty(M^2)$ and $J \in \mathscr{O}(C^\infty(M^2))^+[[\hbar]]$) allows one to control the support of $W_{i,k}(\Phi, J)$ in terms of the support of J and that of Φ.

This shows that $I[P]$ satisfies the first two properties we want to verify. It remains to check that $I[P]$ has smooth first derivative. This follows from the fact that all local functionals have smooth first derivative, and that the property of having smooth first derivative is preserved under the renormalization group flow.

□

Now, we need to prove the converse. This follows from the following lemma.

LEMMA 12.0.2. *Let $\{I_{(r,s)}[P]\}$ and $\{I'_{(r,s)}[P]\}$ be two parametrix theories, defined for all $(r,s) \le (I,K)$. Suppose that $I_{(r,s)}[P] = I'_{(r,s)}[P]$ if $(r,s) < (I,K)$. Then,*

$$J_{(I,K)} = I_{(I,K)}[P] - I'_{(I,K)}[P]$$

is a local functional, that is, an element of $\mathscr{O}_{loc}(C^\infty(M))$.

PROOF. Note that the renormalization group equation implies that the functional $J_{(I,K)}$ is independent of P. The locality axiom implies that $J_{(I,K)}$ is supported on the small diagonal of M^K. Further, $J_{(I,K)}$ has smooth first derivative. Any distribution $J \in \mathcal{D}(M^K)_{S_K}$ which is supported on the small diagonal and which has smooth first derivative is a local functional. □

13. Vector-bundle valued field theories

We would like to have a bijection between theories and Lagrangians for a more general class of field theories. The most general set-up we will need is when the fields are sections of some vector bundle on a manifold; and the interactions depend smoothly on some additional supermanifold. In this section we will explain how to do this on a compact manifold.

DEFINITION 13.0.1. *A* nilpotent graded manifold *is the following data:*
  (1) A smooth manifold with corners X,
  (2) A sheaf A of commutative superalgebras over the sheaf of algebras C_X^∞,

satisfying the following properties:
  (1) A is locally free of finite rank as a C_X^∞-module. In other words, A is the sheaf of sections of some super vector bundle on X.
  (2) A is equipped with an ideal I such that $A/I = C_X^\infty$, and $I^k = 0$ for some $k > 0$. The ideal I, its powers I^l, and the quotient sheaves A/I^l, are all required to be locally free sheaves of C_X^∞-modules.

The algebra $\Gamma(X,A)$ of C^∞ global sections of A will be denoted by $\mathscr{A}$.

Everything in this section will come in families, parameterized by a nilpotent graded manifold (X,A).

13.1. We are interested in vector-valued theories on a compact manifold M. As in the case of scalar field theories, we will fix the data of the free theory, which gives us our propagator; and then consider possible interacting theories which deform this.

The following definition aims to be broad enough to capture all of the free field theories used in this book, and in future applications. Unfortunately, it is not particularly transparent.

DEFINITION 13.1.1. *A* free theory *on a manifold M consists of the following data.*

(1) A super vector bundle E over the field $\mathbb{R}$ or $\mathbb{C}$ on M, equipped with a direct sum decomposition $E = E_1 \oplus E_2$ into the spaces of propagating and non-propagating fields, respectively. We will denote the space of smooth global sections of E or E_i by $\mathscr{E}, \mathscr{E}_i$ respectively.

We will let
$$\mathscr{E}_1^! = \Gamma(M, E_1^\vee \otimes \mathrm{Dens}(M)).$$
There is an inclusion
$$\mathscr{E}_1^! \subset \mathscr{E}_1^\vee.$$

(2) An even, $\mathscr{A}$-linear, order two differential operator
$$\mathrm{D}_{\mathscr{E}_1} : \mathscr{E}_1 \otimes \mathscr{A} \to \mathscr{E}_1 \otimes \mathscr{A}$$
(where the tensor product is the completed projective tensor product).

$\mathrm{D}_{\mathscr{E}_1}$ must be a generalized Laplacian, which means that the symbol
$$\sigma(\mathrm{D}_{\mathscr{E}_1}) \in \Gamma(T^*M, \mathrm{Hom}(E, E)) \otimes \mathscr{A}$$
must be the identity on E times a smooth family of Riemannian metrics
$$g \in C^\infty(T^*M) \otimes C^\infty(X).$$
(Recall that $C^\infty(X) \subset \mathscr{A}$ is a subalgebra, as $\mathscr{A}$ is the global sections of a bundle of algebras on X).

(3) A differential operator
$$D' : \mathscr{E}_1^! \to \mathscr{E}_1.$$
This operator is required to be symmetric: the formal adjoint
$$(D')^* : \mathscr{E}_1^! \to \mathscr{E}_1$$
is required to be equal to D'.

(4) Let
$$\mathrm{D}_{\mathscr{E}_1}^* : \mathscr{E}_1^! \to \mathscr{E}_1^!$$
be the formal adjoint of $\mathrm{D}_{\mathscr{E}_1}$. We require that
$$D' \mathrm{D}_{\mathscr{E}_1}^* = \mathrm{D}_{\mathscr{E}_1} D'.$$

We will abuse notation and refer to the entirely of the data of a free theory on M as $\mathscr{E}$.

The most basic example of this definition is the free scalar field theory, as considered in chapter 2. There, the space $\mathscr{E}_1$ of fields is $C^\infty(M)$. The space $\mathscr{E}_2$ of non-interacting fields is 0. The operator $\mathrm{D}_{\mathscr{E}_1}$ is the usual positive-definite Laplacian operator $C^\infty(M) \to C^\infty(M)$. The operator D' is the identity, where we have used the Riemannian volume element to trivialize the bundle of densities on M, and so to identify $\mathscr{E}_1$ with $\mathscr{E}_1^!$.

More interesting examples will be presented in chapter 5, when we consider the Batalin-Vilkovisky formalism. The use of graded vector bundles will be essential in the BV formalism.

13.2. The space $\mathscr{E}_2$ of non-propagating fields is introduced into this definition with an eye to future applications: none of the examples treated in this book will have non-propagating fields. Thus, the reader will lose nothing by ignoring the space of non-propagating fields.

For those who are interested, however, let me briefly explain the reason for considering non-propagating fields. Let us consider a free scalar field theory on a Riemannian manifold M with metric g_0. Let us consider perturbing the metric to $g_0 + h$. The action of the scalar field theory is given by $S(\phi) = \int \phi \triangle_{g_0+h} \phi$, as usual. Notice that the action is not just local as a function of the field ϕ, but also local as a function of the perturbation h of the metric. However, h is *not* treated as a quantum field, only as a classical field: we do not consider integration over the space of metrics.

In this situation, the tensor h is said to be a background field, or (in the terminology adopted here) a non-propagating field.

In this example, the space of fields is
$$\mathscr{E} = C^\infty(M) \oplus \Gamma(M, \operatorname{Sym}^2 TM)$$
where the space $\mathscr{E}_1$ of propagating fields is $C^\infty(M)$, and the space $\mathscr{E}_2$ of non-propagating fields is $\Gamma(M, \operatorname{Sym}^2 TM)$.

The action of the theory is $S(\phi) = \int_M \phi \triangle_{g_0+h} \phi$. This action can, as usual, be split into quadratic and interacting parts:
$$S(\phi) = \int_M \phi \triangle_{g_0} \phi + \int_M \phi \left(\triangle_{g_0+h} - \triangle_{g_0}\right) \phi.$$

The quadratic part of the action is the only part relevant to the definition of a free field theory as presented above. As we see, the quadratic part only depends on the propagating field ϕ, and not on h. However, the interaction term depends both on ϕ and h. It is a general feature of the interaction terms that they must have some dependence on ϕ; they canot be functions just of the non-propagating field h. We will build this into our definition of interactions in the presence of non-propagating fields shortly.

If we take the free theory associated to this example – given by discarding the interacting terms in the action S – we fit it into the general definition as follows. The operator $D_{\mathscr{E}_1}$ is the Laplacian $\triangle_{g_0} : C^\infty(M) \to C^\infty(M)$. The operator D' is the identity map $C^\infty(M) \to C^\infty(M)$.

13.3. Now that we have the general definition of a free field theory, we can start to define the concept of effective interaction in this context. First, we have to define the heat kernel.

If the manifold M is compact, there is a unique heat kernel
$$K_t \in \mathscr{E}_1^! \otimes \mathscr{E}_1 \otimes C^\infty(\mathbb{R}_{>0}) \otimes \mathscr{A}$$
for the operator D_{E_1}.

Composing with the operator D' gives an element
$$D' K_t \in \mathscr{E}_1 \otimes \mathscr{E}_1 \otimes C^\infty(\mathbb{R}_{>0}) \otimes \mathscr{A}.$$
We will view this as an element of $\mathscr{E} \otimes \mathscr{E} \otimes C^\infty(\mathbb{R}_{>0})$.

74   2. THEORIES, LAGRANGIANS AND COUNTERTERMS

The adjointness properties of the differential operators D', $D_{\mathscr{E}_1}$ imply that $D'K_t$ is symmetric.

The propagator for the theory is
$$P(\varepsilon, L) = \int_\varepsilon^L D'K_t \in \mathscr{E}^{\otimes 2} \otimes \mathscr{A}$$
which is again symmetric.

Note that unless we impose additional positivity conditions on the operator D_{E_1}, the heat kernel K_t may not exist at $t = \infty$; thus, the propagator $P(\varepsilon, \infty)$ may not exist. In almost all examples, however, the operator $D_{\mathscr{E}_1}$ is positive, and so the heat kernel K_∞ does exist.

If we specialize the case of the free scalar field theory on a Riemannian manifold (M, g_0), then, as we have seen, $\mathscr{E}_1 = C^\infty(M)$, $D_{\mathscr{E}_1} = \triangle_{g_0}$ is the non-negative Laplacian for the metric g_0. In this example the operator D' is the identity. Thus, the propagator prescribed by this general definition coincides with the propagator presented in our earlier analysis of the free field theory.

13.4. As before, we can define the algebra
$$\mathscr{O}(\mathscr{E}, \mathscr{A}) = \prod \mathrm{Hom}(\mathscr{E}^{\otimes n}, \mathscr{A})_{S_n}$$
of all functionals on $\mathscr{E}$ with values in $\mathscr{A}$. Here Hom denotes the space of continuous linear maps. The properties of the symmetric monoidal category of nuclear spaces, as detailed in Appendix 2, show that
$$\mathscr{O}(\mathscr{E}, \mathscr{A}) = \left(\prod_n \mathrm{Sym}^n \mathscr{E}^\vee\right) \otimes \mathscr{A}.$$
There is a subspace
$$\mathscr{O}_l(\mathscr{E}, \mathscr{A}) \subset \mathscr{O}(\mathscr{E}, \mathscr{A})$$
of $\mathscr{A}$-valued local action functionals, defined as follows.

DEFINITION 13.4.1. *A functional $\Phi \in \mathscr{O}(\mathscr{E}, \mathscr{A})$ is a local action functional if, when we expand Φ as a sum $\Phi = \sum \Phi_n$ of its homogeneous components, each*
$$\Phi_n : \mathscr{E}^{\otimes n} \to \mathscr{A}$$
can be written in the form
$$\Phi_n(e_1, \ldots, e_n) = \sum_{j=1}^k \int_M (D_{1,j} e_1) \cdots (D_{n,j} e_n) \mathrm{d}\mu$$
where
$$\mathrm{d}\mu \in \mathrm{Densities}(M)$$
is some volume element on M, and each
$$D_{i,j} : \mathscr{E} \otimes \mathscr{A} \to C^\infty(M) \otimes \mathscr{A}$$
is an $\mathscr{A}$-linear differential operator.

Note that $\mathcal{O}_l(\mathcal{E}, \mathcal{A})$ is *not* a closed subspace. However, as we will see in Appendix 2, $\mathcal{O}_{loc}(\mathcal{E}, \mathcal{A})$ has a natural topology making it into a complete nuclear space, and a module over $\mathcal{A}$ in the symmetric monoidal category of nuclear spaces.

Our interactions will be elements of
$$\mathcal{O}_l(\mathcal{E}, \mathcal{A})[[\hbar]].$$
We would like to allow our interactions to have quadratic and linear terms modulo $\hbar$. However, we require that these quadratic terms are accompanied by elements of the nilpotent ideal
$$\mathscr{I} = \Gamma(X, I) \subset \mathcal{A}$$
(recall that $\mathcal{A}/\mathscr{I} = C^\infty(X)$). If we don't impose this condition, we will encounter infinite sums.

Thus, let us denote by
$$\mathcal{O}^+(\mathcal{E}, \mathcal{A})[[\hbar]] \subset \mathcal{O}(\mathcal{E}, \mathcal{A})[[\hbar]]$$
the subset of those functionals which are at least cubic modulo the ideal generated by $\mathscr{I}$ and $\hbar$.

Then, the renormalization group operator
$$W(P(\varepsilon, L), I)$$
$$= \hbar \log \left(\exp(\hbar \partial_{P(\varepsilon, L)}) \exp(I/\hbar) \right) : \mathcal{O}^+(\mathcal{E}, \mathcal{A})[[\hbar]] \to \mathcal{O}^+(\mathcal{E}, \mathcal{A})[[\hbar]]$$
is well-defined.

Because we now allow quadratic and linear interaction terms modulo $\hbar$, the Feynman graph expansion of this expression involves univalent and bivalent genus 0 vertices. However, each such vertex is accompanied by an element of the ideal $\mathscr{I}$ of $\mathcal{A}$. Since this ideal is nilpotent, there is a uniform bound on the number of such vertices that can occur, so there are no infinite sums.

DEFINITION 13.4.2. *A theory is given by a collection of even elements*
$$I[L] \in \mathcal{O}^+(\mathcal{E}, C^\infty((0, \infty)_L) \otimes \mathcal{A})[[\hbar]],$$
such that

(1) *The renormalization group equation*
$$I[L'] = W\left(P(L, L'), I[L]\right)$$
holds.

(2) *Each $I_{(i,k)}[L]$ has a small L asymptotic expansion*
$$I_{(i,k)}[L](e) \simeq \sum \Psi_r(e) f_r(L)$$
where $\Psi_r \in \mathcal{O}_l(\mathcal{E})$ are local action functionals.

Let $\mathscr{T}^{(\infty)}(\mathcal{E})$ denote the space of such theories, and let $\mathscr{T}^{(n)}(\mathcal{E})$ denote the space of theories defined modulo $\hbar^{n+1}$, so that $\mathscr{T}^{(\infty)}(\mathcal{E}) = \varprojlim \mathscr{T}^{(n)}(\mathcal{E})$.

Let me explain more precisely what I mean by saying there is a small L asymptotic expansion

$$I_{i,k}[L] \simeq \sum_{j \in \mathbb{Z}_{\geq 0}} g_r(L) \Phi_r.$$

Without loss of generality, we can require that the local action functionals Φ_r appearing here are homogeneous of degree k in the field $e \in \mathscr{E}$.

Recall that $\mathscr{A}$ is the global sections of some bundle of algebras A on a manifold with corners X. Let A_x denote the fibre of A at $x \in X$. For every element $\alpha \in \mathscr{A}$, let $\alpha_x \in A_x$ denote the value of α at x.

The statement that there is such an asymptotic expansion means that there is a non-decreasing sequence $d_R \in \mathbb{Z}$, tending to infinity, such that for all R, for all fields $e \in \mathscr{E}$, for all $x \in X$,

$$\lim_{L \to 0} L^{-d_R} \alpha_x \left(I_{i,k}[L](e) - \sum_{r=0}^{R} g_r(L) \Phi_r(e) \right) = 0$$

in the finite dimensional vector space A_x.

Then, as before, the theorem is:

THEOREM 13.4.3. *The space $\mathscr{T}^{(n+1)}(\mathscr{E})$ has the structure of a principal $\mathscr{O}_{loc}(\mathscr{E}, \mathscr{A})$ bundle over $\mathscr{T}^{(n)}(\mathscr{E})$, in a canonical way. Further, $\mathscr{T}^{(0)}(\mathscr{E})$ is canonically isomorphic to the space $\mathscr{O}^+_{loc}(\mathscr{E}, \mathscr{A})$ of $\mathscr{A}$-valued local action functionals on $\mathscr{E}$ which are at least cubic modulo the ideal $\mathscr{I} \subset \mathscr{A}$.*

Further, the choice of renormalization scheme gives rise to a section $\mathscr{T}^{(n)}(\mathscr{E}) \to \mathscr{T}^{(n+1)}(\mathscr{E})$ of each torsor, and so a bijection between $\mathscr{T}^{(\infty)}(\mathscr{E})$ and the space

$$\mathscr{O}^+_{loc}(\mathscr{E}, \mathscr{A})[[\hbar]]$$

of local action functionals with values in $\mathscr{A}$, which are at least cubic modulo $\hbar$ and modulo the ideal $\mathscr{I} \subset \mathscr{A}$.

PROOF. The proof is essentially the same as before. The extra difficulties are of two kinds: working with an auxiliary parameter space X introduces extra analytical difficulties, and working with quadratic terms in our interaction forces us to use Artinian induction with respect to the powers of the ideal $\mathscr{I} \subset \mathscr{A}$.

For simplicity, I will only give the proof when the effective interactions $I[L]$ are all at least cubic modulo $\hbar$. The argument in the general case is the same, except that we also must perform Artinian induction with respect to the powers of the ideal $\mathscr{I} \subset \mathscr{A}$.

As before, we will prove the renormalization scheme dependent version of the theorem, saying that there is a bijection between $\mathscr{T}^{(\infty)}(\mathscr{E})$ and $\mathscr{O}^+_{loc}(\mathscr{E}, \mathscr{A})[[\hbar]]$. The renormalization scheme independent formulation is an easy corollary.

Let us start by showing how to construct a theory associated to a local interaction

$$I = \sum \hbar^i I_{(i,k)} \in \mathscr{O}_{loc}(\mathscr{E}, \mathscr{A})[[\hbar]].$$

We will assume that $I_{(0,k)} = 0$ if $k < 3$.

The argument is essentially the same as the argument we gave earlier. We will perform induction on the set $\mathbb{Z}_{\geq 0} \times \mathbb{Z}_{\geq 0}$ with the lexicographical order: $(i,k) < (r,s)$ if $i < r$ or if $i = r$ and $k < s$.

Suppose, by induction, we have constructed counterterms
$$I^{CT}_{(i,k)}(\varepsilon) \in \mathcal{O}_{loc}(\mathcal{E}, C^\infty((0,\infty)_\varepsilon) \otimes \mathcal{A})$$
for all $(i,k) < (I,K)$. The $I^{CT}_{(i,k)}$ are supposed, by induction, to have the following properties:

(1) Each $I^{CT}_{(i,k)}(\varepsilon)$ is homogeneous of degree k as a function of the field $e \in \mathcal{E}$.

(2) Each $I^{CT}_{(i,k)}(\varepsilon)$ is required to be a finite sum
$$I^{CT}_{(i,k)}(\varepsilon) = \sum g_r(\varepsilon) \Phi_r$$
where $g_r(\varepsilon) \in C^\infty((0,\infty)_\varepsilon)$ and $\Phi_r \in \mathcal{O}_{loc}(\mathcal{E}, \mathcal{A})$. Each $g_r(\varepsilon)$ is required to have a finite order pole at 0; that is, $\lim_{\varepsilon \to 0} \varepsilon^k g_r(\varepsilon) = 0$ for some $k > 0$.

(3) Recall that $\mathcal{A}$ is the space of global sections of a vector bundle A on X. For any element $\alpha \in A$, let $\alpha_x \in A_x$ denote its value at $x \in X$.

We require that, for all $L \in (0, \infty)$ and all $x \in X$, the limit
$$\lim_{\varepsilon \to 0} W_{(r,s)} \left(P(\varepsilon, L), I - \sum_{(i,k) \leq (r,s)} \hbar^i I^{CT}_{(i,k)}(\varepsilon) \right)_x$$
exists in the topological vector space $\mathrm{Hom}(\mathcal{E}^{\otimes r}, \mathbb{R}) \otimes A_x$. Here, $\mathrm{Hom}(\mathcal{E}^{\otimes r}, \mathbb{R})$ is given the strong topology (i.e. the topology of uniform convergence on bounded subsets).

Now we need to construct the next counterterm $I^{CT}_{(I,K)}(\varepsilon)$. We would like to define
$$I^{CT}_{(I,K)}(\varepsilon) = \mathrm{Sing}_\varepsilon W_{(I,K)} \left(P(\varepsilon, L), I - \sum_{(r,s) < (I,K)} \hbar^r I_{(r,s)}^{CT}(\varepsilon) \right).$$

In order to be able to define the singular part like this, we need to know that
$$W_{(I,K)} \left(P(\varepsilon, L), I - \sum_{(r,s) < (I,K)} \hbar^r I_{(r,s)}^{CT}(\varepsilon) \right)$$
has a nice small ε asymptotic expansion. The required asymptotic expansion is provided by the following theorem, proved in Appendix 1.

THEOREM 13.4.4. *For all graphs γ, and all $I \in \mathcal{O}_{loc}(\mathcal{E}, \mathcal{A})[[\hbar]]$, there exist local action functionals $\Phi_r \in \mathcal{O}_{loc}(\mathcal{E}, \mathcal{A} \otimes C^\infty((0,\infty)_L)$ and functions g_r*

in the space $P((0,\infty)_\varepsilon) \subset C^\infty((0,\infty)_\varepsilon)$ of periods, such that for all $l \in \mathbb{Z}_{\geq 0}$, there is a small ε asymptotic expansion

$$\frac{\partial^l}{\partial^l L} w_\gamma(P(\varepsilon, L), I) \simeq \frac{\partial^l}{\partial^l L} \sum_{r \geq 0} g_r(\varepsilon) \Phi_r.$$

This means that for each l and each $x \in X$, there exists a non-decreasing sequence $d_R \in \mathbb{Z}$, tending to ∞, such that

$$\lim_{\varepsilon \to 0} \varepsilon^{-d_R} \frac{\partial^l}{\partial^l L} \left(w_\gamma(P(\varepsilon, L), I) - \sum_{r \geq 0} g_r(\varepsilon) \Phi_r \right)_x = 0$$

in the topological vector space $\mathrm{Hom}(\mathscr{E}^{\otimes T(\gamma)}, \mathbb{R}) \otimes A_x$. Here, $\mathrm{Hom}(\mathscr{E}^{\otimes T(\gamma)}, \mathbb{R})$ is given the strong topology (i.e. the topology of uniform convergence on bounded subsets).

Also, each g_r has a finite order pole at $\varepsilon = 0$. This means that the limit $\lim_{\varepsilon \to 0} \varepsilon^k g_r(\varepsilon)$ is 0 for some $k > 0$.

Further, each

$$\Phi_r(L, e) \in \mathrm{Hom}(\mathscr{E}^{\otimes T(\gamma)}, C^\infty((0, \infty)_L))$$

has a small L asymptotic expansion

$$\Phi_r \simeq \sum h_s(L) \Psi_{r,s}$$

where each $\Psi_{r,s} \in \mathscr{O}_{loc}(\mathscr{E}, \mathscr{A})$. The definition of small L asymptotic expansion is in the same sense as before: there exists a non-decreasing sequence $d_S \in \mathbb{Z}_{\geq 0}$, tending to ∞, such that, for all $x \in X$, and all $S \in \mathbb{Z}_{\geq 0}$,

$$\lim_{L \to 0} L^{-d_S} \left(\Phi_r(e) - \sum_{s=0}^{S} h_s(L) \Psi_{r,s}(e) \right)_x = 0$$

in the topological vector space $\mathrm{Hom}(\mathscr{E}^{\otimes T(\gamma)}, \mathbb{R}) \otimes A_x$.

It follows from this theorem that it makes sense to define the next counterterm $I^{CT}_{(I,K)}$ simply by

$$I^{CT}_{(I,K)}(\varepsilon, L) = \mathrm{Sing}_\varepsilon \left(W_{(I,K)} \left(P(\varepsilon, L), I - \sum_{(r,s) < (I,K)} \hbar^r I^{CT}_{(r,s)}(\varepsilon) \right) \right).$$

We would like to show the following properties of $I^{CT}_{(I,K)}(\varepsilon, L)$.

(1) $I^{CT}_{(I,K)}(\varepsilon, L)$ is independent of L.
(2) $I^{CT}_{(I,K)}(\varepsilon)$ is local, that is, it is an element of

$$\mathscr{O}_{loc}(\mathscr{E}, \mathscr{A}) \otimes_{alg} C^\infty((0,\infty)_\varepsilon).$$

(3) For all $L \in (0, \infty)$ and all $x \in X$, the limit

$$\lim_{\varepsilon \to 0} W_{(I,K)} \left(P(\varepsilon, L), I - \sum_{(i,k) \le (I,K)} \hbar^i I^{CT}_{(i,k)}(\varepsilon) \right)_x$$

exists in the topological vector space $\mathrm{Hom}(\mathscr{E}^{\otimes K}, \mathbb{R}) \otimes A_x$.

If we can prove these three properties, we can continue the induction.

The third property is immediate: it follows from the small ε asymptotic expansion of Theorem 13.4.4.

For the first property, observe that

$$\frac{\partial}{\partial L} \mathrm{Sing}_\varepsilon W_{(I,K)} \left(P(\varepsilon, L), I - \sum_{(r,s) < (I,K)} \hbar^r I^{CT}_{(r,s)}(\varepsilon) \right)$$

$$= \mathrm{Sing}_\varepsilon \frac{\partial}{\partial L} W_{(I,K)} \left(P(\varepsilon, L), I - \sum_{(r,s) < (I,K)} \hbar^r I^{CT}_{(r,s)}(\varepsilon) \right).$$

This follows from the fact that the small ε asymptotic expansion proved in Theorem 13.4.4 commutes with taking L derivatives.

Thus, to show that $I^{CT}_{(I,K)}$ is independent of L, it suffices to show that, for all L, all $x \in X$, and all $e \in \mathscr{E}$,

$$\frac{\partial}{\partial L} W_{(I,K)} \left(P(\varepsilon, L), I - \sum_{(r,s) < (I,K)} \hbar^r I^{CT}_{(r,s)}(\varepsilon) \right)_x (e) \in A_x$$

has an $\varepsilon \to 0$ limit. This is immediate by induction, using the renormalization group equation.

The small L asymptotic expansion in Theorem 13.4.4 now implies that $I^{CT}_{(I,K)}(\varepsilon)$ is local.

Thus, $I^{CT}_{(I,K)}(\varepsilon)$ satisfies all the required properties, and we can continue our induction, to construct all the counterterms.

So far, we have shown how to construct the effective interactions

$$I[L] = \lim_{\varepsilon \to 0} W \left(P(\varepsilon, L), I - \sum_{(r,s)} \hbar^r I^{CT}_{(r,s)}(\varepsilon) \right).$$

These effective interactions satisfy the renormalization group equation. It is immediate from Theorem 13.4.4 that the $I[L]$ satisfy the locality axiom, so that they define a theory.

Now, we need to prove the converse. This is again an inductive argument. Suppose we have a theory, given by effective interactions

$$I[L] \in \mathscr{O}^+(\mathscr{E}, \mathscr{A})[[\hbar]].$$

Suppose that we have a local action functional
$$J = \sum_{(i,k)<(I,K)} \hbar^i J_{(i,k)} \in \mathscr{O}^+_{loc}(\mathscr{E},\mathscr{A})[[\hbar]]$$
with associated effective interactions $J[L]$.

Suppose, by induction, that
$$J_{(i,k)}[L] = I_{(i,k)}[L]$$
for all $(i,k) < (I,K)$.

We need to find some $J'_{(I,K)}$ such that, if we set $J' = J + \hbar^I J'_{(I,K)}$, then
$$J'_{(I,K)}[L] = I_{(I,K)}[L].$$
We simply let
$$J'_{(I,K)} = I_{(I,K)}[L] - J_{(I,K)}[L].$$
The renormalization group equation implies that $J'_{(I,K)}$ is independent of L. It is automatic that
$$J'_{(I,K)}[L] = I_{(I,K)}[L].$$
Finally, the fact that both $J_{(I,K)}[L]$ and $J_{(I,K)}[L]$ satisfy the small L asymptotics axiom of a theory implies that $J'_{(I,K)}$ is local.

□

14. Field theories on non-compact manifolds

A second generalization is to non-compact manifolds. On non-compact manifolds, we don't just have ultraviolet divergences (arising from small scales) but infrared divergences, which arise when we try to integrate over the non-compact manifold.

However, by imposing a suitable infrared cut-off, we will find a notion of theory on a non-compact manifold; and a bijection between theories and Lagrangians.

The infrared cut-off we impose is rather brutal: we multiply the propagator by a cut-off function so that it becomes supported on a small neighbourhood of the diagonal in M^2. However, the notion of theory is independent of the cut-off chosen.

We will also show that there are restriction maps, allowing one to restrict a theory on a non-compact manifold M to any open subset U. This allows one to define a sheaf of theories on any manifold. If we are on a compact manifold, global sections of this sheaf are theories in the sense we defined before.

The sheaf-theoretic statement of our main theorem asserts that this sheaf is isomorphic to the sheaf of local action functionals. As always, this isomorphism depends on the choice of a renormalization scheme. The renormalization scheme independent statement of the theorem is that the sheaf of theories defined modulo $\hbar^{n+1}$ is a torsor over the sheaf of theories defined modulo $\hbar^n$, for the sheaf of local action functionals on M.

14. FIELD THEORIES ON NON-COMPACT MANIFOLDS

14.1. Now let us start defining the notion of theory on a possibly non-compact manifold M.

As in Section 13, we will fix a nilpotent graded manifold (X, A), and a family of free field theories on M parameterized by (X, A). The free field theory is given by a super vector bundle E on M, whose space of global sections will be denoted by $\mathscr{E}$; together with various auxiliary data detailed in Definition 13.1.1.

We will use the following notation. We will let $\mathscr{E}$ denote the space of all smooth sections of E, $\mathscr{E}_c$ the space of compactly supported sections, $\overline{\mathscr{E}}$ the space of distributional sections and $\overline{\mathscr{E}}_c$ the space of compactly supported distributional sections. The bundle $E \otimes \mathrm{Dens}_M$ will be denoted $E^!$. We will use the notation $\mathscr{E}^!$, $\mathscr{E}_c^!$, $\overline{\mathscr{E}}^!$ and $\overline{\mathscr{E}}_c^!$ to denote spaces of smooth, compactly supported, distributional and compactly supported distributional sections of the bundle $E^!$. sections of We will let $\overline{\mathscr{E}}$ denote the space of distributional sections of E, and $\overline{\mathscr{E}}_c$ denote the compactly supported distributional sections. With this notation, $\mathscr{E}^\vee = \overline{\mathscr{E}}_c^!$, $\mathscr{E}_c^\vee ee = \overline{\mathscr{E}}^!$, and so on.

14.2.

DEFINITION 14.2.1. *Let M, X be topological spaces. A subset $C \subset M^n \times X$ is called* proper *if each of the projection maps $\pi_i : M^n \times X \to M \times X$ is proper when restricted to C.*

We say that a section of a vector bundle on $M^n \times X$ has proper support if its support is a proper subset of $M^n \times X$.

Recall that we can identify the space $\mathscr{O}(\mathscr{E}_c, \mathscr{A})$ of $\mathscr{A}$-valued functions on $\mathscr{E}_c$ (modulo constants) with the completed symmetric algebra

$$\mathscr{O}(\mathscr{E}_c, \mathscr{A}) = \prod_{n>0} \mathrm{Sym}^n(\overline{\mathscr{E}}^!) \otimes \mathscr{A},$$

where we have identified $\overline{\mathscr{E}}^!$ – the space of distributional sections of the bundle $E^!$ on M – with $(\mathscr{E}_c)^\vee$. We will let

$$\mathscr{O}_p(\mathscr{E}_c, \mathscr{A}) \subset \mathscr{O}(\mathscr{E}_c, \mathscr{A})$$

be the subset consisting of those functionals Φ each of whose Taylor components

$$\Phi_n \in \mathrm{Sym}^n \overline{\mathscr{E}}^!$$

have proper support. We are only interested in functions on $\mathscr{E}_c$ modulo constants.

Note that $\mathscr{O}_p(\mathscr{E}, \mathscr{A})$ is *not* an algebra; the direct product of two properly supported distributions does not necessarily have proper support.

Note also that every $\mathscr{A}$-valued local action functional $I \in \mathscr{O}_l(\mathscr{E}, \mathscr{A})$ is an element of $\mathscr{O}_p(\mathscr{E}, \mathscr{A})$.

2. THEORIES, LAGRANGIANS AND COUNTERTERMS

14.3. Recall that the super vector bundle E on M has additional structure, as described in Definition 13.1.1. This data includes a decomposition $E = E_1 \oplus E_2$ and a generalized Laplacian $\mathrm{D}_{\mathscr{E}_1} : \mathscr{E}_1 \to \mathscr{E}_1$.

We are interested in the heat kernel for the Laplacian Δ_{E_1}. On a compact manifold M, this is unique, and is an element of

$$K_t \in \mathscr{E}_1^! \otimes \mathscr{E}_1 \otimes C^\infty(\mathbb{R}_{>0}) \otimes \mathscr{A}.$$

On a non-compact manifold, there are many heat kernels, corresponding to various boundary conditions. In addition, such heat kernels may grow on the boundary of the non-compact manifold in ways which are difficult to control.

To remedy this, we will introduce the concept of *fake heat kernel*. A fake heat kernel is something which solves the heat equation but *only up to the addition of a smooth kernel*.

DEFINITION 14.3.1. *A* fake heat kernel *is a smooth section*

$$K_t \in \mathscr{E}_1^! \otimes \mathscr{E}_1 \otimes C^\infty(\mathbb{R}_{>0}) \otimes \mathscr{A}$$

with the following properties.

(1) *K_t extends, at $t = 0$ to a distribution. Thus, K_t extends to an element of*

$$\overline{\mathscr{E}}_1^! \otimes \overline{\mathscr{E}}_1 \otimes C^\infty(\mathbb{R}_{\geq 0}) \otimes \mathscr{A}$$

Further, K_0 is the kernel for the identity map $\mathscr{E}_1 \to \mathscr{E}_1$.

(2) *The support*

$$\operatorname{Supp} K_t \subset M \times M \times \mathbb{R}_{>0} \times X$$

is proper. Recall that this means that both projection maps from $\operatorname{Supp} K_t$ to $M \times \mathbb{R}_{>0} \times X$ are proper.

(3) *The heat kernel K_t satisfies the heat equation up to exponentially small terms in t. More precisely, $\frac{\mathrm{d}}{\mathrm{d}t} K_t + \mathrm{D}_{\mathscr{E}_1} K_t$ extends to a smooth section*

$$\frac{\mathrm{d}}{\mathrm{d}t} K_t + \mathrm{D}_{\mathscr{E}_1} K_t \in \mathscr{E}_1^! \otimes \mathscr{E}_1^! \otimes C^\infty(\mathbb{R}_{\geq 0}) \otimes \mathscr{A}$$

which vanishes at $t = 0$, with all derivatives in t and on M, faster than any power of t.

(4) *The heat kernel K_t admits a small t asymptotic expansion which can be written, in normal coordinates x, y near the diagonal of M, in the form*

$$K_t \simeq t^{-\dim M/2} e^{-\|x-y\|^2/t} \sum_{i \geq 0} t^i \Phi_i(x,y).$$

Let me explain more carefully about what I mean by a small t asymptotic expansion. We will let

$$K_t^N = \psi(x,y) t^{-\dim M/2} e^{-\|x-y\|^2/t} \sum_{i=0}^{N} t^i \Phi_i(x,y)$$

be the Nth partial sum of this asymptotic expansion (where we have introduced a cut-off $\psi(x,y)$ so that K_t^N is zero outside of a small neighbourhood of the diagonal).

Then, we require that for all compact subsets $C \subset M \times M \times X$,
$$\left\| \partial_t^k \left(K_t - K_t^N \right) \right\|_{C,l,m} = O(t^{N - \dim M/2 - l/2 - k}).$$
where $\|-\|_{l,m}$ refers to the norm on $\Gamma(M, E^\vee \otimes \text{Dens}(M)) \otimes \mathscr{E} \otimes \mathscr{A}$ where we differentiate l times on $M \times M$, m times on X, and take the supremum over the compact subset C.

These estimates are the same as the ones satisfied by the actual heat kernel on a compact manifold M, as detailed in (BGV92).

The fake heat kernel K_t satisfies the heat equation up to a function which vanishes faster than any power of t. This implies that the asymptotic expansion
$$t^{-\dim M/2} e^{-\|x-y\|^2/t} \sum_{i \geq 0} t^i \Phi_i(x, y).$$
of K_t must be a formal solution to the heat equation (in the sense described in (BGV92), Section 2.5). This characterizes the functions $\Phi_i(x, y)$ (defined in a neighbourhood of the diagonal) uniquely.

LEMMA 14.3.2. *Let K_t, $\widetilde{K}_t$ be two fake heat kernels. Then $K_t - \widetilde{K}_t$ extends across $t = 0$ to a smooth kernel, that is,*
$$K_t - \widetilde{K}_t \in \mathscr{E}^! \otimes \mathscr{E} \otimes \mathscr{A} \otimes C^\infty(\mathbb{R}_{\geq 0}).$$
Further, $K_t - \widetilde{K}_t$ vanishes to all orders at $t = 0$.

PROOF. This is clear from the existence and uniqueness of the small t asymptotic expansion. □

LEMMA 14.3.3. *A fake heat kernel always exists.*

PROOF. The techniques of (BGV92) allow one to construct a fake heat kernel by approximating it with the partial sums of the asymptotic expansion. □

14.4. Let us suppose that M is an open subset of a compact manifold N, and that the free field theory $\mathscr{E}$ on M is restricted from one, say $\mathscr{F}$, on N.

Then, the restriction of the heat kernel for $\mathscr{F}$ to M is an element
$$\widetilde{K}_t \in \mathscr{E}_1^! \otimes \mathscr{E}_1 \otimes C^\infty(\mathbb{R}_{>0}) \otimes \mathscr{A},$$
which satisfies all the axioms of a fake heat kernel except that of requiring proper support.

If $\Psi \in C^\infty(M \times M)$ is a smooth function, with proper support, which takes value 1 on a neighbourhood of the diagonal, then
$$\Psi \widetilde{K}_t \in \mathscr{E}_1^! \otimes \mathscr{E}_1 \otimes C^\infty(\mathbb{R}_{>0}) \otimes \mathscr{A}$$
is a fake heat kernel.

14.5. The bundle E_1 of propagating fields is a direct summand of the bundle E of all fields. Thus, the fake heat kernel K_t can be viewed as an element of the space
$$\mathscr{E}^! \otimes \mathscr{E} \otimes C^\infty(\mathbb{R}_{>0}) \otimes \mathscr{A}.$$
The operator $D' : \mathscr{E}_1^! \to \mathscr{E}_1$ extends to an operator $\mathscr{E}^! \to \mathscr{E}$. For $0 < \varepsilon < L < \infty$, let us define the fake propagator by
$$P(\varepsilon, L) = \int_\varepsilon^L \sigma\left((D' \otimes 1)K_t\right) \mathrm{d}t \in \mathrm{Sym}^2 \mathscr{E} \otimes \mathscr{A},$$
where $\sigma : \mathscr{E}^{\otimes 2} \to \mathrm{Sym}^2 \mathscr{E}$ is the symmetrization map.

We would like to define a theory, for the fake heat kernel K_t, to be a collection of effective interactions
$$I[L] \in \mathscr{O}_p^+(\mathscr{E}_c, \mathscr{A})[[\hbar]]$$
satisfying the renormalization group equation defined using the fake propagator $P(\varepsilon, L)$. Recall that the subscript p in the expression $\mathscr{O}_p(\mathscr{E}_c, \mathscr{A})$ indicates that we are looking at functionals which are distributions with proper support. A function in $\mathscr{O}(\mathscr{E}_c, \mathscr{A})$ is in $\mathscr{O}_p(\mathscr{E}_c, \mathscr{A})$ if its Taylor components, which are $\mathscr{A}$-valued distributions on M^n, have support which is a proper subset of $M^n \times X$.

In order to do this, we need to know that the renormalization group flow is well defined. Thus, we need to check that if we construct the weights attached to graphs using the propagators $P(\varepsilon, L)$ and interactions $I \in \mathscr{O}_p(\mathscr{E}_c, \mathscr{A})^+[[\hbar]]$.

LEMMA 14.5.1. *Let γ be a connected graph with at least one tail. Let $P \in \mathscr{E} \otimes \mathscr{E} \otimes \mathscr{A}$ have proper support. Suppose for each vertex v of γ we have a continuous linear map*
$$I_v : \mathscr{E}_c^{\otimes H(v)} \to \mathscr{A}$$
which has proper support.

Then,
$$w_\gamma(P, \{I_v\}) : \mathscr{E}_c^{\otimes T(\gamma)} \to \mathscr{A}$$
is well defined, and is a continuous linear map with proper support.

PROOF. Let $f \in \mathscr{E}_c^{\otimes T(\gamma)}$. The expression $w_\gamma(P, \{I_v\})(f)$ is defined by contracting the tensor
$$f \otimes_{e \in E(\gamma)} P_e \in \mathscr{E}^{\otimes H(\gamma)}$$
given by putting a propagator P on each edge of γ and f at the tails of γ, with the distribution
$$\otimes_{v \in V(\gamma)} I_v : \mathscr{E}_c^{\otimes H(\gamma)} \to \mathscr{A}.$$
Neither quantity has compact support. However, the restrictions we placed on the supports of f, P and each I_v means that the intersection of the support of $\otimes I_v$ with that of $f \otimes P_e$ is a compact subset of $M^{H(\gamma)}$. Thus, we can contract $\otimes I_v$ with $f \otimes P_e$ to get an element of $\mathscr{A}$.

The resulting linear map
$$\mathscr{E}^{\otimes T(\gamma)} \to \mathscr{A}$$
$$f \mapsto w_\gamma(P, \{I_v\})(f)$$
is easily seen to have proper support. □

Note that this lemma is false if the graph γ has no tails. This is the reason why we only consider functionals on $\mathscr{E}$ modulo constants, or equivalently, without a constant term.

COROLLARY 14.5.2. *The renormalization group operator*
$$\mathscr{O}_p^+(\mathscr{E}_c, \mathscr{A})[[\hbar]] \to \mathscr{O}_p^+(\mathscr{E}_c, \mathscr{A})[[\hbar]]$$
$$I \to W(P(\varepsilon, L), I)$$
is well-defined.

As always, $\mathscr{O}_p^+(\mathscr{E}_c, \mathscr{A})[[\hbar]]$ refers to the subspace of $\mathscr{O}_p(\mathscr{E}_c, \mathscr{A})[[\hbar]]$ consisting of elements which are at least cubic modulo $\hbar$ and the ideal $\Gamma(X, m) \subset \mathscr{A}$.

PROOF. The renormalization group operator is defined by
$$W(P(\varepsilon, L), I) = \sum_\gamma \frac{1}{\|\mathrm{Aut})\gamma)\|} \hbar^{g(\gamma)} w_\gamma(P(\varepsilon, L), I).$$

The sum is over connected stable graphs; and, as we are working with functionals on $\mathscr{E}$ modulo constants, we only consider graphs with at least one tail. Lemma 14.5.1 shows that each $w_\gamma(P(\varepsilon, L), I)$ is well-defined. □

14.6. Now we can define the notion of theory on the manifold M, using the fake propagator $P(\varepsilon, L)$.

DEFINITION 14.6.1. *A theory is a collection* $\{I[L] \mid L \in \mathbb{R}_{>0}\}$ *of elements of* $\mathscr{O}_p^+(\mathscr{E}_c, \mathscr{A})[[\hbar]]$ *which satisfy*

(1) The renormalization group equation,
$$I[L] = W(P(\varepsilon, L), I[\varepsilon])$$

(2) The asymptotic locality axiom: there a small L asymptotic expansion
$$I[L] \simeq \sum f_i(L) \Psi_i$$
in terms of local action functionals $\Psi_i \in \mathscr{O}_{loc}^+(\mathscr{E}_c, \mathscr{A})[[\hbar]]$. *We will assume that the functions* $f_i(L)$ *appearing in this expansion have at most a finite order pole at* $L = 0$; *that is, we can find some* n *such that* $\lim_{L \to 0} L^n f_i(L) = 0$.

Let $\mathscr{T}^{(n)}(\mathscr{E}, \mathscr{A})$ denote the set of theories defined modulo $\hbar^{n+1}$, and let $\mathscr{T}^{(\infty)}(\mathscr{E}, \mathscr{A})$ denote the set of theories defined to all orders in $\hbar$.

14.7. One can ask how the definition of theory depends on the choice of fake heat kernel. It turns out that there is no dependence. Let $K_t, \widetilde{K}_t$ be two heat fake heat kernels, with associated propagators $P(\varepsilon, L), \widetilde{P}(\varepsilon, L)$. We have seen that $K_t - \widetilde{K}_t$ vanishes to all orders at $t = 0$. It follows that $P(0, L) - \widetilde{P}(0, L)$ is smooth, that is, an element of $\mathscr{E} \otimes \mathscr{E} \otimes \mathscr{A}$. Also, $P(0, L) - \widetilde{P}(0, L)$ has proper support. Further, as $L \to 0$, $P(0, L)$ and all of its derivatives vanish faster than any power of L.

Thus, the renormalization group operator

$$W\left(P(0,L) - \widetilde{P}(0,L), -\right) : \mathscr{O}_p^+(\mathscr{E}_c, \mathscr{A})[[\hbar]] \to \mathscr{O}_p^+(\mathscr{E}_c, \mathscr{A})[[\hbar]]$$

is well-defined.

LEMMA 14.7.1. *Let $\{\widetilde{I}[L]\}$ be a collection of effective interactions defining a theory for the propagator $\widetilde{P}(\varepsilon, L)$. Then,*

$$I[L] = W\left(P(0,L) - \widetilde{P}(0,L), \widetilde{I}[L]\right)$$

defines a theory for the propagator $P(\varepsilon, L)$.

Further, the small L asymptotic expansion of $I[L]$ is the same as that of $\widetilde{I}[L]$.

PROOF. The renormalization group equation

$$W(P(\varepsilon, L), I[\varepsilon]) = I[L]$$

is a corollary of the general identity,

$$W(P, W(P', I)) = W(P + P', I)$$

for any $P, P' \in \mathscr{E} \otimes \mathscr{E} \otimes \mathscr{A}$ of proper support, and any $I \in \mathscr{O}_p^+(\mathscr{E}, \mathscr{A})[[\hbar]]$. The statement about the small L asymptotics of $I[L]$ – and hence the locality axiom which says that $I[L]$ defines a theory – follows from the locality axiom for $\widetilde{I}[L]$ and the fact that $P(0,L)$ and all its derivatives tend to zero faster than any power of L, as $L \to 0$. □

14.8. Now let $U \subset M$ be any open subset. The free field theory on M – defined by the vector bundle E, together with certain differential operators on it – restricts to define a free field theory on $U \subset M$. One can ask if a theory on M will restrict to one on U as well.

The following proposition, which will be proved later, shows that one can do this.

PROPOSITION 14.8.1. *Let $U \subset M$ be an open subset.*
Given any theory

$$\{I[L] \in \mathscr{O}_p^+(\mathscr{E}, \mathscr{A})[[\hbar]]\}$$

on M (defined using any fake heat kernel on M), there is a unique theory

$$\{I_U[L] \in \mathscr{O}_p^+(\Gamma(U, E\mid_U), \mathscr{A})[[\hbar]]\}$$

14. FIELD THEORIES ON NON-COMPACT MANIFOLDS

again defined using any fake heat kernel on U, with the property that the small L asymptotic expansion of $I_U[L]$ is the restriction to U of the small L asymptotic expansion of $I[L]$.

The existence of this restriction map shows that there is a presheaf on M which assigns to an open subset $U \subset M$ the set $\mathscr{T}^{(n)}(\Gamma(U, E_U)), \mathscr{A})$ of theories on U. We will denote this presheaf by $\mathscr{T}^{(n)}(\mathscr{E}, \mathscr{A})$.

14.9. Now we are ready to state the main theorem.

THEOREM 14.9.1.   (1) *The presheaves $\mathscr{T}^{(n)}(\mathscr{E}, \mathscr{A})$ (of theories defined modulo $\hbar^{n+1}$) and $\mathscr{T}^{(\infty)}(\mathscr{E}, \mathscr{A})$ (of theories defined to all order in $\hbar$) on M are sheaves.*
 (2) *There is a canonical isomorphism between the sheaf $\mathscr{T}^{(0)}(\mathscr{E}, \mathscr{A})$ and the sheaf of local action functionals $I \in \mathscr{O}_{loc}(\mathscr{E}_c, \mathscr{A})$ which are at least cubic modulo the ideal $\Gamma(X, m) \subset \mathscr{A}$.*
 (3) *For $n > 0$, $\mathscr{T}^{(n)}(\mathscr{E}, \mathscr{A})$ is, in a canonical way, a torsor over $\mathscr{T}^{(n-1)}(\mathscr{E}, \mathscr{A})$ for the sheaf of abelian groups $\mathscr{O}_{loc}(\mathscr{E}_c, \mathscr{A})$, in a canonical way.*
 (4) *Choosing a renormalization scheme leads to a map*
$$\mathscr{T}^{(n)}(\mathscr{E}, \mathscr{A}) \to \mathscr{T}^{(n+1)}(\mathscr{E}, \mathscr{A})$$
 of sheaves for each n, which is a section. Thus, the choice of a renormalization scheme leads to an isomorphism of sheaves
$$\mathscr{T}^{(\infty)}(\mathscr{E}, \mathscr{A}) \cong \mathscr{O}_{loc}^+(\mathscr{E}_c, \mathscr{A})[[\hbar]]$$
on M.

14.10. The proof of the theorem, and of Proposition 14.8.1, is along the same lines as before, using counterterms.

PROPOSITION 14.10.1. *Let us fix a renormalization scheme, and a fake heat kernel K_t on M.*
Let
$$I = \sum \hbar^i I_{i,k} \in \mathscr{O}_{loc}^+(\mathscr{E}_c, \mathscr{A})[[\hbar]].$$
Then:
 (1) *there is a unique series of counterterms*
$$I^{CT}(\varepsilon) = \sum \hbar^i I_{i,k}^{CT}(\varepsilon)$$
 where
$$I_{i,k}^{CT}(\varepsilon) \in \mathscr{O}_{loc}(\mathscr{E}, \mathscr{A}) \otimes_{alg} \mathscr{P}((0, \infty))_{<0}$$
 is purely singular as a function of ε, with the property that the limit
$$\lim_{\varepsilon \to 0} W\left(P(\varepsilon, L), I - I^{CT}(\varepsilon)\right)$$
 exists, for all L.

(2) This limit defines a collection of elements $I[L] \in \mathscr{O}_p^+(\mathscr{E}, \mathscr{A})[[\hbar]]$, satisfying the renormalization group equation and locality axiom, and so defines a theory.
(3) The counterterms $I_{i,k}^{CT}(\varepsilon)$ do not depend on the choice of a fake heat kernel.
(4) If $U \subset M$ is an open subset, then the counterterms for I restricted to U are the restrictions to U of the counterterms $I_{i,k}^{CT}(\varepsilon)$ for I. Thus, counterterms define a map of sheaves

$$\mathscr{O}_{loc}^+(\mathscr{E}_c, \mathscr{A})[[\hbar]] \to \mathscr{O}_{loc}^+(\mathscr{E}_c, \mathscr{A} \otimes_{alg} \mathscr{P}((0,\infty)_\varepsilon)_{<0})[[\hbar]].$$

PROOF. The proof of the first two statements is identical to the proof of the corresponding statement on a compact manifold, and so is mostly omitted. One point is worth mentioning briefly, though: the counterterms are defined to be the singular parts of the small ε asymptotic expansion of the weight $u_\gamma(P(\varepsilon, L), I)$ attached to a graph γ. One can ask whether such asymptotic expansions exist when we use a fake heat kernel rather than an actual heat kernel.

The asymptotic expansion, as constructed in Appendix 1, only relies on the small t expansion of the heat kernel K_t, and thus exists whenever the propagator is constructed from a kernel with such a small t expansion.

The third clause is proved using the uniqueness of the counterterms, as follows. Suppose that $K_t, \widetilde{K}_t$ are two fake heat kernels, with associated fake propagators $P(\varepsilon, L), \widetilde{P}(\varepsilon, L)$. Let $I^{CT}(\varepsilon), \widetilde{I}^{CT}(\varepsilon)$ denote the counterterms associated to the two different fake heat kernels. We need to show that they are the same.

Note that

$$\lim_{\varepsilon \to 0} W\left(\widetilde{P}(0, L) - P(0, L), W\left(P(\varepsilon, L), I - I^{CT}(\varepsilon)\right)\right)$$

exists. We can write the expression inside the limit as

$$W\left(P(0, \varepsilon) - \widetilde{P}(0, \varepsilon), W\left(\widetilde{P}(\varepsilon, L), I - I^{CT}(\varepsilon)\right)\right).$$

Note that $P(0, \varepsilon) - \widetilde{P}(0, \varepsilon)$ tends to zero, with all derivatives, faster than any power of ε. Also, $W\left(\widetilde{P}(\varepsilon, L), I - I^{CT}(\varepsilon)\right)$ has a small ε asymptotic expansion in terms of functions of ε of polynomial growth at the origin.

From these two facts it follows that

$$\lim_{\varepsilon \to 0} W\left(\widetilde{P}(\varepsilon, L), I - I^{CT}(\varepsilon)\right)$$

exists.

Uniqueness of the counterterms implies that $I^{CT}(\varepsilon) = \widetilde{I}^{CT}(\varepsilon)$.

The fourth clause can be proved easily using independence of the counterterms of the fake heat kernel. □

14. FIELD THEORIES ON NON-COMPACT MANIFOLDS

14.11. Now we can prove Proposition 14.8.1 and Theorem 14.9.1. The proof is easy once we know that counterterms exist.

Let us start by regarding the set of theories $\mathscr{T}^{(\infty)}(\mathscr{E}, \mathscr{A})$ as just a set, and not as arising from a sheaf on M. (After all, we have not yet proved Proposition 14.8.1, so we do not know that we have a presheaf of theories).

LEMMA 14.11.1. *Let us choose a renormalization scheme, and a fake heat kernel. Then the map of sets*
$$\mathscr{O}_{loc}^+(\mathscr{E}_c, \mathscr{A})[[\hbar]] \to \mathscr{T}^{(\infty)}(\mathscr{E}, \mathscr{A})$$
$$I \mapsto \{I[L] = \lim_{\varepsilon \to 0} W\left(P(\varepsilon, L), I - I^{CT}(\varepsilon)\right)\}$$
is a bijection.

PROOF. This is proved by the usual inductive argument. □

Now we can prove Proposition 14.8.1, which we restate here for convenience.

PROPOSITION 14.11.2. *Let $U \subset M$ be an open subset.*
Given any theory
$$\{I[L] \in \mathscr{O}_p^+(\mathscr{E}, \mathscr{A})[[\hbar]]\}$$
on M, for any fake heat kernel, U, there is a unique theory
$$\{I_U[L] \in \mathscr{O}_p^+(\Gamma(U, E\mid_U), \mathscr{A})[[\hbar]]\}$$
with the property that the small L asymptotics of $I_U[L]$ is the restriction to U of the small L asymptotics of $I[L]$.

PROOF. Uniqueness is obvious, as any two theories on U with the same small L asymptotic expansions must coincide.

For existence, we will use the bijection between theories and Lagrangians which arises from the choice of a renormalization scheme. We can assume that the theory $\{I[L]\}$ on M arises from a local action functional $I \in \mathscr{O}_{loc}^+(\mathscr{E}_c, \mathscr{A})[[\hbar]]$. Then, we define $I_U[L]$ to be the theory on U associated to the restriction of I to U.

It is straightforward to check that, with this definition, $I[L]$ and $I_U[L]$ have the same small L asymptotics. □

It follows from the proof of this lemma that the map
$$\mathscr{O}_{loc}^+(\mathscr{E}_c, \mathscr{A})[[\hbar]] \to \mathscr{T}^{(\infty)}(\mathscr{E}, \mathscr{A})$$
of sets actually arises from a map of presheaves on M. Since $\mathscr{O}_{loc}^+(\mathscr{E}_c, \mathscr{A})[[\hbar]]$ is actually a sheaf on M, it follows that $\mathscr{T}^{(\infty)}(\mathscr{E}, \mathscr{A})$ is also a sheaf, and similarly for $\mathscr{T}^{(n)}(\mathscr{E}, \mathscr{A})$.

The remaining statements of Theorem 14.9.1 are proved in the same way as the corresponding statements on a compact manifold.

CHAPTER 3

Field theories on $\mathbb{R}^n$

This chapter shows how the main theorem of Chapter 2 works on $\mathbb{R}^n$. However, we will only deal with translation-invariant theories on $\mathbb{R}^n$.

Working on $\mathbb{R}^n$ presents difficulties not present when dealing with theories on compact manifolds. In this situation, the finite dimensional integrals one attaches to Feynman graphs are now over products of copies of $\mathbb{R}^n$, and as such may not converge. Divergences of this form are called infrared divergences.

We have seen in Section 14 how to deal with field theory on non-compact manifolds in general. This involved choosing a "fake heat kernel", which is a kernel only satisfying the heat equation in an approximate way, and which has certain conditions on its support. In this section we will verify that, on $\mathbb{R}^n$, there is no need to choose a fake heat kernel. As long as we only consider effective interactions $I[L]$ when $L < \infty$, we can use the ordinary heat kernel on $\mathbb{R}^n$. It will take a certain amount of work, however, to prove this statement, and to prove the bijection between theories and Lagrangians in this context.

The sheaf-theoretic nature of quantum field theories, as stated in 14.9.1, implies that any theory on $\mathbb{R}^n$ which is invariant under $\mathbb{R}^n \ltimes SO(n)$ defines a theory on any manifold with a flat metric.

0.1. We would like to say that a theory on $\mathbb{R}^n$ is given by a collection of effective interactions $I[\varepsilon]$ satisfying the renormalization group equation

$$I[L] = W\left(P(\varepsilon, L), I[\varepsilon]\right).$$

The possible presence of infrared divergences makes it difficult to define this renormalization group flow. The renormalization group operator $I \mapsto W(P(\varepsilon, L), I)$ is only defined for functionals I which satisfy certain growth conditions: roughly, the component distributions $I_{(i,k)}$ of I must be tempered distributions on $\mathbb{R}^{nk}$ which are of rapid decay away from the small diagonal $\mathbb{R}^n \subset \mathbb{R}^{nk}$.

In Section 1 we give a definition of the space of distributions on $\mathbb{R}^{nk}$ of rapid decay away from the small diagonal. The condition is, roughly, that the distribution D decays at ∞ as fast as $e^{-b\|x\|^2}$ for some $b > 0$.

0.2. In Section 2, we prove the main theorem for scalar field theories on $\mathbb{R}^n$. We define a theory to be a collection of effective interactions $I[L]$, satisfying the growth conditions defined in section 1, the renormalization group equation, and a locality axiom. The main theorem, as before, is as follows.

THEOREM 0.2.1. *Let $\mathscr{T}^{(m)}$ be the space of theories defined modulo $\hbar^{n+1}$. Then $\mathscr{T}^{(m+1)} \to \mathscr{T}^{(m)}$ is a torsor for the abelian group of translation invariant local action functionals on $\mathbb{R}^n$.*

Moreover, $\mathscr{T}^{(0)}$ is canonically isomorphic to the space of translation invariant local action functionals on $\mathbb{R}^n$ which are at least cubic modulo $\hbar$.

The choice of a renormalization scheme yields a section of each torsor $\mathscr{T}^{(m+1)} \to \mathscr{T}^{(m)}$, and so an isomorphism between the space $\mathscr{T}^{(\infty)}$ of theories and the space $\mathscr{O}_{loc}^+(\mathscr{S}(\mathbb{R}^n))[[\hbar]]$ of series in $\hbar$ whose coefficients are translation invariant local action functionals on $\mathbb{R}^n$, and which are at least cubic modulo $\hbar$.

The proof is along the same lines as the proof of the corresponding statement on compact manifolds. We first prove the renormalization scheme independent version of the statement, that there is a bijection between theories and local action functionals. Given a local action functional I, we construct the effective interactions $I[L]$ by introducing local counterterms $I^{CT}(\varepsilon)$ and defining

$$I[L] = \lim_{\varepsilon \to 0} W\left(P(\varepsilon, L), I - I^{CT}(\varepsilon)\right).$$

Some work is required to show that the functionals $I[L]$ satisfy the requisite growth conditions; this is the main point where the proof of the theorem on $\mathbb{R}^n$ differs from that on a compact manifold.

The more canonical renormalization scheme independent version of the statement is a straightforward corollary.

0.3. Section 3 proves the main result for a more general class of theories on $\mathbb{R}^n$, where the fields are sections of some vector bundle $\mathscr{E}$, and where everything may depend on an auxiliary ring $\mathscr{A}$. As with scalar field theories, we only treat the case where everything is translation invariant. The proof of the bijection between theories and local action functionals in this more general context is essentially the same as the proof for scalar field theories.

1. Some functional analysis

Before we discuss theories on $\mathbb{R}^n$, we need to develop a little functional analysis. We need to construct a space of distributions on $\mathbb{R}^{nk}$ which are of rapid decay away from the small diagonal $\mathbb{R}^n \subset \mathbb{R}^{nk}$. The effective interactions which appear in the definition of a theory will lie in these spaces of distributions of rapid decay.

This section should probably be skipped on first reading, and referred back to as necessary.

1. SOME FUNCTIONAL ANALYSIS

1.1. Let us fix throughout this section an auxiliary manifold with corners X, and a (possibly graded) vector bundle A on X. We will assume that A has the structure of super-commutative algebra in the category of vector bundles on X. Let $\mathscr{A} = \Gamma(X, A)$ denote the space of global sections of A.

The first thing to do is introduce some of the basic spaces of functions on $\mathbb{R}^n$. Let $\mathscr{S}(\mathbb{R}^n)$ be the space of Schwartz functions on $\mathbb{R}^n$. This is a nuclear Fréchet space. Let $\mathscr{S}(\mathbb{R}^n) \otimes \mathscr{A}$ denote the completed projective tensor product of $\mathscr{S}(\mathbb{R}^n)$ and $\mathscr{A}$. We can think of $\mathscr{S}(\mathbb{R}^n) \otimes \mathscr{A}$ as the space of Schwartz functions on $\mathbb{R}^n$ with values in $\mathscr{A}$.

Let $\mathcal{D}(\mathbb{R}^n, \mathscr{A})$ denote the space of continuous linear maps $\mathscr{S}(\mathbb{R}^n) \to \mathscr{A}$. Thus, $\mathcal{D}(\mathbb{R}^n, \mathscr{A})$ is the space of $\mathscr{A}$-valued tempered distributions on $\mathbb{R}^n$.

There is an $\mathscr{A}$-bilinear direct product map

$$\mathcal{D}(\mathbb{R}^n, \mathscr{A}) \times \mathcal{D}(\mathbb{R}^k, \mathscr{A}) \to \mathcal{D}(\mathbb{R}^{n+k}, \mathscr{A})$$
$$(\Psi, \Phi) \mapsto \Psi \boxtimes \Phi.$$

The direct product $\Psi \boxtimes \Phi$ is uniquely determined by the property that for all Schwartz functions $f \in \mathscr{S}(\mathbb{R}^n)$, $g \in \mathscr{S}(\mathbb{R}^k)$,

$$(\Psi \boxtimes \Phi)(f \boxtimes g) = \Psi(f) \Phi(g).$$

Here the product on the right hand side is taken in the algebra $\mathscr{A}$, and $f \boxtimes g \in \mathscr{S}(\mathbb{R}^{n+k})$ is the usual exterior product of functions.

1.2. We are interested in spaces of distributions of rapid decay on $\mathbb{R}^n$. By "rapid decay" we will mean, roughly, that they decay as fast as $e^{-b\|x\|^2}$ for some $b > 0$.

Such distributions will be continuous linear maps on spaces of functions whose growth is bounded by all $e^{b\|x\|^2}$. We will first introduce these function spaces.

DEFINITION 1.2.1. *Let V, W be finite dimensional vector spaces over $\mathbb{R}$. For all $a \in \mathbb{Z}_{\geq 0}$, $b \in \mathbb{R}_{>0}$ and $l \in \mathbb{Z}_{\geq 0}$, let us define a norm $\|-\|_{a,b,l}$ on $\mathscr{S}(V \oplus W)$ by the formula*

$$\|f\|_{a,b,l} = \sum_{|I| \leq l} \mathrm{Sup}_{(v,w) \in V \oplus W} \left| (1 + \|v\|^2)^a e^{-b\|w\|^2} \partial_I f \right|.$$

Let us extend this to a map $C^\infty(V \oplus W) \to [0, \infty]$ by the same formula. Let

$$\mathscr{T}(V, W) \subset C^\infty(V \oplus W)$$

be the subspace of those functions such that, for all a, b and l,

$$\|f\|_{a,b,l} < \infty.$$

Let us give $\mathscr{T}(V, W)$ the topology induced by the norms $\|-\|_{a,b,l}$. This is the coarsest linear topology containing, as open neighbourhoods of zero, the sets $\{f \mid \|f\|_{a,b,l} < 1\}$. In this topology, a sequence $f_i \to 0$ if and only if $\|f_i\|_{a,b,l} \to 0$ for all a, b, l.

If the space $V = 0$, we will use the notation
$$\mathscr{T}(W) = \mathscr{T}(0, W).$$

DEFINITION 1.2.2. *A continuous linear map*
$$\Phi : \mathscr{S}(V \oplus W) \to \mathscr{A}$$
is of rapid decay along W if it extends to a continuous linear map
$$\Phi : \mathscr{T}(V, W) \to \mathscr{A}.$$

Let $K \subset X$ be a compact subset, and let $D : \mathscr{A} \to \mathscr{A}$ be a differential operator. Let $\|a\|_{K,D}$ denote the norm on $\mathscr{A}$ given by taken the supremum over K of Da. The topology on $\mathscr{A}$ is defined by these norms, as K and D vary.

LEMMA 1.2.3.   *(1) $\mathscr{S}(V \oplus W) \subset \mathscr{T}(V, W)$ is dense.*
(2) A continuous linear map $\Phi : \mathscr{S}(V \oplus W) \to \mathscr{A}$ extends to a continuous linear map $\mathscr{T}(V, W) \to \mathscr{A}$ if and only if, for all compact subsets $K \subset X$ and all differential operators $D : \mathscr{A} \to \mathscr{A}$, there exist some a, b, l and C such that
$$\|\Phi(f)\|_{K,D} \le C \|f\|_{a,b,l}.$$
(If this extension exists, it is of course unique.)
(3) Let $\Phi : \mathscr{T}(V, W) \to \mathscr{A}$ and $\Psi : \mathscr{T}(V', W') \to \mathscr{A}$ be continuous linear maps. Then the direct product
$$\Phi \boxtimes \Psi : \mathscr{S}(V \oplus V' \oplus W \oplus W') \to \mathscr{A}$$
extends to a continuous linear map
$$\mathscr{T}(V \oplus V', W \oplus W') \to \mathscr{A}.$$

PROOF. The proof is straightforward and omitted. □

1.3. Now we are ready to introduce our main objects, which are *good distributions*. These are certain distributions on $\mathbb{R}^{nk}$ which are of rapid decay away from the small diagonal.

We will view $\mathbb{R}^{nk}$ as a configuration space of k points on $\mathbb{R}^n$; in this way, it inherits an $\mathbb{R}^n$ action. When we refer to translation invariant objects on $\mathbb{R}^{nk}$ we will always mean objects invariant under this $\mathbb{R}^n$ action.

DEFINITION 1.3.1. *A tempered distribution $\Phi : \mathscr{S}(\mathbb{R}^{nk}) \to \mathscr{A}$ is good if it has the following two properties.*
(1) Φ is translation invariant.
(2) If we write $\mathbb{R}^{nk}$ as an orthogonal direct sum $\mathbb{R}^n \oplus \mathbb{R}^{n(k-1)}$, where $\mathbb{R}^n \subset \mathbb{R}^{nk}$ is the small diagonal, then Φ has rapid decay along $\mathbb{R}^{n(k-1)}$. That is, Φ extends to a continuous linear map
$$\Phi : \mathscr{T}(\mathbb{R}^n, \mathbb{R}^{n(k-1)}) \to \mathscr{A}.$$

1.4. The main result of this section is that we can use Feynman graphs to contract good distributions, and that the result is again a good distribution.

Let γ be a connected graph. Let $H(\gamma), T(\gamma), V(\gamma)$ and $E(\gamma)$ refer to the sets of half-edges, tails, vertices and internal edges of γ. For every $v \in V(\gamma)$, let $H(v) \subset H(\gamma)$ refer to the set of half-edges adjoining v. For every $e \in E(\gamma)$, let $H(e) \subset H(\gamma)$ be the pair of half-edges forming e.

Suppose that we have the following data:

(1) For every $v \in V(\gamma)$, we have a good distribution
$$I_v \in \mathcal{D}_g(\mathbb{R}^{nH(v)}, \mathscr{A}).$$
In particular, I_v is an $\mathscr{A}$-valued tempered distribution on $\mathbb{R}^{nH(v)}$.

(2) Suppose that for each edge $e \in E(\gamma)$, we have a function
$$P_e \in C^\infty(\mathbb{R}^{nH(e)}).$$
Let h_1, h_2 denote the two half edges of e, and let $x_{h_i} : \mathbb{R}^{nH(e)} \to \mathbb{R}^n$ be the corresponding linear maps. Let us assume that P_e is invariant under the $\mathbb{R}^n$ action on $\mathbb{R}^{nH(e)}$; this amounts to saying that P_e is independent of $x_{h_1} + x_{h_2}$. Let us further assume that for any multi-index I, there exists b such that
$$|\partial_I P_e| \leq e^{-b\|x_{h_1} - x_{h_2}\|^2}.$$

Then we can attempt to form the distribution $w_\gamma(\{I_v\}, \{P_e\})$ on $\mathbb{R}^{nT(\gamma)}$ by contracting the distributions I_v with the functions P_e, according to a combinatorial rule given by γ, as before.

The result is that this procedure works.

THEOREM 1.4.1. *The distribution $w_\gamma(\{I_v\}, \{P_e\})$ is well-defined, and is a good distribution on $\mathbb{R}^{nT(\gamma)}$.*

"Well-defined" means that $w_\gamma(\{I_v\}, \{P_e\})$ is defined by contracting tempered distributions which are of rapid decrease in some directions with a smooth function which is of bounded growth in the corresponding directions. A more precise statement is given in the proof.

1.5. Proof of the theorem. In fact we will prove something a little more general. We will relax the assumption that each I_v is a good distribution. Instead, we will simply assume that I_v is an $\mathscr{A}$-valued tempered distribution on $\mathbb{R}^{nH(v)}$, of rapid decay along $\mathbb{R}^{nH(v)}/\mathbb{R}^n$.

More precisely: let us write
$$\mathbb{R}^{nH(v)} = \mathbb{R}^n \oplus \left(\mathbb{R}^{nH(v)}/\mathbb{R}^n\right),$$
where $\mathbb{R}^n \subset \mathbb{R}^{nH(v)}$ is the small diagonal, and this direct sum is orthogonal. Then, we will assume that I_v is a continuous linear map
$$I_v : \mathscr{T}(\mathbb{R}^n, \mathbb{R}^{nH(v)}/\mathbb{R}^n) \to \mathscr{A}.$$

In particular, I_v is an $\mathscr{A}$-valued tempered distribution on $\mathbb{R}^{nH(v)}$.

Observe that we can write
$$\mathbb{R}^{nH(\gamma)} = \oplus_{v \in V(\gamma)} \mathbb{R}^{nH(v)}$$
$$= \mathbb{R}^{nV(\gamma)} \oplus \left(\oplus_{v \in V(\gamma)} \left(\mathbb{R}^{nH(v)}/\mathbb{R}^n \right) \right).$$

The direct product $\boxtimes_{v \in V(\gamma)} I_v$ is a continuous linear map
$$\boxtimes_{v \in V(\gamma)} I_v : \mathscr{T}\left(\mathbb{R}^{nV(\gamma)}, \oplus_{v \in V(\gamma)} \mathbb{R}^{nH(v)}/\mathbb{R}^n \right) \to \mathscr{A}.$$

Thus, $\boxtimes I_v$ is an $\mathscr{A}$-valued tempered distribution on $\mathbb{R}^{nH(\gamma)}$, of rapid decrease in certain directions.

Recall that for each edge $e \in E(\gamma)$, we have a function
$$P_e \in C^\infty(\mathbb{R}^{nH(e)}).$$
with the property that, for any multi-index I, there exists b such that
$$|\partial_I P_e| \leq e^{-b\|x_{h_1} - x_{h_2}\|^2}.$$

Let us write $\mathbb{R}^{nT(\gamma)}$ as an orthogonal direct sum
$$\mathbb{R}^{nT(\gamma)} = \mathbb{R}^n \oplus \left(\mathbb{R}^{nT(\gamma)}/\mathbb{R}^n \right).$$

Let
$$\phi \in \mathscr{T}(\mathbb{R}^n, \mathbb{R}^{nT(\gamma)}/\mathbb{R}^n).$$
We will view ϕ as a function on $\mathbb{R}^{nH(\gamma)}$ via the map $\mathbb{R}^{nH(\gamma)} \to \mathbb{R}^{nT(\gamma)}$.

Then,
$$\phi \prod_{e \in E(\gamma)} P_e \in C^\infty(\mathbb{R}^{nH(\gamma)}).$$

LEMMA 1.5.1. *(1) The function $\phi \prod_{e \in E(\gamma)} P_e$ is an element of the space*
$$\mathscr{T}\left(\mathbb{R}^{nV(\gamma)}, \oplus_{v \in V(\gamma)} \mathbb{R}^{nH(v)}/\mathbb{R}^n \right).$$
Thus, we can pair $\phi \prod_{e \in E(\gamma)} P_e$ with $\boxtimes I_v$ to yield an element
$$w_\gamma(\{I_v\}, \{P_e\})(\phi) \in \mathscr{A}.$$

(2) The map
$$\phi \to w_\gamma(\{I_v\}, \{P_e\})(\phi)$$
is a continuous linear map
$$w_\gamma(\{I_v\}, \{P_e\}) : \mathscr{T}(\mathbb{R}^n, \mathbb{R}^{nT(\gamma)}/\mathbb{R}^n) \to \mathscr{A}.$$

(3) If we further assume that each I_v is invariant under the $\mathbb{R}^n$ action, so that each I_v is a good distribution, then $w_\gamma(\{I_v\}, \{P_e\})$ is also invariant under the $\mathbb{R}^n$ action, and so is a good distribution.

1. SOME FUNCTIONAL ANALYSIS

PROOF. The third clause is obvious from the first two.
For the first two clauses, it suffices to show that the map

$$\mathscr{T}(\mathbb{R}^n, \mathbb{R}^{nT(\gamma)}/\mathbb{R}^n) \to C^\infty(\mathbb{R}^{nH(\gamma)})$$
$$\phi \mapsto \phi \prod_{e \in E(\gamma)} P_e$$

is a continuous linear map

$$\mathscr{T}(\mathbb{R}^n, \mathbb{R}^{nT(\gamma)}/\mathbb{R}^n) \to \mathscr{T}\left(\mathbb{R}^{nV(\gamma)}, \oplus_{v \in V(\gamma)} \mathbb{R}^{nH(v)}/\mathbb{R}^n\right).$$

Let $\|-\|_{a,b,l}^{H(\gamma)}$ refer to the norm on $\mathscr{T}\left(\mathbb{R}^{nV(\gamma)}, \oplus_{v \in V(\gamma)} \mathbb{R}^{nH(v)}/\mathbb{R}^n\right)$, and let $\|-\|_{c,d,m}^{T(\gamma)}$ refer to the norm on $\mathscr{T}(\mathbb{R}^n, \mathbb{R}^{nT(\gamma)}/\mathbb{R}^n)$.

It suffices to show that for all a, b, l, there exist c, d, m and C such that for all ϕ,

$$\left\| \phi \prod_e P_e \right\|_{a,b,l}^{H(\gamma)} \leq C \|\phi\|_{c,d,m}^{T(\gamma)}.$$

Let us introduce some notation which will allow us to express this condition more explicitly.

For $h \in H(\gamma)$, let $x_h : \mathbb{R}^{nH(\gamma)} \to \mathbb{R}^n$ be the coordinate function.
For every vertex $v \in V(\gamma)$, let

$$c_v = \sum_{h \in H(v)} x_h : \mathbb{R}^{nH(\gamma)} \to \mathbb{R}^n.$$

Let

$$n_v : \mathbb{R}^{nH(\gamma)} \to \mathbb{R}^{nH(v)} \to \mathbb{R}^{nH(v)}/\mathbb{R}^n$$

be the composition of the natural map $\mathbb{R}^{nH(\gamma)} \to \mathbb{R}^{nH(v)}$ with the projection onto the subspace orthogonal to $\mathbb{R}^n \subset \mathbb{R}^{nH(v)}$.

For every edge $e \in E(\gamma)$, corresponding to half-edges h_1, h_2, let

$$d_e = x_{h_1} - x_{h_2}.$$

(Of course, d_e depends on choosing an orientation of the edge e, but this will play no role).

Let us view $T(\gamma)$ as a subset of $H(\gamma)$, and let

$$c_T = \sum_{t \in T(\gamma)} x_t : \mathbb{R}^{nH(\gamma)} \to \mathbb{R}^n.$$

Finally, let

$$n_T : \mathbb{R}^{nH(\gamma)} \to \mathbb{R}^{nT(\gamma)} \to \mathbb{R}^{nT(\gamma)}/\mathbb{R}^n$$

be the composition of the natural map $\mathbb{R}^{nH(\gamma)} \to \mathbb{R}^{nT(\gamma)}$ with the projection onto the orthogonal complement to the subspace $\mathbb{R}^n \subset \mathbb{R}^{nT(\gamma)}$.

To show the inequality we want, it suffices to show the following: for all $a \in \mathbb{Z}_{\geq 0}$, $b \in \mathbb{R}_{>0}$, and $l \in \mathbb{Z}_{\geq 0}$, and all multi-indices I and J with $|I| \leq l$, there exist c, d and C such that for all ϕ,

$$\mathrm{Sup}_{\mathbb{R}^{nH(\gamma)}} \left| \left(1 + \sum_{v \in V(\gamma)} \|c_v\|^2 \right)^a e^{-\sum_{v \in V(\gamma)} b\|n_v\|^2} (\partial_I \phi) \left(\partial_J \prod_{e \in E(\gamma)} P_e \right) \right|$$
$$\leq C \|\phi\|_{c,d,l}^{T(\gamma)}.$$

Now,
$$|\partial_I \phi| \leq C' \|\phi\|_{c,d,l}^{T(\gamma)} (1 + \|c_T\|^2)^{-c} e^{d\|n_T\|^2},$$
for some constant C'; and we can find some b' and C'' such that
$$\partial_J \prod_{e \in E(\gamma)} P_e \leq C' e^{-b' \sum_{e \in E(\gamma)} \|d_e\|^2}.$$

Putting these inequalities together, we see that it suffices to show that, for all $a \in \mathbb{Z}_{\geq 0}$, $b, b' \in \mathbb{R}_{>0}$, there exists $c \in \mathbb{Z}_{\geq 0}$ and $d \in \mathbb{R}_{>0}$ such that

$$\mathrm{Sup}_{\mathbb{R}^{nH(\gamma)}} \left| \left(1 + \sum_{v \in V(\gamma)} \|c_v\|^2 \right)^a e^{-\sum_{v \in V(\gamma)} b\|n_v\|^2} \right.$$
$$\left. e^{-b' \sum_{e \in E(\gamma)} \|d_e\|^2} (1 + \|c_T\|^2)^{-c} e^{d\|n_T\|^2} \right| < \infty.$$

Let us write $\mathbb{R}^{nH(\gamma)}$ as an orthogonal direct sum
$$\mathbb{R}^{nH(\gamma)} = \mathbb{R}^n \oplus \mathbb{R}^{nH(\gamma)}/\mathbb{R}^n,$$
where $\mathbb{R}^n \subset \mathbb{R}^{nH(\gamma)}$ is the small diagonal. Let
$$n_H : \mathbb{R}^{nH(\gamma)} \to \mathbb{R}^{nH(\gamma)}/\mathbb{R}^n$$
be the projection.

All of the maps n_v, for $v \in V(\gamma)$, d_e, for $e \in E(\gamma)$, and n_T factor through the projection $\mathbb{R}^{nH(\gamma)} \to \mathbb{R}^{nH(\gamma)}/\mathbb{R}^n$.

Further, connectedness of the graph γ implies that the quadratic form
$$b \sum_{v \in V(\gamma)} \|n_v\|^2 + b' \sum_{e \in E(\gamma)} \|d_e\|^2$$
is positive definite on $\mathbb{R}^{nH(\gamma)}/\mathbb{R}^n$. Thus, we can find some $\alpha > 0$ so that
$$b \sum_{v \in V(\gamma)} \|n_v\|^2 + b' \sum_{e \in E(\gamma)} \|d_e\|^2 > \alpha \|n_H\|^2.$$

Also, since $n_T : \mathbb{R}^{nH(\gamma)} \to \mathbb{R}^{nT(\gamma)}/\mathbb{R}^n$ factors through $\mathbb{R}^{nH(\gamma)}/\mathbb{R}^n$, we have
$$e^{d\|n_T\|^2} \leq e^{d\|n_H\|^2}.$$

Thus, by choosing d sufficiently small, we can assume that
$$e^{-\sum_{v \in V(\gamma)} b\|n_v\|^2} e^{-b' \sum_{e \in E(\gamma)} \|d_e\|^2} e^{d\|n_T\|^2} \leq e^{-\varepsilon \|n_H\|^2}$$

for some $\varepsilon > 0$.

It remains to show that for all a and all $\varepsilon > 0$, there exists c such that

$$\mathrm{Sup}_{\mathbb{R}^{nH(\gamma)}} \left| \left(1 + \sum_{v \in V(\gamma)} \|c_v\|^2 \right)^a (1 + \|c_T\|^2)^{-c} e^{-\varepsilon \|n_H\|^2} \right| < \infty.$$

For all c and all ε, there exists a constant C such that

$$(1 + \|c_T\|^2)^{-c} e^{-\varepsilon \|n_H\|^2} < C(1 + \|c_T\|^2 + \|n_H\|^2)^{-c}.$$

Thus, it suffices to show that for all a, we can find some c such that

$$\mathrm{Sup}_{\mathbb{R}^{nH(\gamma)}} \left| \left(1 + \sum_{v \in V(\gamma)} \|c_v\|^2 \right)^a (1 + \|c_T\|^2 + \|n_H\|^2)^{-c} \right| < \infty.$$

But this follows immediately from the fact that the quadratic form $\|c_T\|^2 + \|n_H\|^2$ on $\mathbb{R}^{nH(\gamma)}$ is positive definite. $\square$

This completes the proof of the theorem.

2. The main theorem on $\mathbb{R}^n$

In this section we state and prove the main theorem for scalar field theories on $\mathbb{R}^n$.

We would like to give a definition of theory along the same lines as the definition we gave on compact manifolds. To do this, we need to have a definition of the renormalization group flow. The results of Section 1 allow us to construct the renormalization group flow on $\mathbb{R}^n$.

In Section 1, we defined the space $\mathcal{D}_g(\mathbb{R}^{nk})$ of good distributions on $\mathbb{R}^{nk}$. These distributions are invariant under the diagonal $\mathbb{R}^n$ action, and of rapid decay away from the small diagonal $\mathbb{R}^n \subset \mathbb{R}^{nk}$.

We will let

$$\mathscr{O}(\mathscr{S}(\mathbb{R}^n)) = \prod_{k \geq 1} \mathcal{D}_g(\mathbb{R}^{nk})_{S_n}$$

be the space of formal power series on $\mathscr{S}(\mathbb{R}^n)$ whose Taylor components are good distributions. Note that these power series do not have a constant term.

Note that $\mathscr{O}(\mathscr{S}(\mathbb{R}^n))$ is *not* an algebra; the direct product of good distributions is no longer good.

A good distribution

$$\Phi \in \mathcal{D}_g(\mathbb{R}^{nk})$$

will be called *local* if it is supported on the small diagonal $\mathbb{R}^n \subset \mathbb{R}^{nk}$. Translation invariance of good distributions implies that any such local Φ can be written as a finite sum

$$f(x_1, \ldots, x_n) \to \sum \int_{x \in \mathbb{R}^n} (\partial_I f)(x, \ldots, x)$$

where $\partial_I : \mathscr{S}(\mathbb{R}^{nk}) \to \mathscr{S}(\mathbb{R}^{nk})$ are constant-coefficient differential operators corresponding to multi-indices $I \in (\mathbb{Z}_{\geq 0})^{nk}$. Thus, these distributions are local in the sense used earlier.

We will let
$$\mathscr{O}_{loc}(\mathscr{S}(\mathbb{R}^n)) \subset \mathscr{O}(\mathscr{S}(\mathbb{R}^n))$$
denote the subspace of functionals on $\mathscr{S}(\mathbb{R}^n)$ whose Taylor components are local elements of $\mathcal{D}_g(\mathbb{R}^{nk})$.

Finally, we will let
$$\mathscr{O}^+(\mathscr{S}(\mathbb{R}^n))[[\hbar]] \subset \mathscr{O}(\mathscr{S}(\mathbb{R}^n))[[\hbar]]$$
$$\mathscr{O}^+_{loc}(\mathscr{S}(\mathbb{R}^n))[[\hbar]] \subset \mathscr{O}_{loc}(\mathscr{S}(\mathbb{R}^n))[[\hbar]]$$
denote the subspaces of $\mathscr{O}(\mathscr{S}(\mathbb{R}^n))[[\hbar]]$ and $\mathscr{O}_{loc}(\mathscr{S})[[\hbar]]$ of functionals which are at least cubic modulo $\hbar$.

2.1. Now we are ready to define translation invariant theories on $\mathbb{R}^n$. As before, we will always assume that the kinetic part of our actions is $-\frac{1}{2}\langle \phi, (D+m^2)\phi \rangle$ where D is the non-negative Laplacian, and $m \geq 0$ is the mass.

We will let
$$P(\varepsilon, L) = \int_\varepsilon^L e^{-tm^2} K_t dt = \int_\varepsilon^L t^{-n/2} e^{-tm^2} e^{-\|x-y\|^2/t} dt \in C^\infty(\mathbb{R}^n \times \mathbb{R}^n).$$
be the propagator.

As on a compact manifold, for any $I \in \mathscr{O}^+(\mathscr{S}(\mathbb{R}^n))[[\hbar]]$, we will define
$$W(P(\varepsilon, L), I) = \sum_\gamma \hbar^{g(\gamma)} \frac{1}{|\text{Aut}(\gamma)|} w_\gamma(P(\varepsilon, L), I)$$
where the sum, as before, is over stable graphs γ. Theorem 1.4.1 shows that each $w_\gamma(P(\varepsilon, L), I)$ is well-defined.

The definition of quantum field theory on $\mathbb{R}^n$ is essentially the same as that on a compact manifold. The only essential difference is that the effective interactions $I[L]$ are required to be elements of $\mathscr{O}^+(\mathscr{S}(\mathbb{R}^n))[[\hbar]]$, so that the component distributions $I_{(i,k)}[L]$ are good distributions on $\mathbb{R}^{nk}$. In particular, the effective interactions $I[L]$ are translation invariant; we will only discuss translation invariant theories on $\mathbb{R}^n$.

A second difference between the definitions on $\mathbb{R}^n$ and a compact manifold is more technical: the locality axiom on $\mathbb{R}^n$ can be made somewhat weaker. Instead of requiring that $I[L]$ has a small L asymptotic expansion in terms of local action functionals, we instead simply require that $I[L]$ tends to zero away from the diagonals in $\mathbb{R}^{nk}$. Translation invariance guarantees that this weaker axiom is sufficient to prove the bijection between theories and Lagrangians.

Let us now give the definition more formally.

2. THE MAIN THEOREM ON $\mathbb{R}^n$

DEFINITION 2.1.1. *A scalar theory on $\mathbb{R}^n$, with mass m, is given by a collection of functionals*
$$I[L] = \mathcal{O}^+(\mathscr{S}(\mathbb{R}^n), C^\infty((0,\infty)))[[\hbar]]$$
where L is the coordinate on $(0,\infty)$, such that:

(1) *The renormalization group equation*
$$W\left(P(L,L'), I[L]\right) = I[L']$$
is satisfied, where
$$P(L,L') = \int_L^{L'} e^{-tm^2} K_t \mathrm{d}t$$
is the propagator.

(2) *The following locality axiom holds. Let us regard $I_{i,k}[L]$ as a distribution on $\mathbb{R}^{nk}$, and let $C \subset \mathbb{R}^{nk}$ be a compact subset in the complement of the small diagonal. Then, for all functions $f \in \mathscr{S}(\mathbb{R}^{nk})$ with compact support on f,*
$$\lim_{L\to 0} I_{i,k}[L](f) = 0$$

As before, we will let $\mathscr{T}^{(\infty)}$ denote the set of theories, and let $\mathscr{T}^{(n)}$ define the set of theories defined modulo $\hbar^{n+1}$.

The construction of counterterms in the previous section applies, *mutatis mutandis*, to yield the following theorem.

THEOREM 2.1.2. *The space $\mathscr{T}^{(m+1)}$ is a principal bundle over $\mathscr{T}^{(m)}$ for the group $\mathcal{O}_{loc}(\mathscr{S}(\mathbb{R}^n))$, in a canonical way, and $\mathscr{T}^{(0)}$ is canonically isomorphic to the subset of $\mathcal{O}_{loc}(\mathscr{S}(\mathbb{R}^n))$ of functionals which are at least cubic.*

The choice of a renormalization scheme yields a section of each torsor $\mathscr{T}^{(m+1)} \to \mathscr{T}^{(m)}$, and so a bijection between the set $\mathscr{T}^{(\infty)}$ of theories and the set $\mathcal{O}_{loc}^+(\mathscr{S}(\mathbb{R}^n))[[\hbar]]$ of $\hbar$-dependent translation invariant functionals on $\mathscr{S}(\mathbb{R}^n)$ which are at least cubic modulo $\hbar$.

PROOF. The proof is along the same lines as the proof of the same theorem on compact manifolds. Thus, I will sketch only what needs to be changed.

First, we will show how to construct a theory from a translation-invariant local action functional
$$I = \sum \hbar^i I_{i,k} \in \mathcal{O}_l^+(\mathscr{S}(\mathbb{R}^n))[[\hbar]].$$

We will do this in two steps. The first step will show the result with a modified propagator, and the second step will show how to get a theory for the actual propagator from a theory for this modified propagator.

Let f be any compactly supported function on $\mathbb{R}^n$ which is 1 in a neighbourhood of 0. Let
$$\widetilde{P}(\varepsilon, L)(x,y) = f(x-y) P(\varepsilon, L)(x,y) = f(x-y) t^{-n/2} e^{-\|x-y\|^2/t} \in C^\infty(\mathbb{R}^{2n}).$$

This will be our modified propagator. Note that $\widetilde{P}(\varepsilon, L)(x,y)$ is zero if $x-y$ is sufficiently large, and that $\widetilde{P}(\varepsilon, L) = P(\varepsilon, L)$ if $x-y$ is sufficiently small.

Suppose, by induction, we have constructed translation invariant local counterterms
$$I_{i,k}^{CT}(\varepsilon) \in \mathscr{O}_l(\mathscr{S}(\mathbb{R}^n)) \otimes_{alg} \mathscr{P}((0,\infty)_{\varepsilon}ps),$$
for all $(i,k) < (I,K)$, such that
$$\lim_{\varepsilon \to 0} W_{i,k}\left(\widetilde{P}(\varepsilon, L), I - \sum_{(r,s) \le (i,k)} \hbar^r I_{r,s}^{CT}(\varepsilon)\right)$$
exists for all $(i,k) < (I,K)$.

The results proved in Appendix 1 imply that
$$W_{I,K}\left(\widetilde{P}(\varepsilon, L), I - \sum_{(r,s) \le (i,k)} \hbar^r I_{r,s}^{CT}(\varepsilon)\right)(\phi, L) \simeq \sum f_i(\varepsilon) \Psi_i(\phi, L)$$
where $\simeq$ means there is a small ε asymptotic expansion of this form; the functions $f_i(\varepsilon) \in \mathscr{P}((0,\infty))$ are smooth functions of ε, and are periods, as defined in Section 9 of Chapter 2.

The distributions
$$\Psi_i(\phi, L) : \mathscr{S}(\mathbb{R}^{nk}) \to C^\infty((0,\infty)_L)$$
are translation invariant, and are also supported on the product of the small diagonal $\mathbb{R}^n \subset \mathbb{R}^{nk}$ with a compact subset $K \subset \mathbb{R}^{n(k-1)}$. Thus, they are good distributions, that is
$$\Psi_i \in \mathcal{D}_g(\mathbb{R}^{nk}, C^\infty((0,\infty)_L)).$$

Then, as before, we let
$$I_{(I,K)}^{CT}(\varepsilon) = \operatorname{Sing}_\varepsilon \left(W_{(I,K)}\left(\widetilde{P}(\varepsilon, L), I - \sum_{(r,s) \le (i,k)} \hbar^r I_{(r,s)}^{CT}(\varepsilon)\right)\right).$$

The existence of the small ε asymptotic expansion mentioned above implies that it makes sense to take the singular part. This singular part is an element
$$I_{(I,K)}^{CT}(\varepsilon) \in \mathcal{D}_g(\mathbb{R}^{nk}, C^\infty((0,\infty)_L)) \otimes_{alg} C^\infty((0,\infty)_\varepsilon).$$

As before, this singular part is independent of L; this implies that the counterterm $I_{(I,K)}^{CT}$ is local.

This allows us to define
$$\widetilde{I}[L] = \lim_{\varepsilon \to 0} W\left(\widetilde{P}(\varepsilon, L), I - I^{CT}(\varepsilon)\right)$$
as before. Each distribution
$$\widetilde{I}_{(I,K)}[L] : \mathscr{S}(\mathbb{R}^{nk}) \to C^\infty((0,\infty)_L)$$

is translation invariant, and is supported on the product of the small diagonal $\mathbb{R}^n \subset \mathbb{R}^{nk}$ with a compact neighbourhood $K \subset \mathbb{R}^{n(k-1)}$. Therefore each $\widetilde{I}_{(I,K)}[L]$ is a good distribution.

Thus, the collection of effective interactions
$$\widetilde{I}[L] \in \mathscr{O}(\mathscr{S}(\mathbb{R}^n, C^\infty((0,\infty)_L)))$$
define a theory for the modified propagator $\widetilde{P}(\varepsilon, L)$. It is easy to see that the locality axiom holds: if we view $\widetilde{I}_{(i,k)}[L]$ as a distribution on $\mathbb{R}^{nk}$, and fix any $f \in \mathscr{S}(\mathbb{R}^{nk})$ which has compact support way from the small diagonal, the limit
$$\lim_{L \to 0} \widetilde{I}_{(i,k)}[L](f) = 0.$$

Now we need to show how to get a theory for the actual propagator $P(\varepsilon, L)$ from a theory for the modified propagator $\widetilde{P}(\varepsilon, L)$.

Note that the distribution $P(0, L) - \widetilde{P}(0, L)$ is actually a smooth function, because $P(0, L)$ and $\widetilde{P}(0, L)$ agree in a neighbourhood of the diagonal. In fact,
$$\left(P(0,L) - \widetilde{P}(0,L)\right)(x,y) = (1 - f(x-y)) \int_{t=0}^{L} t^{-n/2} e^{-m^2 t} e^{-\|x-y\|^2/t} dt$$
$$= F(x-y)$$
where
$$F(x) = (1 - f(x)) \int_{t=0}^{L} t^{-n/2} e^{-m^2 t} e^{-\|x\|^2/t} dt.$$
Since $f(x) = 1$ in a neighbourhood of $x = 0$, the function $F(x)$ is Schwartz.

Further, the function F and all of its derivatives decay faster than $e^{-\|x\|^2/2L}$ at ∞.

It follows from this and from the fact that the functionals $\widetilde{I}[L]$ are good that we can apply Theorem 1.4.1, so that the expression
$$W\left(P(0,L) - \widetilde{P}(0,L), \widetilde{I}[L]\right)$$
is well-defined. This expression defines $I[L]$; and, Theorem 1.4.1 implies that $I[L]$ is a good distribution.

It is immediate that $I[L]$ satisfies the renormalization group equation
$$I[L] = W\left(P(\varepsilon, L), I[\varepsilon]\right)$$
and the locality axiom. Thus, $\{I[L]\}$ describes the theory associated to the local functional I.

It remains to show that every theory arises in this way. We will show this by induction, as before. Suppose we have a theory described by a collection of effective interactions $\{I[L]\}$. Suppose we have a local action functional
$$J = \sum_{(i,k) < (I,K)} \hbar^i J_{(i,k)}$$

with associated effective interactions $J[L]$. Suppose that
$$J_{(i,k)}[L] = I_{(i,k)}[L]$$
for all $(i,k) < (I,K)$. We need to find some $J'_{(I,K)}$ such that if we let
$$J' = J + \hbar^I J_{(I,K)}$$
then
$$J'_{(I,K)}[L] = I_{(I,K)}[L].$$

Let
$$J'_{(I,K)} = I_{(I,K)}[L] - J_{(I,K)}[L].$$
The renormalization group equation implies that $J_{(I,K)}$ as so defined is independent of L; and it is immediate that, if we define $J' = J + \hbar^I J_{(I,K)}$, then
$$J'_{(I,K)}[L] = I_{(I,K)}[L].$$
It remains to check that $J_{(I,K)}$ is local.

The locality axiom satisfied by $I_{(I,K)}[L]$ and $J_{(I,K)}[L]$ implies that $J_{(I,K)}$ is a distribution on $\mathbb{R}^{nk}$ supported on the small diagonal. Any such distribution which is translation invariant is a local action functional. □

3. Vector-bundle valued field theories on $\mathbb{R}^n$

We would also like to work theories on $\mathbb{R}^n$ where the space of fields is the space of sections of some vector bundle. As with scalar field theories, we are only interested in translation invariant theories.

As Section 13 of Chapter 2, everything will depend on some auxiliary manifold with corners X, equipped with a sheaf A of commutative graded algebras over the sheaf of algebras C_X^∞. The space of global sections of A will be denoted, as before, by $\mathscr{A}$.

The data we need to state our main theorem in this case is the following.

(1) A finite dimensional graded vector space E. We let
$$\mathscr{E} = E \otimes \mathscr{S}(\mathbb{R}^n).$$

Thus, $\mathscr{E}$ is the space of Schwartz sections of the trivial vector bundle $E \times \mathbb{R}^n$.

(2) A degree 0 symmetric element
$$K_l \in C^\infty(\mathbb{R}^n \times \mathbb{R}^n) \otimes E^{\otimes 2} \otimes C^\infty((0,\infty)_l) \otimes \mathscr{A},$$
such that, in some basis e_i of E, K_l can be written
$$K_l = \sum P_{i,j}(x-y, l^{\frac{1}{2}}, l^{-\frac{1}{2}}) e^{-\|x-y\|^2/l} e_i \otimes e_j$$
where each
$$P_{i,j} \in \mathscr{A}[x_k - y_k, l^{\frac{1}{2}}, l^{-\frac{1}{2}}]$$

is a polynomial in the variables $x_k - y_k$ and $l^{\pm \frac{1}{2}}$, with coefficients in $\mathscr{A}$.

The kernel K_l will, in all examples, be a kernel obtained from the heat kernel for some elliptic operator by differentiating in the variables $x_k - y_k$ some number of times.

The propagator will be of the form
$$P(\varepsilon, L) = \int_\varepsilon^L K_l \mathrm{d}l.$$

3.1. Space of functions. As for scalar field theories, there are various spaces of functionals which are relevant. Recall that
$$\mathscr{D}_g(\mathbb{R}^{nk}, \mathscr{A}) \subset \mathscr{D}(\mathbb{R}^{nk}, \mathscr{A})$$
as defined in Section 1 is the space of good distributions, which, heuristically, are those $\mathscr{A}$-valued translation invariant distributions of rapid decay away from the diagonal.

We will let
$$\mathscr{O}(\mathscr{E}, \mathscr{A}) = \prod_{k>0} \left(\mathscr{D}_g(\mathbb{R}^{nk}, \mathscr{A}) \otimes (E^\vee)^{\otimes k} \right)_{S_k}.$$
This is the space of $\mathscr{A}$-valued functionals on $\mathscr{E}$ whose Taylor components are given by good distributions. Since good distributions are defined to be translation invariant, every element of $\mathscr{O}(\mathscr{E}, \mathscr{A})$ is translation invariant.

A good distribution $\Phi \in \mathscr{D}_g(\mathbb{R}^{nk}, \mathscr{A})$ is called *local* if it is supported on the small diagonal $\mathbb{R}^n \subset \mathbb{R}^{nk}$. Any local good distribution can be written as a finite sum of distributions of the form
$$f(x_1, \ldots, x_k) \mapsto \sum_I \int_{x \in \mathbb{R}^n} a_I (\partial_I f)(x, \ldots, x)$$
where $a_I \in \mathscr{A}$ and $I \in (\mathbb{Z}_{\geq 0})^{nk}$ are multi-indices.

We let
$$\mathscr{O}_{loc}(\mathscr{E}, \mathscr{A}) \subset \mathscr{O}(\mathscr{E}, \mathscr{A})$$
denote the subspace of functionals whose Taylor components are all local.

We will let
$$\mathscr{O}^+(\mathscr{E}, \mathscr{A})[[\hbar]] \subset \mathscr{O}(\mathscr{E}, \mathscr{A})[[\hbar]]$$
be the subspace of functionals which are at least cubic modulo the ideal in $\mathscr{A}[[\hbar]]$ generated by $\mathscr{I}$ and $\hbar$. The subspace
$$\mathscr{O}_{loc}^+(\mathscr{E}, \mathscr{A})[[\hbar]] \subset \mathscr{O}_{loc}(\mathscr{E}, \mathscr{A})[[\hbar]]$$
is defined in the same way.

Theorem 1.4.1 allows us to define the renormalization group flow
$$\mathscr{O}_{loc}^+(\mathscr{E}, \mathscr{A})[[\hbar]] \to \mathscr{O}_{loc}^+(\mathscr{E}, \mathscr{A})[[\hbar]]$$
$$I \mapsto W\left(P(\varepsilon, L), I\right).$$

Now we are ready to define the notion of theory in this context.

DEFINITION 3.1.1. *A family of theories on $\mathscr{E}$, over the ring $\mathscr{A}$, is a collection*
$$\{I[L] \in \mathscr{O}^+(\mathscr{E}, \mathscr{A})[[\hbar]]\}$$
of translation invariant effective interactions such that

(1) *Each $I[L] \in \mathscr{O}^+(\mathscr{E}, \mathscr{A})[[\hbar]]$ is of degree 0.*
(2) *The renormalization group equation*
$$I[L] = W\left(P(\varepsilon, L), I[\varepsilon]\right)$$
is satisfied.
(3) *The following locality axiom holds. Let us expand, as usual,*
$$I[L] = \sum \hbar^i I_{(i,k)}[L]$$
where each $I_{(i,k)}[L]$ is an S_k-invariant map $\mathscr{E}^{\otimes k} \to \mathscr{A}$. Then, we require that for all elements $f \in \mathscr{E}^{\otimes k}$ which are compactly supported away from the small diagonal in $\mathbb{R}^{nk}$, and for all $x \in X$,
$$\lim_{L \to 0} I_{(i,k)}[L](f)_x = 0.$$
This limit is taken in the finite dimensional graded vector space A_x. The subscript x denotes restricting an element of $\mathscr{A} = \Gamma(X, A)$ to its value at the fibre A_x of A above $x \in X$.

We let $\mathscr{T}^{(\infty)}(\mathscr{E})$ denote the set of theories, and $\mathscr{T}^{(n)}(\mathscr{E})$ denote the set of theories defined modulo $\hbar^{n+1}$.

As before, the main theorem is:

THEOREM 3.1.2. *The space $\mathscr{T}^{(m+1)} \to \mathscr{T}^{(m)}$ is a torsor for the space $\mathscr{O}^0_{loc}(\mathscr{E}, \mathscr{A})[[\hbar]]$ of local action functionals of degree 0 on $\mathscr{E}$. Further, $\mathscr{T}^{(0)}$ is canonically isomorphic to the space of degree 0 local action functionals on $\mathscr{E}$ which are at least cubic.*

If we choose a renormalization scheme, then we find a section of each torsor $\mathscr{T}^{(m-1)} \to \mathscr{T}^{(m)}$. Thus, the choice of renormalization scheme yields a bijection
$$\mathscr{T}^{(\infty)}(\mathscr{E}, \mathscr{A}) \cong \mathscr{O}^{+,0}_{loc}(\mathscr{E}, \mathscr{A})[[\hbar]]$$
between the set of theories and the set of (translation-invariant) local action functionals on $\mathscr{E}$ which are of degree 0, and which are at least cubic modulo the ideal in $\mathscr{A}[[\hbar]]$ generated by $\mathscr{I}$ and $\hbar$.

PROOF. As before, we will prove the renormalization scheme dependent version of the theorem; the renormalization scheme independent version is a simple corollary.

There are two steps to the proof. First, we have to construct the theory associated to a local action functional $I \in \mathscr{O}_{loc}(\mathscr{E}, \mathscr{A})$. The second step is to show that every theory arises in this way.

The proof of the first step is essentially the same as that of the corresponding statement for scalar field theories, Theorem 2.1.2. The fact that we are dealing with sections of a vector bundle rather than functions presents no extra difficulties. The fact that we are dealing with a family of theories, parameterized by the algebra $\mathscr{A}$, makes the analysis slightly harder. However, the results in Appendix 1, as well as those in Section 1, are all stated and proved in a context where everything depends on an auxiliary ring $\mathscr{A}$. Thus, the argument of Theorem 2.1.2 applies with basically no changes.

The converse requires slightly more work. Suppose we have a theory, given by a collection of effective interactions $I_{(i,k)}[L] \in \mathscr{O}^+(\mathscr{E}, \mathscr{A} \otimes C^\infty((0,\infty)_L))[[\hbar]]$. Suppose, by induction, that we have a local action functional

$$J = \sum_{(i,k)<(I,K)} \hbar^i J_{(i,k)} \in \mathscr{O}^+_{loc}(\mathscr{E}, \mathscr{A})[[\hbar]]$$

with associated effective interactions $J[L]$, such that

$$J_{(i,k)}[L] = I_{(i,k)}[L] \text{ for all } (i,k) < (I,K).$$

As usual, we need to find some $J'_{(I,K)} \in \mathscr{O}_{loc}(\mathscr{E}, \mathscr{A})$, homogeneous of degree K, such that if we set

$$J' = J + \hbar^I J'_{(I,K)}$$

then

$$J'_{(I,K)}[L] = I_{(I,K)}[L].$$

Since

$$J'_{(I,K)}[L] = J_{(I,K)}[L] + J_{(I,K)},$$

we must have

$$J_{(I,K)} = I_{(I,K)}[L] - J_{(I,K)}[L].$$

The right hand side of this equation is independent of L. Thus, the distribution $J_{(I,K)}$ is an element

$$J_{(I,K)} \in (E^\vee)^{\otimes K} \otimes \mathrm{Hom}(\mathscr{S}(\mathbb{R}^{nK}), \mathscr{A}).$$

We need to show that it is local.

Since both $I[L]$ and $J[L]$ satisfy the locality axiom defining a theory, the distribution $J_{(I,K)}$ is supported on the small diagonal $\mathbb{R}^n \subset \mathbb{R}^{nK}$. In addition, it is translation invariant.

Every element of $\mathrm{Hom}(\mathscr{S}(\mathbb{R}^{nK}), \mathscr{A})$ which is supported on the small diagonal, and which is translation invariant, is local. □

4. Holomorphic aspects of theories on $\mathbb{R}^n$

In this section we will show that the Fourier transforms of the effective interactions describing a theory are entire holomorphic functions on products of complexified momentum space. The results in this section will not be used in the rest of this book. However, the holomorphic nature of the Fourier transform of the effective interactions is worth noting, as it lends some insight into the problem of analytically continuing to Lorentzian signature.

4.1. Recall that Fourier transform is a continuous linear isomorphism $\mathscr{F} : \mathscr{S}(\mathbb{R}^l) \to \mathscr{S}(\mathbb{R}^l)$ for any l. Continuity of Fourier transform means that it acts on spaces associated to $\mathscr{S}(\mathbb{R}^l)$, such as spaces of tempered distributions.

Suppose we have a translation invariant vector-valued field theory on $\mathbb{R}^n$, described by a family of effective interactions

$$I[L] \in \mathscr{O}^+(\mathscr{E}, \mathscr{A} \otimes C^\infty((0,\infty)_L))[[\hbar]].$$

The Taylor components $I_{(i,k)}[L]$ of $I[L]$ are good distributions

$$I_{(i,k)}[L] \in \mathcal{D}_g(\mathbb{R}^{nk}, \mathscr{A} \otimes C^\infty((0,\infty)_L)).$$

In particular, they are tempered distributions

$$I_{(i,k)}[L] : \mathscr{S}(\mathbb{R}^{nk}) \to \mathscr{A} \otimes C^\infty((0,\infty)_L).$$

Thus, we can take the Fourier transform

$$\mathscr{F}(I_{(i,k)}[L]) : \mathscr{S}(\mathbb{R}^{nk}) \to \mathscr{A} \otimes C^\infty((0,\infty)_L),$$

defined by composing $I_{(i,k)}[L]$ with the Fourier transform on $\mathscr{S}(\mathbb{R}^{nk})$. Continuity of the Fourier transform implies that $\mathscr{F}(I_{(i,k)}[L])$ is continuous.

We would like to describe the form taken by the Fourier transform of $I_{(i,k)}[L]$. It turns out that it is a holomorphic function of a certain kind.

We will let $\mathbb{R}^{n(k-1)} \subset \mathbb{R}^{nk}$ denote the orthogonal complement to the small diagonal $\mathbb{R}^n \subset \mathbb{R}^{nk}$. We will let $\text{Hol}(\mathbb{C}^l)$ denote the space of entire holomorphic functions on any $\mathbb{C}^l$. This is a nuclear Fréchet space; we will let $\text{Hol}(\mathbb{C}^l) \otimes \mathscr{A}$ denote the projective tensor product. Note that $\text{Hol}(\mathbb{C}^l) \otimes \mathscr{A}$ is the subspace of $C^\infty(\mathbb{C}^l) \otimes \mathscr{A}$ of those elements $\phi(z_1, \ldots, z_l, x)$ such that $\frac{\partial}{\partial \overline{z}_i} \phi = 0$ for $i = 1, \ldots, l$.

The following proposition describes the form taken by the Fourier transform of $I_{(i,k)}[L]$.

PROPOSITION 4.1.1. *There exists some*

$$\Psi_{(i,k)}[L] \in \text{Hol}(\mathbb{C}^{n(k-1)}) \otimes \mathscr{A} \otimes C^\infty((0,\infty)_L)$$

such that, for all $f \in \mathscr{S}(\mathbb{R}^{nk})$,

$$I_{(i,k)}[L](\mathscr{F}(f)) = \int_{\substack{x_1,\ldots,x_n \in \mathbb{R}^n \\ \sum x_i = 0}} \Psi_{(i,k)}[L](x_1, \ldots, x_{n-1}) f(x_1, \ldots, x_n).$$

In other words, the Fourier transform of $I_{(i,k)}[L]$ is the operation which takes a function f on $\mathbb{R}^{nk}$, restricts it to the subspace $\mathbb{R}^{n(k-1)}$, and then integrates against $\Psi_{(i,k)}[L]$.

A corollary of this proposition, and of Proposition 1.3.2 in Chapter 4, is the following.

COROLLARY 4.1.2. *The functions $\Psi_{(i,k)}[L]$ appearing above can be written as finite sums*

$$\Psi_{(i,k)}[L](x_1,\ldots,x_{n-1}) = \sum \Gamma^{r,s}_{i,k}(\sqrt{L}x_1,\ldots,\sqrt{L}x_{n-1})L^{r/2}(\log L)^s$$

where $r \in \mathbb{Z}$, $s \in \mathbb{Z}_{\geq 0}$, and $\Gamma^{r,s}_{i,k} \in \mathrm{Hol}(\mathbb{C}^{n(k-1)})$.

This corollary shows that the $\Psi_{(i,k)}[L]$ have very tightly constrained behaviour as functions of L.

4.2. Proposition 4.1.1 follows immediately from a general lemma about the holomorphic nature of the Fourier transform of a distribution of rapid decay.

Recall from Section 1 that

$$\mathscr{T}(\mathbb{R}^m) \subset C^{\infty}(\mathbb{R}^m)$$

refers to the space of functions f all of whose derivatives are bounded by $e^{b\|x\|^2}$, for all b. The space $\mathscr{T}(\mathbb{R}^m)$ has a sequence of norms $\|-\|_{b,l}$, for $b \in \mathbb{R}_{>0}$ and $l \in \mathbb{Z}_{\geq 0}$, defined by

$$\|f\|_{b,l} = \mathrm{Sup}_{x \in \mathbb{R}^m} \sum_{|I| \leq l} \left| e^{b\|x\|^2} \partial_I f \right|$$

where the sum is over multi-indices $I = (I_1,\ldots,I_l) \in (\mathbb{Z}_{\geq 0})^m$, and $|I| = \sum I_i$.

The space $\mathscr{T}(\mathbb{R}^m)$ is given the topology defined by the family of norms $\|-\|_{b,l}$. A continuous linear function

$$\Phi : \mathscr{T}(\mathbb{R}^m) \to \mathscr{A}$$

is an $\mathscr{A}$-valued distribution of rapid decay.

Since the inclusion $\mathscr{S}(\mathbb{R}^m) \hookrightarrow \mathscr{T}(\mathbb{R}^m)$ is continuous, any $\mathscr{A}$-valued distribution of rapid decay is in particular an $\mathscr{A}$-valued tempered distribution. Thus, we can take its Fourier transform.

LEMMA 4.2.1. *Let $\Phi : \mathscr{T}(\mathbb{R}^m) \to \mathscr{A}$ be an $\mathscr{A}$-valued distribution of rapid decay. Then there exists an entire holomorphic function*

$$\widehat{\Phi} \in \mathrm{Hol}(\mathbb{C}^m) \otimes \mathscr{A}$$

such that, for all $f \in \mathscr{S}(\mathbb{R}^m)$,

$$\Phi(\mathscr{F}(f)) = \int_{x \in \mathbb{R}^m} \widehat{\Phi}(x) f(x).$$

PROOF. Observe that the map

$$\mathbb{C}^m \to \mathscr{T}(\mathbb{R}^m)$$
$$x \mapsto e^{i\langle y, x\rangle}$$

is holomorphic (meaning that it is smooth and satisfies the Cauchy-Riemann equation).

Define a function
$$\widehat{\Phi} : \mathbb{C}^m \to \mathscr{A}$$
by
$$\widehat{\Phi}(x) = \Phi_y(e^{i\langle y,x\rangle})$$
(where the subscript in Φ_y means that Φ is acting on the y variable).

Since $\widehat{\Phi}$ is the composition of a holomorphic map $\mathbb{C}^m \to \mathscr{T}(\mathbb{R}^m)$ with a continuous linear map $\mathscr{T}(\mathbb{R}^m) \to \mathscr{A}$, $\widehat{\Phi}$ is holomorphic. Thus,
$$\widehat{\Phi} \in \mathrm{Hol}(\mathbb{C}^m) \otimes \mathscr{A}.$$

Further, the map
$$\mathrm{Hom}(\mathscr{T}(\mathbb{R}^m), \mathscr{A}) \to \mathrm{Hol}(\mathbb{C}^m) \otimes \mathscr{A}$$
$$\Phi \mapsto \widehat{\Phi}$$
is continuous (this follows from the fact that $e^{i\langle y,x\rangle}$, and any number of its x and y derivatives, can be bounded by some $e^{b\|y\|^2}$ uniformly for x in a compact set).

It remains to show that $\widehat{\Phi}$ is indeed the Fourier transform of Φ. This is true for all Φ in
$$C_c^\infty(\mathbb{R}^m) \otimes \mathscr{A} \subset \mathrm{Hom}(\mathscr{T}(\mathbb{R}^m), \mathscr{A}),$$
where $C_c^\infty(\mathbb{R}^m)$ denotes the space of smooth functions with compact support.

The former subspace is dense in the latter; continuity of $\Phi \mapsto \widehat{\Phi}$ implies that $\widehat{\Phi}$ must be the Fourier transform for all Φ. $\square$

COROLLARY 4.2.2. *Let $\Phi \in \mathcal{D}_g(\mathbb{R}^{nk}, \mathscr{A})$ be a good distribution. Then there exists some $\widehat{\Phi} \in \mathrm{Hol}(\mathbb{C}^{n(k-1)}) \otimes \mathscr{A}$ such that, for all $f \in \mathscr{S}(\mathbb{R}^{nk})$,*
$$\Phi(\mathscr{F}(f)) = \int_{\substack{x_1,\ldots,x_k \in \mathbb{R}^n \\ \sum x_i = 0}} \widehat{\Phi}(x_1,\ldots,x_{k-1}) f(x_1,\ldots,x_k).$$

PROOF. Let us write $\mathbb{R}^{nk} = \mathbb{R}^n \times \mathbb{R}^{n(k-1)}$, where $\mathbb{R}^n \subset \mathbb{R}^{nk})$ is the small diagonal, and $\mathbb{R}^{n(k-1)}$ is its orthogonal complement.

Translation invariance of Φ implies that it can be written as a direct product
$$\Phi = 1 \boxtimes \Phi'$$
where 1 is the distribution on $\mathbb{R}^n$ given by integrating, and Φ' is an $\mathscr{A}$-valued tempered distribution on $\mathbb{R}^{n(k-1)}$. The rapid decay conditions satisfied by a good distribution mean that Φ' gives a continuous linear map
$$\Phi' : \mathscr{T}(\mathbb{R}^{n(k-1)}) \to \mathscr{A}.$$

Fourier transform commutes with direct product of distributions, and the Fourier transform of the distribution $\int$ is the δ function at the origin. Thus,
$$\mathscr{F}(\Phi) = \delta_{x=0} \boxtimes \mathscr{F}(\Phi').$$

The previous lemma implies that $\mathscr{F}(\Phi')$ is an entire holomorphic function on $\mathbb{C}^{n(k-1)}$, as desired. □

This completes the proof of the proposition.

CHAPTER 4

Renormalizability

1. The local renormalization group flow

1.1. One point of view on renormalizability says that a theory is renormalizable if there are finitely many counterterms. The usual justification for this condition is that it selects, in the infinite dimensional moduli space of theories, a finite dimensional space of well-behaved theories. Finite dimensionality means that renormalizable theories are predictive: they have only finitely many free parameters, which can be fixed by performing finitely many experiments.

Any natural condition which selects a finite dimensional subspace of the space of theories can be regarded as a criterion for renormalizability. The more natural the condition, of course, the better.

The philosophy of this book is that the counterterms themselves have no intrinsic importance; they are simply a tool in the construction of the bijection between theories and local action functionals. The choice of a different renormalization scheme will lead to a different set of counterterms associated to a theory. Thus, we must reject the criterion asking for finitely many counterterms as unnatural.

We will use a renormalization criterion arising from the Wilsonian point of view. This criterion selects, in a natural way, a finite dimensional space of renormalizable theories in the infinite dimensional space of all theories.

The idea is that a theory given by a set of effective interactions $\{I[L]\}$ is well-behaved if $I[L]$ doesn't grow too fast as $L \to 0$, when measured in units appropriate to the length scale L. We will take "not too large" to mean they grow at most logarithmically as $L \to 0$. A theory will be called renormalizable if it satisfies this growth condition and if, in addition, the space of deformations of the theory which also satisfy this growth condition is finite dimensional. This means that one can always specify a renormalizable theory by a finite number of parameters.

1.2. Notation. Throughout this chapter, we will only consider translation invariant theories on $\mathbb{R}^n$. We will use the notation from Chapter 3; thus, $\mathscr{O}(\mathscr{S}(\mathbb{R}^n))$ will refer to the space of translation invariant functionals on $\mathbb{R}^n$, satisfying the technical conditions described in Chapter 3.

Local action functionals will also be translation invariant. We will let

$$\mathscr{O}_{loc}(\mathscr{S}(\mathbb{R}^n)) \subset \mathscr{O}(\mathscr{S}(\mathbb{R}^n))$$

be the space of translation invariant local action functionals on $\mathscr{S}(\mathbb{R}^n)$.

Theories on $\mathbb{R}^n$ which are invariant under $\mathbb{R}^n \ltimes SO(n)$ yield theories on any manifold with a flat metric.

1.3. Now we will start to make the definition of renormalizability precise. First, we need to introduce the local renormalization group flow. This is a flow on the space of theories which is a combination of the renormalization group flow we already know, a rescaling on $\mathbb{R}^n$, and a rescaling of the field ϕ. These rescalings are the change of units mentioned above.

Let us define an operation
$$R_l : \mathscr{S}(\mathbb{R}^n) \to \mathscr{S}(\mathbb{R}^n)$$
$$R_l(\phi)(x) = l^{n/2-1} \phi(lx)$$
for $x \in \mathbb{R}^n$. If $I \in \mathscr{O}(\mathscr{S}(\mathbb{R}^n))$ is a functional on $\mathscr{S}(\mathbb{R}^n)$, define
$$R_l^*(I)(\phi) = I(R_{l^{-1}}\phi).$$
Thus
$$R_l^*(I)(\phi(x)) = I(l^{1-n/2}\phi(l^{-1}x))$$
The reason for this definition is that then the pairing between functionals and fields is invariant:
$$I(\phi) = (R_l^* I)(R_l \phi).$$

DEFINITION-LEMMA 1.3.1. *The local renormalization group flow*
$$\mathcal{RG}_l : \mathscr{T}^{(\infty)} \to \mathscr{T}^{(\infty)}$$
on the space of theories is defined as follows. If $\{I[L]\}$ is a collection of effective interactions defining a theory, then
$$\mathcal{RG}_l(\{I[L]\}) = \{\mathcal{RG}_l(I[L])\}$$
is defined by
$$\mathcal{RG}_l(I[L]) = R_l^* \left(I[l^2 L] \right).$$
This collection of effective interactions $I[L] \in \mathscr{O}(\mathscr{S}(\mathbb{R}^n), C^\infty((0,\infty)_l) \otimes C^\infty((0,\infty)_L)$ defines a smooth family of theories parameterized by l, in the sense of Definition 3.1.1.

Before we proceed, I should comment on the choice of the power $l^{n/2-1}$ which appears in the definition of the local renormalization group flow. Let us suppose we have a classical field theory described by a local action functional S. If $c \in \mathbb{R}^\times$, then we can define a new action functional S_c by $S_c(\phi) = S(c\phi)$. The action functionals S and S_c describe equivalent classical field theories: they are related by an automorphism of the vector bundle of fields covering the identity on the space-time manifold.

The local renormalization group flow is defined by a certain change of coordinates on the space of fields, of the form $\phi(x) \mapsto c(l)\phi(lx)$, where $c(l) \in \mathbb{R}^\times$. Because changes of coordinates of the form $\phi(x) \to c\phi(x)$ take a field theory to an equivalent field theory, it doesn't matter what function

$c(l)$ we choose (as long as we consider how these changes of coordinates act on *actions* instead of interactions).

We have been dealing throughout with interactions, and suppressing the quadratic term in the action. Thus, it is convenient to choose $c(l)$ so that the action for the massless free field is preserved under this change of coordinates. Taking $c(l) = l^{n/2-1}$ is the unique way to do this.

Now let us turn to the proof of the definition-lemma stated above.

PROOF. In order to check that $\mathcal{RG}_l(I[L])$ defines a theory, we need to check that the locality axiom and the renormalization group equation are satisfied.

The locality axiom of is immediate. We need to check that $\mathcal{RG}_l(I[L])$ satisfies the renormalization group equation. Now,

$$P(\varepsilon, L)(x, y) = \int_\varepsilon^L t^{-n/2} e^{-\|x-y\|^2/t} \mathrm{d}t$$

so that

$$R_l P(\varepsilon, L) = l^{n-2} \int_\varepsilon^L t^{-n/2} e^{-l^2 \|x-y\|^2/t} \mathrm{d}t$$
$$= \int_{l^{-2}\varepsilon}^{l^{-2}L} u^{-n/2} e^{-\|x-y\|^2/u} \mathrm{d}u \text{ where } u = l^{-2} t$$
$$= P(l^{-2}\varepsilon, l^{-2} L).$$

The way we defined the action R_l^* on functionals $\mathcal{O}(\mathcal{S}(\mathbb{R}^n))$ means that, for all $P \in C^\infty(M^2)$, and all $I \in \mathcal{O}(\mathcal{S}(\mathbb{R}^n))$,

$$R_l^*(\partial_P I) = \partial_{R_l P} R_l^* I.$$

The renormalization group equation for $I[L]$ says that

$$I[l^2 L] = \hbar \log \left\{ \exp\left(\hbar \partial_{P(l^2 \varepsilon, l^2 L)}\right) \exp\left(\hbar^{-1} I[l^2 \varepsilon]\right) \right\}.$$

Applying R_l^* to both sides, we find

$$\mathcal{RG}_l(I[L]) = \hbar \log \left\{ \exp\left(\hbar \partial_{P(\varepsilon, L)}\right) \exp\left(\hbar^{-1} R_l^* I[l^2 \varepsilon]\right) \right\}$$
$$= \hbar \log \left\{ \exp\left(\hbar \partial_{P(\varepsilon, L)}\right) \exp\left(\hbar^{-1} \mathcal{RG}_l(I[\varepsilon])\right) \right\}$$
$$= W\left(P(\varepsilon, L), \mathcal{RG}_l(I[\varepsilon])\right)$$

as desired.

□

PROPOSITION 1.3.2. *Let $\{I[L]\}$ be a theory on $\mathbb{R}^n$ (as always, translation invariant). Then,*

$$\mathcal{RG}_l(I[L]) \in \mathcal{O}^+(\mathcal{S}(\mathbb{R}^n))[[\hbar]] \otimes \mathbb{C}[\log l, l, l^{-1}].$$

In other words, each $\mathcal{RG}_l(I_{i,k}[L])$, as a function of l, is a polynomial in l, l^{-1} and $\log l$.

We will prove this later.

A term in $\mathcal{RG}_l(I[L])$ is called
(1) *relevant* if it varies as $l^k(\log l)^r$ for some $k \geq 0$ and some $r \in \mathbb{Z}_{\geq 0}$.
(2) *irrelevant* if it varies as $l^k(\log l)^r$ for some $k < 0$ and some $r \in \mathbb{Z}_{\geq 0}$.
(3) *marginal* if it varies as $(\log l)^r$ for some $r \in \mathbb{Z}_{\geq 0}$.

Thus, marginal terms are a subset of relevant terms. Logarithmic terms only appear because of quantum effects; at the classical level (modulo $\hbar$) all terms scale by l^k for some $k \in \frac{1}{2}\mathbb{Z}$.

DEFINITION 1.3.3. *A theory $\{I[L]\}$ is* relevant *if, for each L, $\mathcal{RG}_l(I[L])$ consists entirely of relevant terms; or, in other words, if*
$$\mathcal{RG}_l(I[L]) \in \mathscr{O}^+(\mathscr{S}(\mathbb{R}^n))[[\hbar]] \otimes \mathbb{C}[\log l, l].$$
A theory is marginal *if, for each L, each $\mathcal{RG}_l(I[L])$ consists entirely of marginal terms; that is,*
$$\mathcal{RG}_l(I[L]) \in \mathscr{O}^+(\mathscr{S}(\mathbb{R}^n))[[\hbar]] \otimes \mathbb{C}[\log l].$$
Let
$$\mathscr{R}^{(n)} \subset \mathscr{T}^{(n)}$$
(respectively,
$$\mathscr{M}^{(n)} \subset \mathscr{T}^{(n)})$$
denote the subset of the space $\mathscr{T}^{(n)}$ of theories consisting of relevant (respectively, marginal) theories defined modulo $\hbar^n$. Let $\mathscr{R}^{(\infty)} = \varprojlim \mathscr{R}^{(n)}$ and let $\mathscr{M}^{(\infty)} = \varprojlim \mathscr{M}^{(n)}$ be the spaces of relevant and marginal theories defined to all orders in $\hbar$.

In this definition, and throughout this chapter, we consider only translation invariant theories.

Passing to small length scales (that is, high energy) corresponds to sending $l \to 0$. A theory which is well-behaved at small length scales will have only relevant terms, and a theory which is well-behaved at large length scales will have only marginal and irrelevant terms. A theory which is well-behaved at all length scales will have only marginal terms.

DEFINITION. *A theory on $\mathbb{R}^n$ is* renormalizable *if it is relevant and, term by term in $\hbar$, has only finitely many relevant deformations.*

In other words, a theory $\{I[L]\} \in \mathscr{T}^{(\infty)}$ is renormalizable if $\{I[L]\} \in \mathscr{R}^{(\infty)}$ and, for all finite n, $T_{\{I[L]\}}\mathscr{R}^{(n)}$ is finite dimensional.

A theory is strictly renormalizable *if it is renormalizable and it is marginal.*

A theory is strongly renormalizable *it it is strictly renormalizable, and all of its relevant deformations are marginal; in other words, if $\{I[L]\} \in \mathscr{M}^{(\infty)}$ and $T_{\{I[L]\}}\mathscr{R}^{(\infty)} = T_{\{I[L]\}}\mathscr{M}^{(\infty)}$.*

1.4. The choice of a renormalization scheme leads to a bijection
$$\mathscr{T}^{(\infty)} \cong \mathscr{O}_{loc}^{+}(\mathscr{S}(\mathbb{R}^n))[[\hbar]]$$
between the space of theories and the space of local action functionals. Both theories and functionals are, as always in this chapter, translation invariant.

The local renormalization group flow translates, under this bijection, to an $\mathbb{R}_{>0}$ action on the space of local action functionals. If $I \mathscr{O}_{loc}^{+}(\mathscr{S}(\mathbb{R}^n))[[\hbar]]$ we will let $\mathcal{RG}_l(I)$ be the family of local action functionals arising by the action of the local renormalization group. It follows from the fact that $\mathcal{RG}_l(\{I[L]\})$ is a smooth family of theories, and from Theorem 3, that $\mathcal{RG}_l(I)$ is a smooth family of local action functionals, that is, an element
$$\mathcal{RG}_l(I) \in \mathscr{O}_{loc}^{+}(\mathscr{S}(\mathbb{R}^n), C^\infty((0,\infty)_l))[[\hbar]].$$
Modulo $\hbar$, the action of the local renormalization group on local action functionals is simple. If
$$I \in \mathscr{O}_{loc}^{+}(\mathscr{S}(\mathbb{R}^n))[[\hbar]],$$
then
$$\mathcal{RG}_l(I) = R_l^*(I) \bmod \hbar.$$
At the quantum level more subtle things happen, as we will see shortly.

1.5. Every local action functional $I \in \mathscr{O}_{loc}(\mathscr{S}(\mathbb{R}^n))$ which is homogeneous of degree k with respect to the field $\phi \in \mathscr{S}(\mathbb{R}^n)$ can be written as a finite sum of functionals of the form
$$(\dagger) \qquad \phi \mapsto \int_{\mathbb{R}^n} \phi \, (\partial_{I_2} \phi) \cdots (\partial_{I_k} \phi)$$
where the I_j are multi-indices, and
$$\partial_{I_j} = \left(\frac{\partial}{\partial x_1}\right)^{I_{j,1}} \cdots \left(\frac{\partial}{\partial x_n}\right)^{I_{j,n}}$$
are the corresponding translation invariant differential operators.

The operator R_l^* acts diagonally on the space of functionals, with eigenvalues l^k for $k \in \frac{1}{2}\mathbb{Z}$. We say a functional I has *dimension k* if
$$R_l^* I = l^k I.$$
For example, if I is the functional in equation ($\dagger$), then
$$R_l^*(I) = l^{n+k\left(1-\frac{1}{2}n\right) - \sum |I_j|} I$$
so that I is of dimension $n + k\left(1 - \frac{1}{2}n\right) - \sum |I_j|$.

Let
$$\mathscr{O}_{loc,k}(\mathscr{S}(\mathbb{R}^n)) \subset \mathscr{O}_{loc}(\mathscr{S}(\mathbb{R}^n))$$
be the space of local action functionals of dimension k.

Let
$$\mathscr{O}_{loc,\geq 0}(\mathscr{S}(\mathbb{R}^n)) \subset \mathscr{O}_{loc}(\mathscr{S}(\mathbb{R}^n))$$

be the subspace of functionals which are sums of functionals of non-negative dimension. There are projection maps
$$\mathscr{O}_{loc}(\mathscr{S}(\mathbb{R}^n)) \to \mathscr{O}_{loc,\geq 0}(\mathscr{S}(\mathbb{R}^n))$$
and
$$\mathscr{O}_{loc}(\mathscr{S}(\mathbb{R}^n)) \to \mathscr{O}_{loc,0}(\mathscr{S}(\mathbb{R}^n)).$$

The main theorem proved in this chapter is the following.

THEOREM 1.5.1. *The space $\mathscr{R}^{(m+1)}$ is, in a canonical way, a torsor over $\mathscr{R}^{(m)}$ for the abelian group $\mathscr{O}_{loc,\geq 0}(\mathscr{S}(\mathbb{R}^n))$ of local action functionals of non-negative dimensions. Also, $\mathscr{R}^{(0)}$ is canonically isomorphic to the subspace of $\mathscr{O}_{loc,\geq 0}(\mathscr{S}(\mathbb{R}^n))$ of functionals which are at least cubic.*

In a similar way, $\mathscr{M}^{(m+1)}$ is a torsor over $\mathscr{M}^{(m)}$ for the abelian group $\mathscr{O}_{loc,0}(\mathscr{S}(\mathbb{R}^n))$ of local action functionals of dimension zero. Also, $\mathscr{M}^{(0)}$ is canonically isomorphic to the subspace of $\mathscr{O}_{loc,\,0}(\mathscr{S}(\mathbb{R}^n))$ of functionals of degree 0.

The choice of a renormalization scheme leads to sections of the torsors $\mathscr{M}^{(m+1)} \to \mathscr{M}^{(m)}$ and $\mathscr{R}^{(m+1)} \to \mathscr{R}^{(m)}$, and thus to bijections
$$\mathscr{R}^{(\infty)} \cong \mathscr{O}_{loc,\geq 0}^{+}(\mathscr{S}(\mathbb{R}^n))[[\hbar]]$$
and
$$\mathscr{M}^{(\infty)} \to \mathscr{O}_{loc,0}^{+}(\mathscr{S}(\mathbb{R}^n))[[\hbar]].$$

The following is an immediate corollary.

COROLLARY 1.5.2. *Let us choose a renormalization scheme.*

Then, we find a bijection between (translation invariant) strictly renormalizable scalar field theories on $\mathbb{R}^n$ and Lagrangians of the form:
(1) The ϕ^3 theory on $\mathbb{R}^6$.
(2) The ϕ^4 theory on $\mathbb{R}^4$.
(3) The ϕ^6 theory on $\mathbb{R}^3$.
(4) The free field theory on $\mathbb{R}^n$ when $n = 5$ or $n > 6$.

More precisely: there is a bijection between strictly renormalizable theories on $\mathbb{R}^3$ and interactions of the form
$$f(\hbar) \int_{\mathbb{R}^3} \phi \, \mathrm{D} \, \phi + g(\hbar) \int_{\mathbb{R}^3} \phi^6$$
where $f \in \hbar\mathbb{R}[[\hbar]]$ and $g \in \mathbb{R}[[\hbar]]$; and similarly for $\mathbb{R}^4$ and $\mathbb{R}^6$. The point is that, on $\mathbb{R}^3$, the space of local action functionals of dimension zero is spanned by $\int_{\mathbb{R}^3} \phi \, \mathrm{D} \, \phi$ and $\int_{\mathbb{R}^3} \phi^6$; and similarly on $\mathbb{R}^4$ and $\mathbb{R}^6$.

On $\mathbb{R}^n$ where $n = 5$ or $n > 6$, the only local action functional of dimension zero is $\int_{\mathbb{R}^n} \phi \, \mathrm{D} \, \phi$. Thus, renormalizable theories in these dimensions correspond to purely quadratic interactions of the form $f(\hbar) \int_{\mathbb{R}^3} \phi \, \mathrm{D} \, \phi$. These are free theories.

There are no renormalizable scalar field theories on $\mathbb{R}$ or $\mathbb{R}^2$, as any relevant theory in these dimensions has infinitely many relevant deformations.

1. THE LOCAL RENORMALIZATION GROUP FLOW

1.6. Now let us consider the quantum effects which appear in the local renormalization group.

Let us fix a renormalization scheme $\mathbf{RS}_0$. Recall that $\mathscr{P}((0,1)) \subset C^\infty((0,1))$ is the algebra of smooth functions on $(0,1)$ which are periods. A renormalization scheme is defined by a subspace

$$\mathbf{RS}_0 = \mathscr{P}((0,1))_{<0} \subset \mathscr{P}((0,1))$$

of purely singular periods; this subspace is required to be complementary to the subspace $\mathscr{P}((0,1))_{\geq 0}$ of periods which admit an $\varepsilon \to 0$ limit.

Let

$$\mathbf{RS}_l = \{f \in \mathscr{P}((0,1)) \mid f(l^{-2}\varepsilon) \in \mathbf{RS}_0\}.$$

In other words: the action of $\mathbb{R}_{>0}$ on itself induces an action of $\mathbb{R}_{>0}$ on the space of renormalization schemes, and we will denote by $\mathbf{RS}_l$ the renormalization scheme obtained by applying $l^{-2} \in \mathbb{R}_{>0}$ to the original renormalization scheme $\mathbf{RS}_0$.

There is a change of renormalization scheme map

$$\Phi_{l,0} = \Phi_{\mathbf{RS}_l \to \mathbf{RS}_0} : \mathscr{O}^+_{loc}(\mathscr{S}(\mathbb{R}^n))[[\hbar]] \to \mathscr{O}^+_{loc}(\mathscr{S}(\mathbb{R}^n))[[\hbar]].$$

By definition, a theory associated to the action I and the renormalization scheme $\mathbf{RS}_l$ is equivalent to the theory associated to the action $\Phi_{l,0}(I)$ and the renormalization scheme $\mathbf{RS}_0$.

Throughout, we will use the renormalization scheme $\mathbf{RS}_0$ to identify the space $\mathscr{T}^{(\infty)}$ with the space $\mathscr{O}^+_{loc}(\mathscr{S}(\mathbb{R}^n))[[\hbar]]$ of local action functionals using the renormalization scheme $\mathbf{RS}_0$. Once we have made this identification, the local renormalization group flow $\mathscr{RG}_l$ is a map

$$\mathscr{RG}_l : \mathscr{O}^+_{loc}(\mathscr{S}(\mathbb{R}^n))[[\hbar]] \to \mathscr{O}^+_{loc}(\mathscr{S}(\mathbb{R}^n))[[\hbar]].$$

LEMMA 1.6.1. *The local renormalization group flow $\mathscr{RG}_l$ is the composition*

$$\mathscr{RG}_l = \Phi_{l,0} \circ R_l^* : \mathscr{O}^+_{loc}(\mathscr{S}(\mathbb{R}^n))[[\hbar]] \to \mathscr{O}^+_{loc}(\mathscr{S}(\mathbb{R}^n))[[\hbar]].$$

PROOF. The theory associated to a local action functional I is defined by the effective interactions

$$W^R(P(0,L),I) = \lim_{\varepsilon \to 0} W\left(P(\varepsilon,L), I - I^{CT}(\varepsilon)\right).$$

By definition of the local renormalization group flow,

$$W^R(P(0,L), \mathscr{RG}_l(I)) = R_l^* W^R(P(0,l^2L), I)$$

$$= \lim_{\varepsilon \to 0} R_l^* W\left(P(\varepsilon, l^2L), I - I^{CT}(\varepsilon)\right)$$

$$= \lim_{\varepsilon \to 0} W\left(R_l P(\varepsilon, l^2L), R_l^* I - R_l^* I^{CT}(\varepsilon)\right)$$

$$= \lim_{\varepsilon \to 0} W\left(P(l^{-2}\varepsilon, L), R_l^* I - R_l^* I^{CT}(\varepsilon)\right)$$

$$= \lim_{\varepsilon \to 0} W\left(P(\varepsilon, L), R_l^* I - R_l^* I^{CT}(l^2\varepsilon)\right)$$

where in the last step we reparameterize the dummy variable ε. Note that $R_l^* I^{CT}(l^2\varepsilon)$ is purely singular for the renormalization scheme $\mathbf{RS}_l$. Since the limit exists, it follows that $R_l^* I^{CT}(l^2\varepsilon)$ is the counterterm for $R_l^* I$ with this renormalization scheme.

Thus, the effective interaction
$$\lim_{\varepsilon \to 0} W\left(P(\varepsilon, L), R_l^* I - R_l^* I^{CT}(l^2\varepsilon)\right)$$
defines the theory associated to $R_l^* I$ and the renormalization scheme $\mathbf{RS}_l$. By definition of the change of renormalization scheme map, this is the same as the theory associated to $\Phi_{l,0} R_l^* I$ and the renormalization scheme $\mathbf{RS}_0$. Thus, we find that
$$W^R\left(P(0,L), \mathcal{RG}_l(I)\right) = W^R\left(P(0,L), \Phi_{l,0} R_l^* I\right)$$
so that
$$\Phi_{l,0} R_l^* I = \mathcal{RG}_l(I)$$
as desired. □

1.7. As we have seen, the choice of renormalization scheme leads to a bijection between $\mathscr{R}^{(\infty)}$ and the space $\mathscr{O}^+_{loc,\geq 0}(\mathscr{S}(\mathbb{R}^n))[[\hbar]]$ of local action functionals of non-negative dimension. Since the local renormalization group flow acts on $\mathscr{R}^{(\infty)}$, via this bijection, it also acts on $\mathscr{O}^+_{loc,\geq 0}(\mathscr{S}(\mathbb{R}^n))[[\hbar]]$. We will continue to use the notation $\mathcal{RG}_l$ to refer to this action:
$$\mathcal{RG}_l : \mathscr{O}^+_{loc,\geq 0}(\mathscr{S}(\mathbb{R}^n))[[\hbar]] \to \mathscr{O}^+_{loc,\geq 0}(\mathscr{S}(\mathbb{R}^n))[[\hbar]].$$

In a similar way, the bijection between the space $\mathscr{M}^{(\infty)}$ of marginal theories and the space $\mathscr{O}^+_{loc,0}(\mathscr{S}(\mathbb{R}^n))[[\hbar]]$ of local action functionals of dimension zero leads to an action of $\mathcal{RG}_l$ on the latter space.

We have to be careful here; the diagram
$$\begin{array}{ccc} \mathscr{O}^+_{loc,\geq 0}(\mathscr{S}(\mathbb{R}^n))[[\hbar]] & \xrightarrow{\mathcal{RG}_l} & \mathscr{O}^+_{loc,\geq 0}(\mathscr{S}(\mathbb{R}^n))[[\hbar]] \\ \downarrow & & \downarrow \\ \mathscr{O}^+_{loc}(\mathscr{S}(\mathbb{R}^n))[[\hbar]] & \xrightarrow{\mathcal{RG}_l} & \mathscr{O}^+_{loc}(\mathscr{S}(\mathbb{R}^n))[[\hbar]] \end{array}$$
does not commute, where the vertical arrow is the naive inclusion. Neither does the corresponding diagram where the inclusion is replaced by the projection
$$\mathscr{O}^+_{loc}(\mathscr{S}(\mathbb{R}^n))[[\hbar]] \to \mathscr{O}^+_{loc,\geq 0}(\mathscr{S}(\mathbb{R}^n))[[\hbar]].$$
However, there is a modified inclusion map
$$\mathscr{O}^+_{loc,\geq 0}(\mathscr{S}(\mathbb{R}^n))[[\hbar]] \hookrightarrow \mathscr{O}^+_{loc}(\mathscr{S}(\mathbb{R}^n))[[\hbar]]$$
which corresponds to the inclusion
$$\mathscr{R}^{(\infty)} \hookrightarrow \mathscr{T}^{(\infty)};$$
this inclusion is, of course, equivariant with respect to the local renormalization group action.

1. THE LOCAL RENORMALIZATION GROUP FLOW

Classically, that is, modulo $\hbar$, the local renormalization group action is trivial on a local action functional of dimension zero. However, there are quantum effects. We can see this, for instance, in the ϕ^4 theory on $\mathbb{R}^4$. Let us use the notation

$$\langle \phi^4 \rangle = \int_{\mathbb{R}^4} \phi^4.$$

LEMMA 1.7.1.

$$\mathcal{RG}_l \left(c \tfrac{1}{4!} \langle \phi^4 \rangle \right) = \tfrac{1}{4!} \langle \phi^4 \rangle \left(c + \hbar c^2 \frac{3}{16\pi^2} \log l \right) \text{ modulo } \hbar^2$$

(This agrees with a standard result in the physics literature).

PROOF. We will fix a renormalization scheme $\mathbf{RS}_0$ with the property that the functions ε^{-1} and $\log \varepsilon$ are purely singular.

In Section 4 of this chapter, the counterterms for the action

$$I = \frac{1}{4!} c \langle \phi^4 \rangle$$

are calculated to one loop. We find (Proposition 4.0.1) that

$$I_{1,2}^{CT}(\varepsilon) = 2^{-6} \pi^{-2} c \varepsilon^{-1} \langle \phi^2 \rangle$$
$$I_{1,4}^{CT}(\varepsilon) = -\pi^{-2} 2^{-8} c^2 \log \varepsilon \langle \phi^4 \rangle$$
$$I_{1,i}^{CT}(\varepsilon) = 0 \text{ if } i \neq 2, 4.$$

Lemma 1.6.1 says that the local renormalization group flow $\mathcal{RG}_l$ is obtained as a composition

$$\mathcal{RG}_l = R_l^* \circ \Phi_{l,0}$$

of the rescaling operator R_l^* and a change of renormalization scheme map $\Phi_{l,0}$. As discussed earlier, $\Phi_{l,0}$ is the change of renormalization scheme map from the renormalization scheme $\mathbf{RS}_l$, whose purely singular functions are of the form $f(l^2 \varepsilon)$ where $f(\varepsilon)$ is purely singular for $\mathbf{RS}_0$, to the renormalization scheme $\mathbf{RS}_0$.

By definition of $\Phi_{l,0}$, the theory associated to action I and renormalization scheme $\mathbf{RS}_l$ is equivalent to the theory associated to action $\Phi_{l,0}(I)$ and renormalization scheme $\mathbf{RS}_0$.

With $I = \frac{1}{4!} c \langle \phi^4 \rangle$, the one-loop counterterms with renormalization scheme $\mathbf{RS}_l$ are

$$I_{1,2}^{CT,\mathbf{RS}_l}(\varepsilon) = 2^{-6} \pi^{-2} c \varepsilon^{-1} \langle \phi^2 \rangle$$
$$I_{1,4}^{CT,\mathbf{RS}_l}(\varepsilon) = -\pi^{-2} 2^{-8} c^2 \left(\log \varepsilon + \log l^2 \right) \langle \phi^4 \rangle$$
$$I_{1,i}^{CT,\mathbf{RS}_l}(\varepsilon) = 0 \text{ if } i \neq 2, 4$$

To see this, observe that the functions ε^{-1} and $\log l^2 \varepsilon = \log \varepsilon + \log l^2$ are purely singular for the renormalization scheme $\mathbf{RS}_l$.

It follows that
$$\Phi_{l,0}(I) = I + \hbar \pi^{-2} 2^{-8} c^2 \log l^2 \langle \phi^4 \rangle \text{ modulo } \hbar^2.$$
Since $R_l^* \langle \phi^4 \rangle = \langle \phi^4 \rangle$, this yields the desired result.
□

2. The Kadanoff-Wilson picture and asymptotic freedom

2.1. The definition of renormalizability given above is closely related to the Kadanoff-Wilson picture.

The Kadanoff-Wilson renormalization criterion says that a theory is renormalizable if

(1) It converges to a fixed point under the local renormalization group flow as $l \to 0$.
(2) The unstable manifold for the local renormalization group at this fixed point is finite dimensional; in other words, all but finitely many directions are attractive.

"Unstable" means that as the local RG parameter l increases, and we tend to low energy, points in the unstable manifold move away from the fixed point. To avoid using the term "renormalizable" for several slightly different concepts, we will refer to a theory satisfying these two conditions as Kadanoff-Wilson (KW) renormalizable. A fixed point satisfying condition (2) will be called a Kadanoff-Wilson (KW) fixed point.

Any real-world measurements one makes of a quantum field theory occur at low energy, that is, as the local renormalization group parameter l becomes large. Suppose we have a theory which is a Kadanoff-Wilson fixed point, and we make a small deformation in the stable direction. Then, as $l \to \infty$, this deformation is drawn back to the fixed point we started with. Thus, deformations in the stable direction make no difference to real world measurements; they are called *irrelevant*.

If the Kadanoff-Wilson criterion holds, there are only a finite number of deformations which make any difference to low-energy measurements; these deformations occur in the unstable manifold. Thus, to specify a theory up to terms which are irrelevant for experimental purposes amounts to specifying a point in the unstable manifold. The KW criterion says that this unstable manifold is finite dimensional, and so a theory is specified by only a finite number of parameters.

2.2. The Kadanoff-Wilson criterion for renormalizability is an ideal non-perturbative definition. The definition we give is a perturbative approximation to this ideal.

Our perturbative definition of renormalizability contains two conditions: firstly, that the coupling constants in the theory grow at most logarithmically as $l \to 0$; and secondly, that the theory has only finite many deformations satisfying this condition.

2. THE KADANOFF-WILSON PICTURE AND ASYMPTOTIC FREEDOM

The first condition says that there are no obvious perturbative obstructions to convergence to a fixed point. If a term in the theory exhibits polynomial growth as $l \to 0$, then one can reasonably conclude that the theory doesn't converge to a fixed point. The second condition says that if the theory does converge to a fixed point, then this fixed point must satisfy the Kadanoff-Wilson criterion.

The presence of terms with logarithmic growth does not preclude convergence to a fixed point at the non-perturbative level. For instance, suppose we had a theory with a single coupling constant c, and that at $c = 0$ the theory is a fixed point (for example, a free theory). Suppose that the coupling constant c changes as
$$c \mapsto cl^{\hbar} = c + \hbar c \log l + \cdots .$$
Since the formal parameter $\hbar$ should be viewed as being greater than zero, if c changes in this way then the theory does converge to a fixed point.

On the other hand, if the coupling constant c changes as
$$c \mapsto cl^{-\hbar} = c - \hbar c \log l + \cdots$$
we would not expect convergence to a fixed point.

Thus, our definition of perturbative renormalizability is not as strong as the ideal non-perturbative definition given above: our definition only excludes the theories which obviously don't converge to a fixed point, but includes theories which don't converge to a fixed point for more subtle reasons.

2.3. A more refined definition of renormalizability would require our theory to be *asymptotically free*; this idea is explained in this subsection.

Let us consider a theory with a single coupling constant c. Since we are always making the $\hbar$ dependence explicit, our coupling constant c depends on $\hbar$,
$$c(\hbar) = \sum_{i \geq 0} \hbar^i c_i.$$
Geometrically, we will think of the coupling constant c as being a map of formal schemes
$$\operatorname{Spec} \mathbb{R}[[\hbar]] \to \operatorname{Spec} \mathbb{R}[c]$$
or equivalently, a section
$$\operatorname{Spec} \mathbb{R}[[\hbar]] \to \operatorname{Spec} \mathbb{R}[c] \times \operatorname{Spec} \mathbb{R}[[\hbar]].$$

The renormalization group flow then acts on the space of coupling constants:
$$c(\hbar) \mapsto \mathcal{RG}_l(c(\hbar)).$$
This action is $\hbar$ multilinear, in the sense that if we Taylor expand $\mathcal{RG}_l$ as a function of $c(\hbar)$, the terms in the Taylor series are $\hbar$ multilinear maps
$$\mathbb{R}[[\hbar]]^{\otimes k} \to \mathbb{R}[[\hbar]].$$

(This multilinearity property follows from the fact that the change of renormalization scheme maps are $\hbar$-multilinear.) Geometrically, this means that $\mathcal{RG}_l$ is an automorphism

$$\mathcal{RG}_l : \operatorname{Spec} \mathbb{R}[c] \times \operatorname{Spec} \mathbb{R}[[\hbar]] \to \operatorname{Spec} \mathbb{R}[c] \times \operatorname{Spec} \mathbb{R}[[\hbar]]$$

compatible with the projection to $\operatorname{Spec} \mathbb{R}[[\hbar]]$.

Since $\mathcal{RG}_{lm} = \mathcal{RG}_l \circ \mathcal{RG}_m$, and since $\mathcal{RG}_l(c(\hbar))$ is smooth as a function of l, there exists a vector field

$$X = f(c,\hbar)\frac{\partial}{\partial c} \in \mathbb{R}[[\hbar]] \otimes \mathbb{R}[c]\frac{\partial}{\partial c}$$

which generates the renormalization group flow $\mathcal{RG}_l$. This vector field X is an infinitesimal automorphism of $\operatorname{Spec} \mathbb{R}[[\hbar]] \times \operatorname{Spec} \mathbb{R}[c]$ over $\operatorname{Spec} \mathbb{R}[[\hbar]]$, or, in other words, an $\hbar$-dependent family of vector fields on the affine line $\operatorname{Spec} \mathbb{R}[c]$. The statement that X generates $\mathcal{RG}_l$ means that

$$\mathcal{RG}_l = e^{(\log l)X}.$$

We will say that a theory is asymptotically free, if, to leading order in c and $\hbar$, the vector field X moves the coupling constant c closer to the origin in the affine line. This means that if we write

$$X = f(c,\hbar)\frac{\partial}{\partial c} = c^k f_k(\hbar)\frac{\partial}{\partial c} + O(c^{k+1}),$$

then

$$f_k(\hbar) = \alpha \hbar^l + O(\hbar^{l+1})$$

where $\alpha > 0$.

What this means is that if we start with a theory with some very small coupling constant c, then the renormalization group flow will make this coupling constant smaller.

In practise, theories are very rarely asymptotically free. For example, Lemma 1.7.1 shows that the renormalization group vector field X for the ϕ^4 interaction is, to leading order,

$$X = c^2 \hbar \frac{\partial}{\partial c}.$$

Thus, the renormalization group flow takes a small positive coupling constant and moves it towards the origin, and a small negative coupling constant and moves it away from the origin. However, our sign conventions mean that the physically relevant coupling constant is negative.

Of course, the famous result of Gross-Wilczek (GW73) and Politzer (Pol73) says that Yang-Mills theory is asymptotically free. Although this book proves renormalizability of Yang-Mills theory, proof of asymptotic freedom in our context will have to wait for future work.

3. Universality

A key feature of the Kadanoff-Wilson picture is the idea of universality. If we are at a KW fixed point, then all small deformations of the theory become equivalent to small deformations lying on the unstable manifold in the large length (low-energy) limit. The theories lying on the unstable manifold are said to be *universal* among all theories near the fixed point.

One can ask whether is a similar picture in the perturbative version of the Kadanoff-Wilson philosophy we are using here. Let us start by making a precise definition of universality.

DEFINITION 3.0.1. *Two theories, given by collections of effective interactions $\{I[L]\}$, $\{I'[L]\}$, are in the same universality class if for all L,*
$$\mathcal{RG}_l\left(I[L]\right) - \mathcal{RG}_l\left(I'[L]\right) \to 0 \text{ as } l \to \infty.$$

Thus, given a renormalizable theory, one can ask whether all small deformations of it are in the same universality class as a renormalizable deformation. The strongest results along these lines can be obtained for theories which satisfy an extra condition.

DEFINITION 3.0.2. *We say a theory is* strongly renormalizable *if*

(1) It is strictly renormalizable, that is, it only contains marginal terms.
(2) All renormalizable deformations of the theory are also strictly renormalizable. In other words, there are no deformations which introduce relevant terms.

The following theorem is very easy to prove[1].

THEOREM 3.0.3. *Any infinitesimal deformation of a strongly renormalizable theory (to any order) is in the same universality class as a renormalizable deformation.*

Unfortunately, scalar field theories are never strongly renormalizable. However, as we will see later, pure Yang-Mills theory in dimension 4 is strongly renormalizable. Thus, the quantization constructed later for pure Yang-Mills theory is universal.

A weaker result holds for theories which are just strictly renormalizable.

THEOREM 3.0.4. *Any first-order deformation of a strictly renormalizable theory is in the same universality class as a renormalizable first-order deformation.*

Unfortunately, this is not true for higher order deformations. If we have a first order deformation which involves both relevant and irrelevant terms, then these can combine at higher order to create extra marginal or relevant terms which are not present in any renormalizable deformation.

---

[1] When we discuss Yang-Mills theory, we will give a detailed proof of a stronger version of this theorem.

4. Calculations in ϕ^4 theory

In this section, we will calculate explicitly the one-loop counterterms for ϕ^4 theory, with interaction
$$I = I_{0,4} = c\frac{1}{4!}\int \phi^4.$$
This allows explicit calculations of the local renormalization group flow to the first non-trivial order (Lemma 1.7.1).

Let us fix a renormalization scheme with the property that the functions ε^{-1} and $\log \varepsilon$ are purely singular.

PROPOSITION 4.0.1. *The one-loop counterterms for the action I are*
$$I_{1,2}^{CT}(\varepsilon)(\phi) = 2^{-6}\pi^{-2}c\varepsilon^{-1}\int_{x\in\mathbb{R}^4}\phi(x)^2$$
$$I_{1,4}^{CT}(\varepsilon)(\phi) = -\pi^{-2}2^{-8}c^2\log\varepsilon\int_{x\in\mathbb{R}^4}\phi(x)^4$$
$$I_{1,i}^{CT}(\varepsilon)(\phi) = 0 \text{ if } i \neq 2, 4.$$

The rest of this section will be devoted to the proof of this proposition.

4.1. The first counterterm arises from the graph

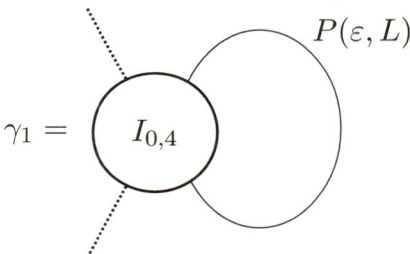

It is easy to see that the weight is
$$w_{\gamma_1}(P(\varepsilon, L), I)(\phi) = c\int_{l=\varepsilon}^{L}\int_{\mathbb{R}^4} K_l(x,x)\phi(x)^2$$
$$= c\int_{l=\varepsilon}^{L}\int_{\mathbb{R}^4}(4\pi l)^{-2}\phi(x)^2$$
$$= c\frac{1}{16\pi^2}\left(\varepsilon^{-1} - L^{-1}\right)\int_{\mathbb{R}^4}\phi^2$$

Recall that we are using a renormalization scheme where ε^{-1} is purely singular. Then, the first counterterm is
$$I_{1,2}^{CT}(\varepsilon) = c\frac{1}{64\pi^2}\varepsilon^{-1}\int_{x\in\mathbb{R}^4}\phi(x)^2.$$
(the extra factor of $\frac{1}{4}$ arises from the automorphism of the graph).

4. CALCULATIONS IN ϕ^4 THEORY

Because $I_{1,2}^{CT}(l\varepsilon)$ is purely singular, this graph does not contribute to the local renormalization group flow.

The next graphs we need to consider are

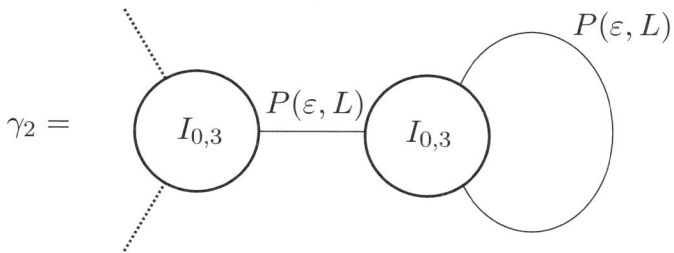

These graphs contribute with opposite signs. The singularity in the graph γ_2 arises from the loop on the second vertex; this singularity is counteracted by the counterterm $I_{1,2}^{CT}(\varepsilon)$ in the graph γ_3. Thus, the sum of contributions of these two graphs is non-singular, so they do not contribute any new counterterms.

4.2. The next graph that needs to be considered is

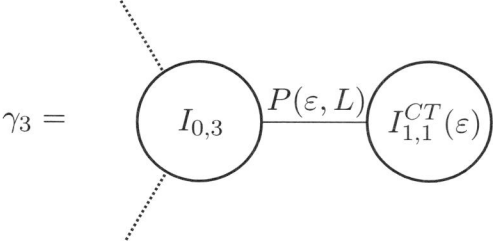

The weight for this graph is

$$\begin{aligned}
w_{\gamma_4}(\phi) =& c^2 \int_{l_1, l_2 \in [\varepsilon, L]} \int_{x_1, x_2 \in \mathbb{R}^4} K_{l_1}(x_1, x_2) K_{l_2}(x_1, x_2) \phi(x_1)^2 \phi(x_2)^2 \\
=& c^2 2^{-8} \pi^{-4} \int_{l_1, l_2 \in [\varepsilon, L]} \int_{x_1, x_2 \in \mathbb{R}^4} \left\{ l_1^{-2} l_2^{-2} \right. \\
& \left. e^{-\|x_1 - x_2\|^2 ((4l_1)^{-1} + (4l_2)^{-1})} \phi(x_1)^2 \phi(x_2)^2 \right\}
\end{aligned}$$

Let us perform a change of variables on $\mathbb{R}^4 \times \mathbb{R}^4$, by letting
$$y = \tfrac{1}{2}(x_1 + x_2)$$
$$z = \tfrac{1}{2}(x_1 - x_2).$$
This introduces a Jacobian factor of 2^4, so that this becomes
$$w_{\gamma_4}(\phi) = c^2 2^{-4} \pi^{-4} \int_{l_1, l_2 \in [\varepsilon, L]} \int_{y, z \in \mathbb{R}^4} l_1^{-2} l_2^{-2} e^{-\|z\|^2 (l_1^{-1} + l_2^{-1})} \phi(y-z)^2 \phi(y+z)^2.$$

We can get an asymptotic expansion for the integral over z with the help of Wick's lemma. Since we're ultimately only interested in the singular part, this asymptotic expansion contains all the information we need.

We find that
$$\int_{z \in \mathbb{R}^4} l_1^{-2} l_2^{-2} e^{-\|z\|^2 (l_1^{-1} + l_2^{-1})} \phi(y-z)^2 \phi(y+z)^2$$
$$= \pi^2 l_1^{-2} l_2^{-2} (l_1^{-1} + l_2^{-1})^{-2} \left(\phi(y)^4 + O\left(\frac{l_1 l_2}{l_1 + l_2}\right) \right)$$
$$= \pi^2 (l_1 + l_2)^{-2} \left(\phi(y)^4 + O\left(\frac{l_1 l_2}{l_1 + l_2}\right) \right)$$

where $O(l_1 l_2 / (l_1 + l_2))$ indicates some polynomial in $l_1 l_2 / (l_1 + l_2)$, where each term in the polynomial is multiplied by a local action functional of ϕ.

The terms like $(l_1 + l_2)^{-2-k} l_1^k l_2^k$ do not contribute to the singular part, because the integral
$$\int_{l_1, l_2 \in [0, L]} (l_1 + l_2)^{-2-k} l_1^k l_2^k \, dl_1 \, dl_2$$
converges absolutely for any $k > 0$.

Thus, we find that
$$\mathrm{Sing}_\varepsilon w_{\gamma_4}(P(\varepsilon, L), I)(\phi) = \mathrm{Sing}_\varepsilon(f(\varepsilon)) \int_{y \in \mathbb{R}^4} \phi(y)^4$$
where
$$f(\varepsilon) = \pi^{-2} 2^{-4} c^2 \int_{l_1, l_2 \in [\varepsilon, L]} (l_1 + l_2)^{-2} \, dl_1 \, dl_2$$
$$= \pi^{-2} 2^{-4} c^2 \left(-\log 2\varepsilon + 2\log(L + \varepsilon) - \log 2L \right).$$

We will choose a renormalization scheme such that $\log \varepsilon$ is purely singular. Then,
$$\mathrm{Sing}_\varepsilon f(\varepsilon) = \pi^{-2} 2^{-4} c^2 \log \varepsilon.$$
The counterterm is
$$I_{1,4}^{CT} = \frac{1}{|\mathrm{Aut}\,\gamma_4|} \mathrm{Sing}_\varepsilon w_{\gamma_4}(P(\varepsilon, L), I)(\phi)$$
$$= -\frac{1}{|\mathrm{Aut}\,\gamma_4|} \pi^{-2} 2^{-4} c^2 \log \varepsilon \int_{y \in \mathbb{R}^4} \phi(y)^4.$$

Now,
$$|\text{Aut}\,\gamma_4| = 2^{-4}$$
so that
$$I_{1,4}^{CT}(\varepsilon)(\phi) = -2^{-8}\pi^{-2}c^2\log\varepsilon \int_{y\in\mathbb{R}^4}\phi(y)^4.$$

4.3. It turns out that all other one-loop counterterms vanish. The calculation which proves this is similar to the calculations we have already seen.

We say that a connected graph is *one-particle reducible*, or 1PR, if it has a separating edge, that is, an edge which when cut leaves a disconnected graph. A graph which is not one-particle reducible is one-particle irreducible, or 1PI. If we cut a one-particle reducible graph along all of its separating edges, we are left with a disjoint union of one-particle irreducible graphs; these are called the irreducible components of the graph.

One-particle reducible graphs do not contribute to the counterterms, because they are already made non-singular by the counterterms for its irreducible components. We have already seen this for the graphs γ_2 and γ_3 in the calculation above; although the contribution of the graph γ_2 is singular, this singularity is cancelled out by the contribution of the graph γ_3. The graph γ_3 contains the counterterms for the irreducible components of γ_2.

Thus, when calculating the remaining one-loop counterterms, we only have to worry about 1PI graphs. Every one-loop 1PI graph is given by a sequence of vertices arranged in a circle, like

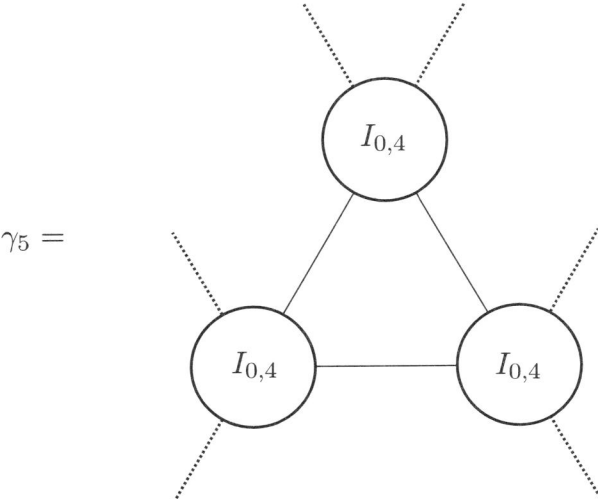

The integral attached to each such graph converges absolutely, and so these graphs do not contribute to the counterterms. We will see this explicitly for γ_5; the calculation for other circular graphs is similar.

Note that

$$w_{\gamma_5}(P(\varepsilon, L), I)(\phi)$$
$$= c^3 \int_{l_1,l_2,l_3 \in [\varepsilon,L]} \int_{x_1,x_2,x_3 \in \mathbb{R}^4} \{K_{l_1}(x_2, x_3) K_{l_2}(x_3, x_1)$$
$$K_{l_3}(x_1, x_2) \phi(x_1)^2 \phi(x_2)^2 \phi(x_3)^2\}$$

Since we are only interested in showing that the limit as $\varepsilon \to 0$ exists, we will not worry about constants appearing in the formulae.

Plugging the expression

$$K_l(x, y) = (4\pi l)^{-2} e^{-\|x-y\|^2/4l}$$

into the above formula, we see that we need to show that the integral

$$\int_{l_1,l_2,l_3 \in [0,L]} \int_{x_1,x_2,x_3 \in \mathbb{R}^4} \{(l_1 l_2 l_3)^{-2} e^{-\|x_2-x_3\|^2/4l_1 - \|x_3-x_1\|^2/4l_2 - \|x_1-x_2\|^2/4l_3}$$
$$\phi(x_1)^2 \phi(x_2)^2 \phi(x_3)^2\}$$

is absolutely convergent.

Let us perform the change of variables

$$z_1 = \tfrac{1}{2}(x_2 - x_3)$$
$$z_2 = \tfrac{1}{2}(x_3 - x_1)$$
$$y = \tfrac{1}{2}(x_1 + x_2 + x_3).$$

The integral we need to show is absolutely convergent is

$$\int_{l_1,l_2,l_3 \in [0,L]} \int_{y,z_1,z_2 \in \mathbb{R}^4} (l_1 l_2 l_3)^{-2} e^{-\|z_1\|^2/l_1 - \|z_2\|^2/l_2 - \|z_1+z_2\|^2/l_3} F(y, z_1, z_2)$$

where F is a Schwartz function on $\mathbb{R}^4 \times \mathbb{R}^4 \times \mathbb{R}^4$. Since F is Schwartz, for all $n > 0$ there exists a constant C_n such that

$$F(y, z_1, z_2) \le C_n (1 + \|y\|)^{-n}$$

for all y, z_1, z_2. Thus, it suffices to show that the integral

$$\int_{l_1,l_2 \in [0,L]} \int_{z_1,z_2 \in \mathbb{R}^4} (l_1 l_2 l_3)^{-2} e^{-\|z_1\|^2/l_1 - \|z_2\|^2/l_2 - \|z_1+z_2\|^2/l_3}$$

converges absolutely.

Note that

$$\int_{z_1,z_2 \in \mathbb{R}^4} e^{-\|z_1\|^2/l_1 - \|z_2\|^2/l_2 - \|z_1+z_2\|^2/l_3} = \pi^4 (\det A)^{-1/2}$$

where A is the matrix of the quadratic form

$$\|z_1\|^2/l_1 + \|z_2\|^2/l_2 + \|z_1 + z_2\|^2/l_3.$$

Since

$$\det A = (l_1^{-1} l_2^{-1} + l_1^{-1} l_3^{-1} + l_2^{-1} l_3^{-1})^4,$$

we are left with the integral
$$\int_{l_1,l_2,l_3\in[0,L]} (l_1 l_2 l_3)^{-2}(l_1^{-1}l_2^{-1}+l_1^{-1}l_3^{-1}+l_2^{-1}l_3^{-1})^{-2} = \int_{l_1,l_2,l_3\in[0,L]} (l_1+l_2+l_3)^{-2}.$$
This integral is easily seen to be absolutely convergent.

5. Proofs of the main theorems

5.1. Before we prove the main results of this chapter, we will need a lemma about the behaviour of the change of renormalization scheme map.

Let $\mathbf{RS}_0$, $\mathbf{RS}_1$ be two renormalization schemes. Let
$$\Phi_{\mathbf{RS}_0 \to \mathbf{RS}_1} : \mathscr{O}^+_{loc}(\mathscr{S}(\mathbb{R}^n))[[\hbar]] \to \mathscr{O}^+_{loc}(\mathscr{S}(\mathbb{R}^n))[[\hbar]]$$
be the change of renormalization scheme map, defined so that for all $I \in \mathscr{O}^+_{loc}(\mathscr{S}(\mathbb{R}^n))[[\hbar]]$ the theory associated to I and $\mathbf{RS}_0$ is the same as the theory associated to $\Phi_{\mathbf{RS}_0 \to \mathbf{RS}_1}(I)$ and $\mathbf{RS}_1$.

LEMMA 5.1.1. $\Phi_{\mathbf{RS}_0 \to \mathbf{RS}_1}(I)$ *is a series of the form*
$$\Phi_{\mathbf{RS}_0 \to \mathbf{RS}_1}(I) = I + \sum_{a>0,\ b\geq 0} \hbar^a P_{a,b}(I)$$
where:

(1) *Each $P_{a,b}(I) \in \mathscr{O}_{loc}(\mathscr{S}(\mathbb{R}^n))$ is homogeneous of degree b as a function on $\mathscr{S}(\mathbb{R}^n)$;*
(2) *$P_{a,b}(I) \in \mathscr{O}_{loc}(\mathscr{S}(\mathbb{R}^n))$ is a polynomial in the $I_{r,s}$ where $2r - s \leq 2a - b$.*

Note that the space $\mathscr{O}_{loc}(\mathscr{S}(\mathbb{R}^n))$ of translation invariant local action functionals on $\mathscr{S}(\mathbb{R}^n)$, homogeneous of degree k, is isomorphic, as a topological vector space, to $\mathbb{R}^\infty = \oplus_{i \in \mathbb{Z}_{\geq 0}} \mathbb{R}$. A polynomial map $\mathbb{R}^\infty \to \mathbb{R}^\infty$ is simply an element of
$$\mathbb{R}^\infty[t_1, t_2, \ldots] = \oplus_{i \in \mathbb{Z}_{\geq 0}} \mathbb{R}[t_1, t_2, \ldots].$$

PROOF. The proof is by induction. As usual, let us write $\Phi_{\mathbf{RS}_0 \to \mathbf{RS}_1}(I)$ as a sum
$$\Phi_{\mathbf{RS}_0 \to \mathbf{RS}_1}(I) = \sum \hbar^a \Phi_{\mathbf{RS}_0 \to \mathbf{RS}_1}(I)_{a,b}$$
where $\Phi_{\mathbf{RS}_0 \to \mathbf{RS}_1}(I)_{a,b}$ is homogeneous of degree b as a function of the field.

Let us assume, by induction, that $\Phi_{\mathbf{RS}_0 \to \mathbf{RS}_1}(I)_{a,b}$ is of the desired form for all $(a,b) < (A,B)$.

Let us write $\mathrm{Sing}_\varepsilon^{\mathbf{RS}_0}$ and $\mathrm{Sing}_\varepsilon^{\mathbf{RS}_1}$ to denote the operator which picks out the singular part of a function of ε using the renormalization schemes $\mathbf{RS}_0$ and $\mathbf{RS}_1$ respectively. In order to keep the notation short, let us write $\widetilde{I}_{a,b}$ for $\Phi_{\mathbf{RS}_0 \to \mathbf{RS}_1}(I)_{a,b}$. Let $\widetilde{I}^{CT}_{a,b}(\varepsilon)$ refer to the counterterms for $\widetilde{I}_{a,b}$, taken with respect to the renormalization scheme $\mathbf{RS}_1$. Similarly, let $I^{CT}_{a,b}(\varepsilon)$ refer to the counterterms for $I_{a,b}$ taken with respect to the renormalization scheme $\mathbf{RS}_1$.

Let us assume, by induction, that each $\widetilde{I}_{a,b}^{CT}(\varepsilon)$ and $I_{a,b}^{CT}(\varepsilon)$ is a polynomial in $I_{(r,s)}$ for $2r - s \leq 2a - b$.

Then, observe that

$$\widetilde{I}_{A,B}^{CT}(\varepsilon) = \mathrm{Sing}_{\varepsilon}^{RS_1} W_{A,B}\left(P(\varepsilon, L), \sum_{(a,b)<(A,B)} \hbar^a \widetilde{I}_{a,b} - \hbar^a \widetilde{I}_{a,b}^{CT}(\varepsilon)\right)$$

Note that, in a stable graph of genus A with B tails, only vertices of type (r,s) can appear if $2r - s \leq 2A - B$. It follows that we can write

$$\widetilde{I}_{A,B}^{CT}(\varepsilon) = \mathrm{Sing}_{\varepsilon}^{RS_1} W_{A,B}\left(P(\varepsilon, L), \sum_{2a-b\leq 2A-B} \hbar^a \widetilde{I}_{a,b} - \hbar^a \widetilde{I}_{a,b}^{CT}(\varepsilon)\right).$$

Since everything on the right hand side of this expression is a polynomial in $I_{a,b}$ where $2a - b \leq 2A - B$, it follows that $\widetilde{I}_{A,B}^{CT}(\varepsilon)$ is such a polynomial.

In a similar way, $I_{A,B}^{CT}(\varepsilon)$ is a polynomial in $I_{a,b}$ for $2a - b \leq 2A - B$.

By definition, we have

$$\widetilde{I}_{A,B} + \lim_{\varepsilon \to 0}\left(W_{A,B}\left(P(\varepsilon, L), \sum_{(a,b)<(A,B)} \hbar^a \widetilde{I}_{a,b} - \hbar^a \widetilde{I}_{a,b}^{CT}(\varepsilon)\right) - \widetilde{I}_{A,B}^{CT}(\varepsilon)\right)$$

$$= I_{A,B} + \lim_{\varepsilon \to 0}\left(W_{A,B}\left(P(\varepsilon, L), \sum_{(a,b)<(A,B)} \hbar^a I_{a,b} - \hbar^a I_{a,b}^{CT}(\varepsilon)\right) - I_{A,B}^{CT}(\varepsilon)\right).$$

Thus,

$$\widetilde{I}_{A,B} = I_{A,B}$$

$$- \lim_{\varepsilon \to 0}\left(W_{A,B}\left(P(\varepsilon, L), \sum_{(a,b)<(A,B)} \hbar^a \widetilde{I}_{a,b} - \hbar^a \widetilde{I}_{a,b}^{CT}(\varepsilon)\right) - \widetilde{I}_{A,B}^{CT}(\varepsilon)\right)$$

$$+ \lim_{\varepsilon \to 0}\left(W_{A,B}\left(P(\varepsilon, L), \sum_{(a,b)<(A,B)} \hbar^a I_{a,b} - \hbar^a I_{a,b}^{CT}(\varepsilon)\right) - I_{A,B}^{CT}(\varepsilon)\right).$$

Everything on the right hand side is a polynomial in $I_{a,b}$ for $2a - b \leq 2A - B$; thus, $\widetilde{I}_{A,B}$ is such a polynomial, as desired. □

Next, we will prove Proposition 1.3.2.

If $I \in \mathscr{O}_{loc}^+(\mathscr{S}(\mathbb{R}^n))[[\hbar]]$ is a local action functional, then the counterterms $I^{CT}(\varepsilon)$ depend in a polynomial way on I. Thus, the effective interactions $W^R(P(0, L), I)$ depend in a polynomial way on I. It follows that to prove Proposition 1.3.2, it suffices to prove the following.

LEMMA 5.1.2. *For any* $I \in \mathcal{O}_{loc}^+(\mathscr{S}(\mathbb{R}^n))[[\hbar]]$,
$$\mathcal{RG}_l(I) \in \mathcal{O}_{loc}^+(\mathscr{S}(\mathbb{R}^n))[[\hbar]] \otimes \mathbb{C}[l, l^{-1}, \log l].$$
In other words, for each (i,k), $\mathcal{RG}_l(I_{i,k})$ *is a polynomial in* l, l^{-1} *and* $\log l$ *when considered as a function of* l.

PROOF. Recall that the family of theories $\mathcal{RG}_l(\{I[L]\})$ is a smooth family of theories parameterized by l, in the sense of Definition 3.1.1 of Chapter 3. It follows from this and from Theorem 3 of Chapter 3 that the local action functionals $\mathcal{RG}_l(I)$ are a smooth family parameterized by l, that is,
$$\mathcal{RG}_l(I) \in \mathcal{O}_{loc}(\mathscr{S}(\mathbb{R}^n), C^\infty((0,\infty)_l)).$$
Recall that
$$\mathcal{RG}_l = \Phi_{l,0} \circ R_l^*$$
where $\Phi_{l,0}$ is the change of renormalization scheme map.

We will prove the result by the usual inductive argument. Fix some $I \in \mathcal{O}_{loc}^+(\mathscr{S}(\mathbb{R}^n))[[\hbar]]$. Let us fix $(a,b) \in \mathbb{Z}_{>0} \times \mathbb{Z}_{\geq 0}$, and let us assume, by induction, that for all $(r,s) < (a,b)$.
$$\mathcal{RG}_l(I)_{r,s} \in \mathcal{O}_{loc}(\mathscr{S}(\mathbb{R}^n)) \otimes \mathbb{C}[l, l^{-1}, \log l].$$

The operator $\Phi_{l,0}$ is "upper triangular", as proved in Theorem 5.1.1 in Chapter 2. We have
$$\mathcal{RG}_{ml}(I) = \mathcal{RG}_m(\mathcal{RG}_l(I)) = \Phi_{m,0} R_m^*(\mathcal{RG}_l(I)).$$
It follows that, for all $m, l \in \mathbb{R}_{>0}$, we have a difference equation of the form
$$\mathcal{RG}_{ml}(I)_{a,b} = R_m^* \mathcal{RG}_l(I)_{a,b} + J$$
where
$$J \in \mathcal{O}_{loc}(\mathscr{S}(\mathbb{R}^n)) \otimes \mathbb{C}[l, l^{-1}, \log l].$$
We are using the induction assumption to specify the form of the dependence of J on l. Of course, J is homogeneous of degree a in the field ϕ. Also, J depends on m, but this dependence is suppressed.

Let us write
$$\mathcal{RG}_l(I)_{a,b} = \sum_c \mathcal{RG}_l(I)_{a,b}^c$$
where $\mathcal{RG}_l(I)_{a,b}^c$ is of dimension $c \in \frac{1}{2}\mathbb{Z}$. Then, we find that
(‡)
$$\mathcal{RG}_{ml}(I)_{a,b}^c = m^c \mathcal{RG}_l(I)_{a,b}^c + J^c$$
for some $J^c \in \mathcal{O}_{loc,c}(\mathscr{S}(\mathbb{R}^n)) \otimes \mathbb{C}[l, l^{-1}, \log l]$ of dimension c.

This is the point at which we use the fact that $\mathcal{RG}_l(I)$ is smooth as a function of l. Because of this, the difference equation (‡) for $\mathcal{RG}_l(I)_{a,b}^c$ implies that $\mathcal{RG}_l(I)_{a,b}^c$ satisfies a differential equation of the form
$$l \frac{\partial}{\partial l} \mathcal{RG}_l(I)_{a,b}^c = c \mathcal{RG}_l(I)_{a,b}^c + \widetilde{J}^c$$
for some $\widetilde{J}^c \in \mathcal{O}_{loc,c}(\mathscr{S}(\mathbb{R}^n)) \otimes \mathbb{C}[l, l^{-1}, \log l]$ of dimension c.

Any solution of this differential equation is an element $\mathscr{O}_{loc,c}(\mathscr{S}(\mathbb{R}^n)) \otimes \mathbb{C}[l, l^{-1}, \log l]$.

□

Next, we can prove Theorem 1.5.1. The following is an equivalent statement of the theorem.

THEOREM 5.1.3. *Let*
$$I \in \mathscr{O}^+_{loc,0}(\mathscr{S}(\mathbb{R}^n))[[\hbar]]$$
be any local action functional of dimension 0. Then, there exists a unique
$$\widetilde{I} \in \hbar\mathscr{O}_{loc,\neq 0}(\mathscr{S}(\mathbb{R}^n))[[\hbar]]$$
which is a sum of terms of non-zero dimension, such that the theory associated to $I + \widetilde{I}$ is marginal, that is, an element of $\mathscr{M}^{(\infty)}$.

Similarly, for each local action functional
$$I \in \mathscr{O}^+_{loc,\geq 0}(\mathscr{S}(\mathbb{R}^n))[[\hbar]]$$
which is a sum of terms of non-negative dimension, there exists a unique
$$\widetilde{I} \in \hbar\mathscr{O}_{loc,\neq 0}(\mathscr{S}(\mathbb{R}^n))[[\hbar]]$$
which is a sum of terms of negative dimension, such that the theory associated to $I + \widetilde{I}$ is relevant, that is, an element of $\mathscr{R}^{(\infty)}$.

PROOF. We will prove the first statement; the proof of the second statement is essentially the same.

We need to show that there is a unique $\widetilde{I}$ which is a sum of terms of non-zero dimension such that
$$\mathcal{RG}_l(I + \widetilde{I}) \in \mathscr{O}^+_{loc}(\mathscr{S}(\mathbb{R}^n))[[\hbar]] \otimes \mathbb{C}[\log l].$$

The proof is similar to the proof of Lemma 5.1.2. Let us assume inductively that there are unique $\widetilde{I}_{r,s}$ for all $(r,s) < (a,b)$ such that:

(1) Each $\widetilde{I}_{r,s}$ is translation invariant, homogeneous of degree r in the field $\phi \in \mathscr{S}(\mathbb{R}^n)$, and is a sum of terms of non-zero dimension.
(2) Let
$$\widetilde{I}_{<(a,b)} = \sum_c \sum_{(r,s)<(a,b)} \hbar^r \widetilde{I}_{r,s}.$$

Then,
$$\mathcal{RG}_l\left(I + \widetilde{I}_{<(a,b)}\right)_{r,s} \in \mathscr{O}^+_{loc}(\mathscr{S}(\mathbb{R}^n))[[\hbar]] \otimes \mathbb{C}[\log l]$$

for all $(r,s) < (a,b)$.

Let $\mathcal{RG}_l \left(I + \widetilde{I}_{<(a,b)} \right)^c_{a,b}$ denote the part of dimension $c \in \frac{1}{2}\mathbb{Z}$. As in the proof of the previous lemma, we can see inductively that $\mathcal{RG}_l \left(I + \widetilde{I}_{<(a,b)} \right)^c_{a,b}$ satisfies a differential equation of the form

$$l\frac{\partial}{\partial l}\mathcal{RG}_l \left(I + \widetilde{I}_{<(a,b)} \right)^c_{a,b} = c\mathcal{RG}_l \left(I + \widetilde{I}_{<(a,b)} \right)^c_{a,b} + \widetilde{J}^c$$

for some $\widetilde{J}^c \in \mathscr{O}_{loc,c}(\mathscr{S}(\mathbb{R}^n)) \otimes \mathbb{C}[\log l]$ of dimension c.

If $c = 0$, we see immediately that

$$\mathcal{RG}_l \left(I + \widetilde{I}_{<(a,b)} \right)^c_{a,b} \in \mathscr{O}_{loc,0}(\mathscr{S}(\mathbb{R}^n)) \otimes \mathbb{C}[\log l]$$

varies only logarithmically with l, as desired.

If $c \neq 0$, this differential equation implies that

$$\mathcal{RG}_l \left(I + \widetilde{I}_{<(a,b)} \right)^c_{a,b} = l^c H + H'$$

for some $H \in \mathscr{O}_{loc,c}(\mathscr{S}(\mathbb{R}^n))$ and $H' \in \mathscr{O}_{loc,c}(\mathscr{S}(\mathbb{R}^n)) \otimes \mathbb{C}[\log l]$, both of conformal weight c.

Now set

$$\widetilde{I}^c_{a,b} = -H.$$

Then,

$$\mathcal{RG}_l \left(I + \widetilde{I}_{<(a,b)} + \hbar^a \widetilde{I}^c_{a,b} \right)^c_{a,b} = \mathcal{RG}_l \left(I + \widetilde{I}_{<(a,b)} \right)^c_{a,b} + \hbar^a l^c \widetilde{I}^c_{a,b}$$
$$= \mathcal{RG}_l \left(I + \widetilde{I}_{<(a,b)} \right)^c_{a,b} - \hbar^a l^c H$$
$$= H'$$
$$\in \mathscr{O}_{loc,c}(\mathscr{S}(\mathbb{R}^n)) \otimes \mathbb{C}[\log l]$$

as desired.

We do this for all c; for all but finitely many c, we find $\widetilde{I}^c_{a,b} = 0$. Thus, we can set

$$\widetilde{I}_{a,b} = \sum_{c \neq 0} \widetilde{I}^c_{a,b}$$

and we can continue the induction. $\square$

6. Generalizations of the main theorems

The main results of this chapter hold in a more general context. This generalization will be needed when we consider Yang-Mills theory in Chapter 6.

In this section, everything will depend on some auxiliary manifold with corners X, equipped with a sheaf A of commutative superalgebras over the sheaf of algebras C^∞_X. The space of global sections of A will be denoted, as before, by $\mathscr{A}$.

Let us assume we are in the situation of Chapter 3, Subsection 3. Thus, we have

(1) A finite dimensional super vector space E. Let
$$\mathscr{E} = E \otimes \mathscr{S}(\mathbb{R}^n).$$
Thinking of E as a trivial vector bundle on $\mathbb{R}^n$, $\mathscr{E}$ is the space of Schwartz sections of E.

(2) An even symmetric element
$$K_l \in C^\infty(\mathbb{R}^n \times \mathbb{R}^n) \otimes E^{\otimes 2} \otimes C^\infty((0,\infty)_l) \otimes \mathscr{A},$$
playing the role of the heat kernel. We assume that in some basis e_i of E, K_l can be written
$$K_l = \sum P_{i,j}(x-y, l^{-1/2}) e^{-\|x-y\|^2/l} e_i \otimes e_j$$
where the
$$P_{i,j} \in \mathscr{A}\left[x-y, l^{\pm 1/2}\right]$$
are polynomials in the variables $x_k - y_k$ and $l^{\pm 1/2}$, with coefficients in $\mathscr{A}$.

Then, we can write the propagator as
$$P(\varepsilon, L) = \int_{l=\varepsilon}^{L} K_l \, \mathrm{d}l$$
as an integral of the kernel K_l.

Let us further assume that E is equipped with a direct sum decomposition
$$E = \bigoplus_{i \in \frac{1}{2}\mathbb{Z}} E_i$$
into super vector spaces. E_i is the space of elements of E of dimension i.

The direct sum decomposition on E induces an $\mathbb{R}_{>0}$ action on $\mathscr{E} = \mathscr{S}(\mathbb{R}^n) \otimes E$, by
$$R_l(f(x) e_i) = f(lx) l^i e_i$$
where $f(x) \in \mathscr{S}(\mathbb{R}^n)$ and $e_i \in E_i$.

We require that the propagator scales as
$$R_l(P(\varepsilon, L)) = P(m^{-2}\varepsilon, m^{-2}L).$$

As explained in Chapter 3, Subsection 3, we have a notion of theory defined using this propagator; let $\mathscr{T}^{(\infty)}(\mathscr{E}, \mathscr{A})$ denote the set of theories, and let $\mathscr{T}^{(n)}(\mathscr{E}, \mathscr{A})$ denote the set of theories defined modulo $\hbar^{n+1}$. The choice of renormalization scheme leads to a bijection
$$\mathscr{T}^{(\infty)}(\mathscr{E}, \mathscr{A}) \cong \mathscr{O}_{loc}^{+,ev}(\mathscr{E}, \mathscr{A})[[\hbar]]$$
between theories and local action functionals $I \in \mathscr{O}_{loc}(\mathscr{E})[[\hbar]]$ which are even and at least cubic modulo $\hbar$. As always, both theories and local action functionals are translation invariant.

There is an action
$$\mathcal{RG}_l : \mathscr{T}^{(\infty)}(\mathscr{E}, \mathscr{A}) \to \mathscr{T}^{(\infty)}(\mathscr{E}, \mathscr{A})$$

of the local renormalization group on the space of theories. And, just as in the case of scalar field theories, we have:

PROPOSITION 6.0.1. *For any theory* $\{I[L]\} \in \left(\mathscr{T}^{(\infty)}(\mathscr{E}, \mathscr{A})\right)$,
$$\mathcal{RG}_l(I[L]) \in \mathscr{O}^+(\mathscr{E}, \mathscr{A})[[\hbar]] \otimes \mathbb{C}[l, l^{-1}, \log l].$$

Then, we define the spaces $\mathscr{R}^{(\infty)}(\mathscr{E}, \mathscr{A})$ of relevant theories to be the space of those theories such that
$$\mathcal{RG}_l(I[L]) \in \mathscr{O}^+(\mathscr{E}, \mathscr{A})[[\hbar]] \otimes \mathbb{C}[l, \log l].$$
Similarly, the space $\mathscr{M}^{(\infty)}(\mathscr{E}, \mathscr{A})$ of marginal theories is defined to be the space of theories such that
$$\mathcal{RG}_l(I[L]) \in \mathscr{O}^+(\mathscr{E}, \mathscr{A})[[\hbar]] \otimes \mathbb{C}[\log l].$$

Let
$$\mathscr{O}_{loc,k}(\mathscr{E}, \mathscr{A}) \subset \mathscr{O}_{loc}(\mathscr{E}, \mathscr{A})$$
be the space of local action functionals which are of dimension $k \in \frac{1}{2}\mathbb{Z}$. Then, there are projection maps
$$\mathscr{O}_{loc}(\mathscr{E}, \mathscr{A}) \to \mathscr{O}_{loc,0}(\mathscr{E}, \mathscr{A})$$
$$\mathscr{O}_{loc}(\mathscr{E}, \mathscr{A}) \to \mathscr{O}_{loc,\geq 0}(\mathscr{E}, \mathscr{A})$$
onto the space of local action functionals of dimension zero and the space of local action functionals which are sums of terms of non-zero dimension.

The main theorem is analogous to the previous theorem:

THEOREM 6.0.2. *The space $\mathscr{R}^{(n)}(\mathscr{E}, \mathscr{A})$ is a torsor over $\mathscr{R}^{(n-1)}(\mathscr{E}, \mathscr{A})$ for the abelian group $\mathscr{O}_{l, \geq 0}(\mathscr{E}, \mathscr{A})$ of local action functionals of non-negative dimension. The space $\mathscr{R}^{(0)}(\mathscr{E}, \mathscr{A})$ is canonically isomorphic to the subspace of $\mathscr{O}_{l, \geq 0}(\mathscr{E}, \mathscr{A})$ of functionals which are at least cubic.*

In a similar way, $\mathscr{M}^{(n)}(\mathscr{E}, \mathscr{A})$ is a torsor over $\mathscr{M}^{(n-1)}(\mathscr{E}, \mathscr{A})$ for the abelian group $\mathscr{O}_{l,0}(\mathscr{E}, \mathscr{A})$ of local action functionals of dimension zero, and $\mathscr{M}^{(0)}(\mathscr{E}, \mathscr{A})$ is canonically isomorphic to the subspace of $\mathscr{O}_{l,0}(\mathscr{E}, \mathscr{A})$ which are sums of terms which are at least cubic.

CHAPTER 5

Gauge symmetry and the Batalin-Vilkovisky formalism

1. Introduction

1.1. The philosophy of this book is that everything one says about a quantum field theory should be said in terms of its effective theories. The bijection between theories and local action functionals allows one to translate statements about effective theories into statements about local action functionals.

We would like to apply this philosophy to understand gauge theories. Naively, one could imagine that to give a gauge theory would be to give an effective gauge theory at every energy level, in a way related by the renormalization group flow.

One immediate problem with this idea is that the space of low-energy gauge symmetries is *not a group*. The product of low-energy gauge symmetries is no longer low-energy; and if we project this product onto its low-energy part, the resulting multiplication on the set of low-energy gauge symmetries is not associative.

For example, if $\mathfrak{g}$ is a Lie algebra, then the Lie algebra of infinitesimal gauge symmetries on a manifold M is $C^\infty(M) \otimes \mathfrak{g}$. The space of low-energy infinitesimal gauge symmetries is then $C^\infty(M)_{<\Lambda} \otimes \mathfrak{g}$ where we use the traditional Wilsonian cut-off – $C^\infty(M)_{<\Lambda}$ is the direct sum of all eigenspaces of the Laplacian with eigenvalue less than Λ. In general, the product of two functions in $C^\infty(M)_{<\Lambda}$ can have arbitrary energy; so that $C^\infty(M)_{<\Lambda} \otimes \mathfrak{g}$ is not closed under the Lie bracket.

This problem is solved by a very natural union of the Batalin-Vilkovisky formalism and the effective interaction philosophy. This effective BV formalism is the second main theme of this book.

1.2. The Batalin-Vilkovisky formalism is widely regarded as being the most powerful and general way to quantize gauge theories. The first step in the BV procedure is to introduce extra fields – ghosts, corresponding to infinitesimal gauge symmetries; anti-fields dual to fields; and anti-ghosts dual to ghosts – and then write down an extended action on this extended space of fields. This extended action encodes the original action, the Lie bracket on the space of infinitesimal gauge symmetries, and the way this Lie algebra acts on the original space of fields. This action is supposed to satisfy the *quantum master equation*, which is a succinct way of encoding the following conditions:

(1) The Lie bracket on the space of infinitesimal gauge symmetries satisfies the Jacobi identity.
(2) This Lie algebra acts in a way preserving the action functional on the space of fields.
(3) The Lie algebra of infinitesimal gauge symmetries preserves the "Lebesgue measure" on the original space of fields. That is, the vector field on the original space of fields associated to every infinitesimal gauge symmetry is divergence-free.
(4) The adjoint action of the Lie algebra on itself also preserves the "Lebesgue measure". Again, this says that the vector field associated to every infinitesimal gauge symmetry is divergence-free.

Unfortunately, the quantum master equation is an ill-defined expression. The 3rd and 4th conditions above are the source of the problem: the divergence of a vector field on the space of fields is a singular expression, involving the same kind of singularities as those appearing in one-loop Feynman diagrams.

1.3. This form of the quantum master equation violates the philosophy of this book: we should always express things in terms of the effective interactions. The QME above is about the original infinite energy / length scale zero action, so we should not be surprised that it doesn't make sense.

The solution to this problem is to combine the BV formalism with the effective interaction philosophy. With Wilson's sharp energy cut-off, this works as follows. We say that to give an effective interaction in the BV formalism is to give a functional $S[\Lambda]$ on the space of *extended* fields (ghosts, fields, anti-fields and anti-ghosts) of energy less than Λ. This energy Λ effective action must satisfy a certain *energy Λ quantum master equation*.

The reason that the effective action philosophy and the BV formalism work well together is the following.

LEMMA. *The renormalization group flow from scale Λ to scale Λ' carries solutions of the energy Λ QME into solutions of the energy Λ' QME.*

Thus, to give a gauge theory in the effective BV formalism is to give a collection of effective actions $S[\Lambda]$ for each Λ, such that $S[\Lambda]$ satisfies the scale Λ QME, and such that $S[\Lambda']$ is obtained from $S[\Lambda]$ by the renormalization group flow.

This picture also solves the problem that the low energy gauge symmetries are not a group. The energy Λ effective action $S[\Lambda]$, satisfying the energy Λ quantum master equation, gives the extended space of low-energy fields a certain homotopical algebraic structure, which has the following interpretation:

(1) The space of low-energy infinitesimal gauge symmetries has a Lie bracket.

(2) This Lie algebra acts on the space of low-energy fields.
  (3) The space of low-energy fields has a functional, invariant under the bracket.
  (4) The action of the Lie algebra on the space of fields, and on itself, preserves the Lebesgue measure.

However, these axioms don't hold on the nose, but hold *up to a sequence of coherent higher homotopies*.

In this book, of course, we prefer to use the proper-time cut-off, based on the heat kernel. Thus, an effective BV theory is given by a set of effective interactions $\{I[L]\}$ on the space of extended fields, such that

  (1) Each $I[L]$ satisfies the scale L quantum master equation.
  (2) The various $I[L]$ are related by the renormalization group equation.
  (3) As $L \to 0$, $I[L]$ is asymptotically local in the same sense as before.

1.4. Renormalizing gauge theories. It is straightforward to generalize the Wilsonian definition of renormalizability given in Chapter 4 to effective BV theories. Thus, to give a renormalizable theory on $\mathbb{R}^n$ in the BV formalism is to give a collection $\{I[L]\}$ of effective interactions, satisfying the quantum master equation and the renormalization group equation as above, such that

  (1) As $L \to 0$, $I[L]$ grows at most logarithmically in L, when measured in units appropriate to length scale L.
  (2) Term by term in $\hbar$, the theory has only finitely many deformations satisfying the first condition. [1]

This chapter provides the foundations for the renormalized BV formalism, and gives a general homological criterion which is sufficient for renormalizability. In Chapter 6, we will use these results to prove:

THEOREM. *Pure Yang-Mills theory on $\mathbb{R}^4$, or on any compact four-manifold with a flat metric, with coefficients in any semi-simple Lie algebra $\mathfrak{g}$, is perturbatively renormalizable.*

This chapter makes extensive use of homological algebra; an excellent reference for this material is the book-in-progress (KS).

2. A crash course in the Batalin-Vilkovisky formalism

In this section I will give an informal overview of the Batalin-Vilkovisky approach to quantizing gauge theories. I will present two different points of view on the BV formalism: one where the BV formalism is viewed as a machine for understanding integrals, and a second where the BV formalism is interpreted from the point of view of derived geometry.

---

[1] It is more natural to weaken this axiom to require that there are only finitely many deformations up to Batalin-Vilkovisky equivalence. Roughly, two theories in the BV formalism are equivalent if they are related by a local symplectic change of coordinates on the space of fields. A precise definition along these lines is given later.

2.1. The BV construction and integrals.
In this section, our vector spaces and manifolds will *always* be finite dimensional. Of course, none of the difficulties of renormalisation are present in this simple case. Many of the expressions we write in the finite dimensional case are ill-defined in the infinite dimensional case.

Let us suppose that we have a finite dimensional vector space V of fields, which will always be treated as a formal manifold near $0 \in V$. Let $\mathfrak{g}$ be a Lie group acting on V, in a possibly non-linear way. Since V is formal, this action integrates to an action of the formal group G associated to $\mathfrak{g}$.

Let $\mathscr{O}(V) = \widehat{\operatorname{Sym}}^* V^\vee$ denote the algebra of formal power series on V. Let $f \in \mathscr{O}(V)$ be a G-invariant function, with the property that $0 \in V$ is a critical point of f.

We will think of this data as defining a finite-dimensional classical gauge theory, with gauge group G, space of fields V and action functional f.

One is interested in making sense of functional integrals of the form

$$\int_{V/G} e^{f/\hbar}$$

over the quotient space V/G. The starting point in the Batalin-Vilkovisky formalism is the BRST construction, which says one should try to interpret this quotient in a homological fashion.

Functions on the quotient V/G are the same as G-invariant functions on V, or, equivalently, $\mathfrak{g}$-invariant functions on V. In homological algebra, it is always better to take the *derived* invariants, rather than the naive invariants. The derived invariants for the action of $\mathfrak{g}$ on the algebra $\mathscr{O}(V)$ of functions on V is the Chevalley-Eilenberg complex

$$C^*(\mathfrak{g}, \mathscr{O}(V)).$$

Thus, we can define the derived quotient of V be G to be the "spectrum" of $C^*(\mathfrak{g}, \mathscr{O}(V))$; that is, the object whose algebra of functions is $C^*(\mathfrak{g}, \mathscr{O}(V))$.

As a graded vector space,

$$C^*(\mathfrak{g}, \mathscr{O}(V)) = \widehat{\operatorname{Sym}}^*(\mathfrak{g}[1] \oplus V)$$

(where symmetric powers are taken, as always, in the graded sense, and [1] refers to a change of degree, so $\mathfrak{g}$ is in degree -1.).

Thus, we can identify the derived quotient of V by G with a differential graded manifold, whose underlying graded manifold is

$$\mathfrak{g}[1] \oplus V.$$

The Chevalley-Eilenberg differential can be thought of as a degree 1 vector field of square 0 on $\mathfrak{g}[1] \oplus V$. Let us denote this vector field by X.

The BRST construction says one should replace the integral over V/G by an integral over this derived quotient, that is, an integral the form

$$\int_{\mathfrak{g}[1] \oplus V} e^{f/\hbar}.$$

2. A CRASH COURSE IN THE BATALIN-VILKOVISKY FORMALISM

2.2. This leaves us in a better situation than before, as the derived quotient $\mathfrak{g}[1] \oplus V$ is a smooth formal dg manifold, and we can attempt to make sense of the integral perturbatively. However, we still have problems; the quadratic part of the functional f is highly degenerate on $\mathfrak{g}[1] \oplus V$. Indeed, f is independent of $\mathfrak{g}[1]$ and is constant on G-orbits on V. Thus, we cannot compute the integral above by a perturbation expansion around the critical points of the quadratic part of f.

This is where the Batalin-Vilkovisky formalism comes in. Let E denote the shifted cotangent bundle of $T^*[-1](\mathfrak{g}[1] \oplus V)$, so that

$$E = \mathfrak{g}[1] \oplus V \oplus V^\vee[-1] \oplus \mathfrak{g}^\vee[-2].$$

The various summands of E are usually given the following names: $\mathfrak{g}[1]$ is the space of ghosts, V is the space of fields, $V^\vee[-1]$ is the space of antifields and $\mathfrak{g}^\vee[-2]$ is the space of antighosts. (We will see shortly that these additional fields also have a natural interpretation from the point of view of derived geometry).

The function f on $\mathfrak{g}[1] \oplus V$ pulls back to a function on E, via the projection $E \to \mathfrak{g}[1] \oplus V$; we continue to call this function f. By naturality, the vector field X on $\mathfrak{g}[1] \oplus V$ induces a vector field on E, which we continue to call X. As $[X, X] = 0$ on $\mathfrak{g}[1] \oplus V$, the same identity holds on E. As X preserves f on $\mathfrak{g}[1] \oplus V$, it continues to preserve f on E.

E is a dg manifold equipped with a symplectic form of degree -1, and X is a degree 1 vector field preserving the symplectic form. Thus, there exists a unique function h_X on E whose Hamiltonian vector field is X, and which vanishes at the origin in E. As X is of degree 1 and the symplectic pairing is of degree -1, the function h_X is of cohomological degree 0.

As E has a symplectic form of cohomological degree -1, the space of functions on E has a Poisson bracket of cohomological degree 1. The statement that $[X, X] = 0$ translates into the equation $\{h_X, h_X\} = 0$. The statement that $Xf = 0$ becomes $\{h_X, f\} = 0$. And, as f is pulled back from $\mathfrak{g}[1] \oplus V$, it automatically satisfies $\{f, f\} = 0$. These identities together tell us that the function $f + h_X$ satisfies the *Batalin-Vilkovisky classical master equation*,

$$\{f + h_X, f + h_X\} = 0.$$

The function $f + h_X$ is the BV action

$$S_{BV}(e) = f(e) + h_X(e).$$

Let us split S_{BV} into "kinetic" and "interacting" parts, by

$$S_{BV}(e) = \frac{1}{2} \langle e, Qe \rangle + I_{BV}(e)$$

where $Q : V \to V$ is a cohomological degree 1 linear map, skew self-adjoint for the degree -1 pairing $\langle -, - \rangle$, and I_{BV} is a function which is at least cubic. The fact that S_{BV} satisfies the classical master equation implies that

$Q^2 = 0$. Also, the identity

$$QI_{BV} + \frac{1}{2}\{I_{BV}, I_{BV}\} = 0$$

holds as a consequence of the classical master equation for $f + h_X$.

Let $L \subset E$ be a small, generic, Lagrangian perturbation of the zero section $\mathfrak{g}[1] \oplus V \subset E$. The Batalin-Vilkovisky formalism tells us to consider the functional integral

$$\int_{e \in L}^{\sim} \exp(S_{BV}(e)/\hbar) = \int_{e \in L} \exp(\frac{1}{2\hbar}\langle e, Qe\rangle + \frac{1}{\hbar}I_{BV}(e))$$

As L is generic, the pairing $\langle e, Qe\rangle$ will have very little degeneracy on L. In fact, if the complex (E, Q) has zero cohomology, then the pairing $\langle e, Qe\rangle$ is non-degenerate on a generic Lagrangian L. This means we can perform the above integral perturbatively, around the critical point $0 \in L$.

2.3. The BV construction is the derived critical scheme. In this subsection we will describe how the BV complex, as described above, can be interpreted as the derived critical scheme of the function f on the derived quotient of V by G. In my opinion, this is the best way to interpret the BV formalism; this point of view is developed extensively in (CG10).

As before, let V be a vector space, viewed as a formal manifold completed near the origin. Let $\mathfrak{g}$ be a Lie algebra, with a possibly non-linear action on V (defined in a formal neighbourhood of the origin). Let G be the formal group associated to $\mathfrak{g}$; G also acts on the formal manifold V. Let f be a $\mathfrak{g}$-invariant function on V.

The BV construction, as described above, associates to this data a formal differential graded supermanifold

$$E = \mathfrak{g}[1] \oplus V \oplus V^\vee[-1] \oplus \mathfrak{g}^\vee[-2].$$

The differential on the algebra $\mathscr{O}(E)$ of formal power series on E is given by bracketing with the BV action S_{BV}.

In this subsection we will show how we can interpret this differential graded supermanifold as the derived critical scheme for the function f on the derived quotient of V by $\mathfrak{g}$.

In general, if we have a scheme M with a function f on M, the critical scheme of f is simply the zero locus of the one-form df. The derived critical scheme of f is the differential graded manifold obtained by forming a Koszul resolution of the equation $df = 0$.

Explicitly, the derived critical scheme of f has, as underlying graded manifold, the shifted cotangent bundle $T^*[-1]M$ of M. The algebra of functions on this space is

$$C^\infty(T^*[-1]M) = \Gamma(M, \wedge^*TX).$$

The differential on this algebra of functions which yields the derived critical scheme of f is given by contracting with the one-form df.

It is clear that $H^0(C^\infty(T^*[-1]M), \mathrm{d}f\vee)$ is the ring of functions on the critical locus of f (that is, the Jacobi ring of f).

We will see that when we apply this construction to the derived quotient of V by $\mathfrak{g}$, we find the BV dg manifold E.

As we have seen above, the derived quotient of V by $\mathfrak{g}$ is the dg manifold $\mathfrak{g}[1] \oplus V$. The algebra of functions on this, $\mathscr{O}(\mathfrak{g}[1] \oplus V)$, has a Chevalley-Eilenberg differential which we have been denoting X.

The function $f \in \mathscr{O}(\mathfrak{g}[1] \oplus V)$ is X-invariant, that is, closed for the differential. Thus, we can form the derived critical scheme for the function f; we find the dg manifold $T^*[-1](\mathfrak{g}[1] \oplus V)$. The differential on this dg manifold has two terms; one arising from the differential X on the base $\mathfrak{g}[1] \oplus V$, and the other from contracting with the one-form $\mathrm{d}f$.

It is straightforward to check that

$$\mathscr{O}(T^*[-1](\mathfrak{g}[1] \oplus V)) = \mathscr{O}(E)$$

where the differential on the left hand side comes from the derived critical scheme construction, and that on the right hand side from bracketing with the BV action function S_{BV}.

2.4. The BV construction as symplectic reduction. Yet another point of view on the BV construction is to interpret it as being obtained by a derived graded symplectic reduction from the derived critical scheme of the function f on V.

The derived critical scheme of f is the dg graded symplectic manifold (with symplectic form of degree -1)

$$T^*[-1]V = V \oplus V^\vee[-1],$$

with differential arising from contracting with the one-form $\mathrm{d}f$.

The Lie algebra $\mathfrak{g}$ acts on the dg graded symplectic manifold $T^*[-1]V$, preserving the differential. The claim is that the dg graded symplectic manifold E constructed above is the homotopical symplectic reduction of the dg graded symplectic manifold $T^*[-1]V$ by $\mathfrak{g}$.

Symplectic reduction has two steps: first, we set the moment map to zero; and then, we form the quotient.

There is a map of dg Lie algebras from $\mathfrak{g}$ to the dg Lie algebra of symplectic vector fields on $T^*[-1]V$. Since symplectic vector fields correspond to Hamiltonian functions, we get a map from $\mathfrak{g}$ to the Lie algebra of Hamiltonian functions on $T^*[-1]V$ of degree -1 which vanish at the origin. This map can be viewed as a map of dg Lie algebras with bracket of degree $+1$,

$$\mathfrak{g}[1] \to \mathscr{O}(T^*[-1]V).$$

This map induces a map of commutative algebras

$$\widehat{\mathrm{Sym}}^*(\mathfrak{g}[1]) = \mathscr{O}(\mathfrak{g}^\vee[-1]) \to \mathscr{O}(T^*[-1]V).$$

This map is in fact a map of graded Poisson algebras (with Poisson bracket of degree 1). The map of formal dg manifolds

$$m : T^*[-1]V \to \mathfrak{g}^\vee[-1]$$

is the moment map.

The first step in the homotopical symplectic reduction is to take the homotopy fibre of this moment map over $0 \in \mathfrak{g}^\vee[-1]$. This homotopy fibre $m^{-1}(0)$ can be defined by a Koszul complex:

$$\mathscr{O}(m^{-1}(0)) = \mathscr{O}(T^*[-1]V) \otimes^{\mathbb{L}}_{\mathscr{O}(\mathfrak{g}^\vee[-1])} \mathbb{C}$$
$$\simeq \mathscr{O}(T^*[-1]V \oplus \mathfrak{g}^\vee[-2])$$

with a certain Koszul differential.

The next step in the symplectic reduction procedure is to take the quotient of the homotopy fibre $m^{-1}(0)$ of the moment map by the action of $\mathfrak{g}$. The quotient is taken in the homotopical sense, yielding the Chevalley-Eilenberg complex

$$\mathbb{R}\operatorname{Hom}_{U\mathfrak{g}}(\mathbb{C}, \mathscr{O}(T^*[-1]V \oplus \mathfrak{g}^\vee[-2])) = \mathscr{O}(\mathfrak{g}[1] \oplus T^*[-1]V \oplus \mathfrak{g}^\vee[-2]).$$

The formal dg graded symplectic manifold $\mathfrak{g}[1] \oplus T^*[-1]V \oplus \mathfrak{g}^\vee[-2]$, with differential arising from the symplectic reduction procedure described above, is the same as the formal dg graded symplectic manifold

$$E = \mathfrak{g}[1] \oplus V \oplus V^\vee[-1] \oplus \mathfrak{g}^\vee[-2]$$

with differential arising from the action S_{BV}.

2.5. Quantum master equation. Let us now turn to a more general situation, where E is a finite dimensional vector space with a symplectic pairing of cohomological degree -1 and $Q : E \to E$ is a operator of square zero and degree 1, which is skew self-adjoint for the pairing. E is not necessarily of the form constructed above.

The symplectic pairing gives an isomorphism

$$E[-1] \cong E^\vee.$$

The symplectic two-form is an element

$$\omega \in \wedge^2 E^\vee$$

of cohomological degree -1. Note that

$$\wedge^2 E^\vee \cong \left(\operatorname{Sym}^2 E\right)[-2].$$

Under this isomorphism, the symplectic form ω becomes an element

$$\omega^\vee \in \operatorname{Sym}^2 E$$

of cohomological degree 1.

Let

$$\Delta : \mathscr{O}(E) \to \mathscr{O}(E)$$

be the linear order two differential operator corresponding to $\omega^\vee$. Thus, Δ is the unique order two differential operator which vanishes on $\operatorname{Sym}^{\le 1} E^\vee$, and which, on $\operatorname{Sym}^2 E^\vee$, is given by pairing with $\omega^\vee$.

Let $I \in \mathcal{O}(E)[[\hbar]]$ be an $\hbar$-dependent function on E, which modulo $\hbar$ is at least cubic. The function I satisfies the *quantum master equation* if
$$(Q + \hbar\Delta)e^{I/\hbar} = 0.$$
This equation is equivalent to the equation
$$QI + \frac{1}{2}\{I, I\} + \hbar\Delta I = 0.$$

The key lemma in the Batalin-Vilkovisky formalism is the following.

LEMMA 2.5.1. *Let $L \subset E$ be a Lagrangian on which the pairing $\langle e, Qe \rangle$ is non-degenerate. (Such a Lagrangian exists if and only if $H^*(E, Q) = 0$.) Suppose that I satisfies the quantum master equation. Then the integral*
$$\int_{e \in L} \exp(\frac{1}{2\hbar}\langle e, Qe \rangle + \frac{1}{\hbar}I(e))$$
is unchanged under deformations of L.

The non-degeneracy of the inner product on L, and the fact that I is at least cubic modulo $\hbar$, means that one can compute this integral perturbatively.

Suppose $E, Q, \langle\,,\,\rangle, I$ are obtained as before from a gauge theory. Then I automatically satisfies the classical master equation $QI + \frac{1}{2}\{I,I\} = 0$. If, in addition, $\Delta I = 0$, then I satisfies the quantum master equation. Thus, we see that we can quantize the gauge theory in a way independent of the choice of L as long as I satisfies the equation $\Delta I = 0$. When I does not satisfy this equation, one looks to replace I by a series $I' = I + \sum_{i>0} \hbar^i I_i$ which does satisfy the quantum master equation $QI' + \frac{1}{2}\{I', I'\} + \hbar\Delta I' = 0$.

2.6. Geometric interpretation of the quantum master equation. The quantum master equation has a geometric interpretation, first described by Albert Schwarz (Sch93). I will give a very brief summary; the reader should refer to this paper for more details.

Let μ denote the unique up to scale translation invariant "measure" on E, that is, section of the Berezinian. The operator Δ can be interpreted as a kind of divergence associated to the measure μ, as follows. As E has a symplectic form of degree -1, the algebra $\mathcal{O}(E)$ has a Poisson bracket of degree 1. Every function $I \in \mathcal{O}(E)$ has an associated vector field X_I, defined by the formula $X_I f = \{I, f\}$.

The operator Δ satisfies the identity
$$\mathcal{L}_{X_I}\mu = (\Delta I)\mu$$
where $\mathcal{L}_{X_I}$ refers to the Lie derivative. In other words, ΔI is the infinitesimal change in volume associated to the vector field X_I.

Thus, the two equations
$$\{I, I\} = 0$$
$$\Delta I = 0$$
say that the vector field X_I has square zero and is measure preserving.

This gives an interpretation of the quantum master equation in the case when I is independent of $\hbar$. When I depends on $\hbar$, the two terms of the quantum master equation do not necessarily hold independently. In this situation, we can interpret the quantum master equation as follows. Let μ_I be the measure on E defined by the formula
$$\mu_I = e^{I/\hbar}\mu.$$
We can define an operator Δ_I on $\mathscr{O}(E)$ by the formula
$$\mathcal{L}_{X_f}\mu_I = (\Delta_I f)\mu_I.$$
This is the divergence operator associated to the measure μ_I, in the same way that Δ is the divergence operator associated to the translation invariant measure μ.

Then, a slightly weaker version of the quantum master equation is equivalent to the statement
$$\Delta_I^2 = 0.$$
Indeed, one can compute that
$$\hbar \Delta_I f = \{I, f\} + \hbar \Delta f$$
so that
$$\hbar^2 \Delta_I^2 f = \tfrac{1}{2}\{\{I, I\}, f\} + \hbar\{\Delta I, f\}.$$
Thus, $\Delta_I^2 = 0$ if and only if $\tfrac{1}{2}\{I, I\} + \hbar \Delta I$ is in the centre of the Poisson bracket, that is, is constant.

This discussion shows that the quantum master equation is the statement that the measure $e^{I/\hbar}\mu$ is compatible in a certain sense with the graded symplectic structure on E.

REMARK. In fact it is better to use half-densities rather than densities in this picture. A solution of the quantum master equation is then given by a half-density which is compatible in a certain sense with the graded symplectic form. As all of our graded symplectic manifolds are linear, we can ignore this subtlety.

2.7. Integrating over isotropic subspaces. As we have described it, the BV formalism only has a chance to work when $H^*(E, Q) = 0$. This is because one cannot make sense of the relevant integrals perturbatively otherwise. However, there is a generalisation of the BV formalism which works when $H^*(E, Q) \neq 0$. In this situation, let $L \subset E$ be an isotropic subspace such that $Q : L \to \operatorname{Im} Q$ is an isomorphism. Let $\operatorname{Ann}(L) \subset E$ be

the set of vectors which pair to zero with any element of L. Then we can identify
$$H^*(E, Q) = \operatorname{Ann}(L) \cap \operatorname{Ker} Q.$$
We thus have a direct sum decomposition
$$E = L \oplus H^*(E, Q) \oplus \operatorname{Im} Q.$$
Note that $H^*(E, Q)$ acquires a degree -1 symplectic pairing from that on E. Thus, there is a BV operator $\Delta_{H^*(E,Q)}$ on functions on $H^*(E, Q)$. We say a function f on $H^*(E, Q)$ satisfies the quantum master equation if $\Delta e^{f/\hbar} = 0$.

The analog of the "key lemma" of the Batalin-Vilkovisky formalism is the following. This lemma is well known to experts in the area (Mnë09; KL09).

LEMMA 2.7.1. *Let $I \in \mathcal{O}(E)[[\hbar]]$ be an $\hbar$-dependent function on E which satisfies the quantum master equation. Then the function on $H^*(E, Q)$ defined by*
$$a \mapsto \hbar \log \left(\int_{e \in L} \exp(\frac{1}{2\hbar} \langle e, Qe \rangle + \frac{1}{\hbar} I(e + a)) \right)$$
(where we think of $H^*(E, Q)$ as a subspace of E) satisfies the quantum master equation.

Further, if we perturb the isotropic subspace L a small amount, then this solution of the QME on $H^*(E, Q)$ is changed to a homotopic solution of the QME.

Note that this integral is the renormalization group flow for the effective interaction.

This integral is an explicit way of writing the homological perturbation lemma for BV algebras, which transfers a solution of the quantum master equation at chain level to a corresponding solution on cohomology. From this observation it's clear (at least philosophically) why the lemma should be true; the choice of the Lagrangian L is essentially the same as the choice of symplectic homotopy equivalence between E and its cohomology.

I need to explain what a "homotopy" of a solution of the QME is. There is a general concept of homotopy equivalence of algebraic objects, which I learned from the work of Deligne, Griffiths, Morgan and Sullivan (DGMS75). Two algebraic objects are homotopic if they are connected by a family of such objects parametrised by the commutative differential algebra $\Omega^*([0, 1])$. In our context, this means that two solutions f_0, f_1 to the quantum master equation on $H^*(E, Q)$ are homotopic if there exists an element $F \in \mathcal{O}(H^*(E, Q)) \otimes \Omega^*([0, 1])[[\hbar]]$ which satisfies the quantum master equation
$$(\mathrm{d}_{DR} + \hbar \Delta_{H^*(E,Q)}) e^{F/\hbar} = 0$$
and which restricts to f_0 and f_1 when we evaluate at 0 and 1. Here d_{DR} refers to the de Rham differential on $\Omega^*([0, 1])$.

The quantum master equation imposed on F is equivalent to the equation

$$\mathrm{d}_{DR} F + \frac{1}{2}\{F,F\} + \hbar \Delta F = 0.$$

If we write $F(t, \mathrm{d}t) = A(t) + \mathrm{d}t B(t)$, then the QME imposed on F becomes the system of equations

$$\frac{1}{2}\{A(t), A(t)\} + \hbar \Delta A(t) = 0$$

$$\frac{\mathrm{d}}{\mathrm{d}t} A(t) + \{A(t), B(t)\} + \hbar \Delta B(t) = 0.$$

The first equation says that $A(t)$ satisfies the ordinary QME for all t, and the second says that the family $A(t)$ is tangent at every point to an orbit of a certain "gauge group" acting on the space of solutions to the QME.

2.8. Operadic interpretation of the quantum master equation.
In this subsection we will show how solving the quantum master equation can be interpreted as solving a certain algebraic quantization problem, expressed in the language of operads. This point of view will not be considered in the remainder of the book; thus, I will not give all the details. However, this operadic perspective is developed extensively in (CG10).

Let (E, Q) be a cochain complex with a symplectic pairing of degree -1, as above. The algebra $\mathscr{O}(E)$ of formal power series on E has a Poisson bracket $\{-,-\}$ of cohomological degree 1, derived from the symplectic pairing.

DEFINITION 2.8.1. *A P_0 algebra is a commutative differential graded equipped with a Poisson bracket of cohomological degree 1. This Poisson bracket must be a derivation in each factor, satisfy the graded Jacobi identity, and must be compatible with the differential.*

We will let P_0 denote the operad (in the category of cochain complexes) whose algebras are P_0 algebras.

REMARK. In general, a P_k algebra is a commutative differential graded algebra with a Poisson bracket of degree $1-k$. Thus, P_1 algebras are ordinary Poisson algebras. The reason for the terminology is that P_k algebras are closely related to E_k algebras.

The algebra $\mathscr{O}(E)$, with differential Q and Poisson bracket arising from the symplectic form, is a P_0 algebra.

If $I \in \mathscr{O}(E)$ solves the quantum master equation, we will let

$$Q_I = Q + \{I, -\} : \mathscr{O}(E) \to \mathscr{O}(E)$$

is a derivation for both the commutative and Poisson brackets. The classical master equation implies that $Q_I^2 = 0$. Thus, $(\mathscr{O}(E), Q_I)$ defines a P_0 algebra.

2. A CRASH COURSE IN THE BATALIN-VILKOVISKY FORMALISM

DEFINITION 2.8.2. *A BD algebra[2] is a flat $\mathbb{R}[[\hbar]]$ module A, equipped with a commutative product, a Poisson bracket of degree 1, and a differential d, all of which are linear over $\hbar$; such that*

$$d(a \cdot b) = (da) \cdot b + (-1)^{|a|} a \cdot db + \hbar\{a, b\}.$$

A BD algebra is not a differential graded algebra, because the differential is not a derivation for the commutative product. However, the failure of the differential to be a derivation is measured by the Poisson bracket.

There is an operad **BD**, in the category of cochain complexes over the ring $\mathbb{R}[[\hbar]]$, whose algebras are **BD** algebras. As a graded operad, **BD** is simply

$$\mathbf{BD} = P_0 \otimes \mathbb{R}[[\hbar]].$$

The differential on **BD** is defined by

$$d(- * -) = \hbar\{-, -\}$$

where $(- * -), \hbar\{-, -\} \in P_0(2)$ denote the product and bracket.

DEFINITION 2.8.3. *A quantization of a P_0 algebra A is a BD algebra $\widetilde{A}$, flat over $\mathbb{R}[[\hbar]]$, with an isomorphism of P_0 algebras*

$$\widetilde{A} \otimes_{\mathbb{R}[[\hbar]]} \mathbb{R} \cong A.$$

The main results of this subsection is that quantizations of the P_0 algebra $\mathscr{O}(E)$ are the same as solutions to the quantum master equation. Here, as before, (E, Q, ω) is a cochain complex with a symplectic pairing of degree -1.

We will state this result as an equivalence of groupoids (a more refined analysis would yield an equivalence of infinity-groupoids). Given a P_0 algebra A, let us define a groupoid $\text{Quant}(A)$ whose objects are quantizations of $\widetilde{A}$ of A, and whose morphisms are $\mathbb{R}[[\hbar]]$-linear isomorphisms $\phi : \widetilde{A} \to \widetilde{A}'$ of **BD** algebras, which are the identity modulo $\hbar$.

Let $I \in \mathscr{O}(E)$ be a solution to the classical master equation. Let

$$\mathscr{O}^{>0}(E) = \mathscr{O}(E)/\mathbb{R}$$

denote the quotient of E by $\mathbb{R}$. Note that we can identify $\mathscr{O}^{>0}(E)[1]$ with the vector space of derivations of $\mathscr{O}(E)$ which preserve the symplectic form.

Let us define a simplicial set $\text{QME}(E, I)$ of solutions to the quantum master equation by saying that the set $\text{QME}(E, I)[n]$ of n-simplices are elements

$$\widehat{I} \in \mathscr{O}^{>0}(E) \otimes \Omega^*(\Delta^n)[[\hbar]]$$

which satisfy the quantum master equation and which agree with I modulo $\hbar$. Here, the quantum master equation includes the differential on the algebra $\Omega^*(\Delta^n)$ of forms on Δ^n. (We will analyze this simplicial set in more detail in Section 10).

---

[2] BD stands for Beilinson-Drinfeld, who considered such algebras in their book (BD04). I use this notation because, unfortunately, the term Batalin-Vilkovisky algebra has been used in the mathematical literature to refer to a different, but closely related object.

152   5. GAUGE SYMMETRY AND THE BATALIN-VILKOVISKY FORMALISM

Let us define a groupoid
$$\Pi\,\mathrm{QME}(E,I)$$
as the fundamental groupoid of $\mathrm{QME}(E,I)$. One can verify easily that $\mathrm{QME}(E,I)$ is a Kan complex. This means that $\Pi\,\mathrm{QME}(E,I)$ can be described, in concrete terms, as the groupoid whose objects are solutions to the quantum master equation $\widehat{I}$ which restrict to I modulo $\hbar$; and whose morphisms are homotopies of such solutions; but these homotopies are themselves only taken up to homotopy.

PROPOSITION 2.8.4. *There is an equivalence of groupoids*
$$\Pi\,\mathrm{QME}(E,I) \simeq \mathrm{Quant}(\mathscr{O}(E), Q + \{I,-\}).$$

This proposition won't be used in the rest of this book.

PROOF. Given any 0-simplex $\widehat{I} \in \mathrm{QME}(E,I)$, let us define a differential
$$Q_{\widehat{I}} = Q + \{\widehat{I},-\} + \hbar\Delta$$
on $\mathscr{O}(E) \otimes \Omega^*(\Delta^n)[[\hbar]]$. The quantum master equation implies that $Q_{\widehat{I}}^2 = 0$.

This shows that each object of $\Pi\,\mathrm{QME}(E,I)$ (that is, each 0-simplex in $\mathrm{QME}(E,I)$) gives an object of the category $\mathrm{Quant}(\mathscr{O}(E), Q + \{I,-\})$.

Next, we need to show that homotopies between solutions of the QME give rise to isomorphisms of quantizations. Suppose that $F(t,\mathrm{d}t) \in \mathscr{O}(E) \otimes \Omega^*([0,1])[[\hbar]]$ is such a homotopy. Then, we can write
$$F(t,\mathrm{d}t) = \widehat{I}(t) + \mathrm{d}t\,\widehat{J}(t)$$
where $\widehat{I}(t)$ is a family of solutions to the QME, depending on t. Let
$$X(t) = -\{\widehat{J}(t),-\}$$
be the derivation given by Poisson bracket with $\widehat{J}(t)$. The statement that $F(t,\mathrm{d}t)$ satisfies the QME implies
$$\frac{\mathrm{d}}{\mathrm{d}t}Q_{\widehat{I}(t)} = [X(t), Q_{\widehat{I}(t)}].$$
(Here, as before, $Q_{\widehat{I}(t)} = Q + \{\widehat{I}(t),-\} + \hbar\Delta$).

Let us define a one-parameter family of automorphisms
$$\Phi(T) : \mathscr{O}(E)[[\hbar]] \to \mathscr{O}(E)[[\hbar]]$$
for $T \in [0,1]$, by requiring that $\Phi(0)$ is the identity, and that $\Phi(T)$ satisfies the differential equation
$$\frac{\mathrm{d}}{\mathrm{d}T}\Phi(T)(f) = X(T)\Phi(T)(f).$$
The fact that $\hbar$ is a formal parameter and that $\mathscr{O}(E)$ is the completed symmetric algebra implies that there is a unique such $\Phi(T)$.

Note that
$$\Phi(T)Q_{\widehat{I}(0)}\Phi(T)^{-1} = Q_{\widehat{I}(T)}.$$

2. A CRASH COURSE IN THE BATALIN-VILKOVISKY FORMALISM

In particular, $\Phi(1)$ defines an isomorphism of quantizations
$$\Phi(1) : (\mathscr{O}(E)[[\hbar]], Q_{\widehat{I}(0)}) \to (\mathscr{O}(E)[[\hbar]], Q_{\widehat{I}(1)}).$$
Thus, every one-simplex in $\mathrm{QME}(E,I)$ defines a morphism in the groupoid $\mathrm{Quant}(\mathscr{O}(E), Q_I)$. It is not hard to check that homotopic one-simplices give the same morphism, so that we have a map of groupoids
$$\Pi\,\mathrm{QME}(E, I) \to \mathrm{Quant}(\mathscr{O}(E), Q_I).$$

Next we will check that this functor is essentially surjective. If A is a quantization of $(\mathscr{O}(E), Q_I)$, then A (without the differential) can be thought of as a family of formal graded symplectic manifolds over $\mathrm{Spec}\,\mathbb{R}[[\hbar]]$ which, modulo $\hbar$, is equipped with an isomorphism to the formal graded symplectic manifold E. The Darboux lemma (which applies in this graded formal situation) shows that there is an isomorphism of graded P_0 algebras over $\mathbb{R}[[\hbar]]$ (without the differential)
$$A \cong \mathscr{O}(E)[[\hbar]].$$
We will let d_A be the differential on A, viewed as an operator on $\mathscr{O}(E)[[\hbar]]$. This operator is not a derivation, but satisfies
$$\mathrm{d}_A(a \cdot b) = (\mathrm{d}_A a) \cdot b + (-1)^{|a|} a \cdot \mathrm{d}_A b + \hbar\{a, b\}.$$
Then,
$$\mathrm{d}_A - Q - \hbar\Delta$$
is a derivation for both the commutative and Poisson structure on $\mathscr{O}(E)[[\hbar]]$.

Now, every such derivation of $\mathscr{O}(E)[[\hbar]]$ is given by bracketing with a Hamiltonian function, defined up to a constant. Thus, let $\widehat{I} \in \mathscr{O}^{>0}(E)[[\hbar]]$ be the unique element such that
$$\{\widehat{I}, -\} = \mathrm{d}_A - Q - \hbar\Delta.$$
That is, $\mathrm{d}_A = Q_{\widehat{I}}$ for some unique $\widehat{I} \in \mathscr{O}(E)^{>0}[[\hbar]]$. The fact that $\mathrm{d}_A^2 = 0$ implies the that $\widehat{I}$ satisfies the quantum master equation.

Thus, we see that the functor
$$\Pi\,\mathrm{QME}(E, I) \to \mathrm{Quant}(\mathscr{O}(E), Q_I)$$
is essentially surjective.

It remains to check that the functor is full and faithful. Thus, suppose that $\widehat{I}_0, \widehat{I}_1$ are solutions to the quantum master equation. Suppose that
$$\Phi : \mathscr{O}(E)[[\hbar]] \to \mathscr{O}(E)[[\hbar]]$$
is an automorphism of P_0 algebra $\mathscr{O}(E)[[\hbar]]$ such that
$$\Phi Q_{\widehat{I}_0} \Phi^{-1} = Q_{\widehat{I}_1},$$
and such that Φ is the identity modulo $\hbar$. Then, we can write Φ as the exponential of some derivation X of $\mathscr{O}(E)[[\hbar]]$. Let $\Phi(t) = e^{tX}$. Define $\widehat{I}(t)$ by
$$Q_{\widehat{I}(t)} = \Phi(t) Q_{\widehat{I}_0} \Phi(t)^{-1}.$$

Let $J \in \mathcal{O}^{>0}(E)[[\hbar]]$ be the unique element such that $-\{J,-\} = X$. Then, we find a homotopy between the solutions $\widehat{I}_0$ and $\widehat{I}_1$ of the quantum master equation by setting

(†) $$F(t, \mathrm{d}t) = \widehat{I}(t) + J \mathrm{d}t$$

The differential equation

$$\frac{\mathrm{d}}{\mathrm{d}t}\widehat{I}(t) + QJ + \hbar \Delta J + \{J, \widehat{I}(t)\} = 0$$

implies that $F(t, \mathrm{d}t)$ satisfies the quantum master equation, and so defines a homotopy.

Thus we have shown that the functor is full. The final step is to verify that the functor is faithful. This amounts to checking that any homotopy between two solutions of the quantum master equation $\widehat{I}_0$ and $\widehat{I}_1$ is homotopic to one of the form (†), that is, one such that the coefficient of $\mathrm{d}t$ is independent of t.

This can be proved by an inductive argument, working term by term in $\hbar$. Let $\mathrm{QME}^{\leq n}(E, I)$ denote the simplicial set of solutions to the quantum master equation defined modulo $\hbar^{n+1}$. One can check that the natural map $\mathrm{QME}^{\leq n+1}(E, I) \to \mathrm{QME}^{\leq n}(E, I)$ is a fibration. Let

$$\widetilde{\mathrm{QME}}^{\leq n}(E, I) \subset \mathrm{QME}^{\leq n}(E, I)$$

be the sub-simplicial set consisting of solutions to the quantum master equation of the form (†).

Let us fix two solutions $\widehat{I}_0$, $\widehat{I}_1$ of the quantum master equation which agree with the original I modulo $\hbar$. Let F be a homotopy between them. Let F_n denote the reduction of F modulo $\hbar^{n+1}$; we assume that $F_0 = 0$. Thus, F_n is a one-simplex in $\mathrm{QME}^{\leq n}(E, I)$. Suppose, by induction, that we have a one-simplex G_n in the subcomplex $\widetilde{\mathrm{QME}}^{\leq n}(E, I)$ which is of the form (†). Suppose in addition that H_n is a two-simplex in $\mathrm{QME}^{\leq n}(E, I))$ which is constant with value I_0 on one edge; and on the other two edges is F_n and G_n.

We have a lift F_{n+1} of F to $\mathrm{QME}^{\leq n+1}(E, I)$. The fact that the map $\mathrm{QME}^{\leq n+1}(E, I) \to \mathrm{QME}^{\leq n}(E, I)$ is a fibration means that the homotopy H_n lifts to a two-simplex H'_{n+1} of $\mathrm{QME}^{\leq n+1}(E, I)$, which on one edge is F_{n+1} and on another is constant with value I_0.

Let H''_{n+1} be another such lift of H_n. Then,

$$H'_{n+1} - H''_{n+1} = \hbar^{n+1} A$$

for some $A \in \mathcal{O}(E) \otimes \Omega^*(\Delta^2, \mathrm{Horn})$ (here $\mathrm{Horn} \subset \Delta_2$ is the union of the first two edges).

Further, the fact that H'_{n+1} and H''_{n+1} satisfies the quantum master equation implies

$$(Q_I + \mathrm{d}_{dR})A = 0.$$

We would like to construct some H_{n+1} whose restriction to the third face of Δ_2 is in the subcomplex $\widetilde{\mathrm{QME}}^{\leq n+1}(E,I)$. This condition amounts to saying that the restriction of H_{n+1} to the third edge of Δ^2 lies in the subcomplex of $\Omega^*([0,1])$ spanned by $1, t$ and $\mathrm{d}t$. This subcomplex is quasi-isomorphic to $\Omega^*([0,1])$. Therefore, standard obstruction theory arguments allow us to find some $A \in \mathscr{O}(E) \otimes \Omega^*(\Delta^2, \text{Horn})$ such that $Q_I A + \mathrm{d}_{dR} A = 0$, and that $\widetilde{H}_{n+1} + \hbar^{n+1} A$ restricts, on the third edge of Δ_2, to an element of this subcomplex. $\square$

3. The classical BV formalism in infinite dimensions

In this section we will explain how the constructions and results of the BV formalism in finite dimensions go through with little change at the classical level in infinite dimensions. This leads to many examples of solutions of the Batalin-Vilkovisky classical master equation in infinite dimensions. At the quantum level, as we will see later, we will need renormalization techniques to even define the quantum master equation.

3.1. Before we start, we need a definition.

DEFINITION 3.1.1. *Let $E_1, \ldots, E_{n+1}$ be vector bundles on a manifold M; let $\mathscr{E}_1, \ldots, \mathscr{E}_n$ denote the spaces of global sections. Let $\mathrm{Diff}(\mathscr{E}_i, C^\infty(M))$ be the space of differential operators from the vector bundle E_i to the trivial vector bundle. We will view $\mathrm{Diff}(\mathscr{E}_i, C^\infty(M))$ as a $C^\infty(M)$-module by multiplication on the left: $(fD)(e) = f(De)$ for $f \in C^\infty(M)$ and $e \in \mathscr{E}_i$.*

The space of polydifferential operators

$$\mathscr{E}_1 \otimes \cdots \otimes \mathscr{E}_n \to \mathscr{E}_{n+1}$$

is the space

$$\mathrm{PolyDiff}(\mathscr{E}_1 \otimes \cdots \otimes \mathscr{E}_n, \mathscr{E}_{n+1})$$
$$= \mathrm{Diff}(\mathscr{E}_1, C^\infty(M)) \otimes_{C^\infty(M)} \cdots \otimes_{C^\infty(M)} \mathrm{Diff}(\mathscr{E}_n, C^\infty(M)) \otimes_{C^\infty(M)} \mathscr{E}_{n+1}.$$

There is a natural injective map

$$\mathrm{PolyDiff}(\mathscr{E}_1 \otimes \cdots \otimes \mathscr{E}_n, \mathscr{E}_{n+1}) \to \mathrm{Hom}(\mathscr{E}_1 \otimes \cdots \otimes \mathscr{E}_n, \mathscr{E}_{n+1})$$

defined by

$$(D_1 \otimes \cdots \otimes D_n \otimes e_{n+1})(e_1 \otimes \cdots \otimes e_n) = e_{n+1}(D_1 e_1)(D_2 e_2) \cdots (D_n e_n)$$

where $D_i \in \mathrm{Diff}(\mathscr{E}_i, C^\infty(M))$ and $e_i \in \mathscr{E}_i$.

3.2. We will describe the construction of solutions of the classical master equation on a compact manifold, for simplicity. The construction on $\mathbb{R}^n$ or on a non-compact manifold is similar.

Let M be a compact manifold and E a super vector bundle on M. Let $\mathscr{E} = \Gamma(M, E)$, and, as usual, let $\mathscr{O}(\mathscr{E}) = \prod_{n \geq 0} \mathrm{Hom}(\mathscr{E}^{\otimes n}, \mathbb{R})_{S_n}$ be the algebra of functionals on $\mathscr{E}$.

The Lie algebra
$$\mathrm{Der}(\mathscr{O}(\mathscr{E})) = \mathscr{O}(\mathscr{E}) \otimes \mathscr{E} = \prod_{n \geq 0} \mathrm{Hom}(\mathscr{E}^{\otimes n}, \mathscr{E})_{S_n}$$
is the Lie algebra of symmetries of $\mathscr{E}$. There is a subalgebra
$$\mathrm{Der}_{loc}(\mathscr{O}(\mathscr{E})) = \prod_{n \geq 0} \mathrm{Hom}_{loc}(\mathscr{E}^{\otimes n}, \mathscr{E})_{S_n}$$
of local symmetries of $\mathscr{E}$, where
$$\mathrm{Hom}_{loc}(\mathscr{E}^{\otimes n}, \mathscr{E})$$
is the space of polydifferential operators.

DEFINITION 3.2.1. *A degree -1 symplectic structure on $\mathscr{E}$ is a map*
$$\phi : E \otimes E \to \mathrm{Dens}(M)$$
of vector bundles on M, where $\mathrm{Dens}(M)$ is the line bundle of densities. This map must be of cohomological degree -1, -symmetric, and fibrewise non-degenerate. This map yields a degree -1 anti-symmetric pairing
$$\omega : \mathscr{E} \otimes \mathscr{E} \to \mathbb{C}$$
$$e_1 \otimes e_2 \mapsto \int_M \phi(e_1 \otimes e_2).$$

Let $X \in \mathrm{Der}(\mathscr{O}(\mathscr{E}))$. This vector field is given by a collection of S_n-invariant maps
$$X_n : \mathscr{E}^{\otimes n} \to \mathscr{E}$$
for $n \geq 0$. Formally, we can say that X is symplectic if the Lie derivative $L_X \omega$ vanishes. It is not completely obvious that this equation is well defined. However, it turns out that $L_X \omega = 0$ if and only if the maps
$$X_n \vee \omega : \mathscr{E}^{\otimes n+1} \to \mathbb{C}$$
$$e_1 \otimes \cdots \otimes e_{n+1} \mapsto \omega(e_1, X_n(e_2, \ldots, e_{n+1}))$$
are invariant under S_{n+1}.

If X is symplectic, then there is a Hamiltonian $H_X \in \mathscr{O}(\mathscr{E})$ whose homogeneous components
$$H_n : \mathscr{E}^{\otimes n} \to \mathbb{C}$$
are defined by
$$H_n = \frac{1}{n} X_{n-1} \vee \omega$$
where $X_{n-1} \vee \omega$ is the map defined above.

Formally, X is the Hamiltonian vector field associated to the function H_X.

DEFINITION 3.2.2. *A function $f \in \mathscr{O}(\mathscr{E})$ is Hamiltonian if it is the Hamiltonian function H_X associated to some symplectic vector field X. If f is Hamiltonian, then there is a unique such vector field X; we call this the Hamiltonian vector field associated to f.*

There is a bijection between Hamiltonian functions up to constants and symplectic vector fields.

Not all functions $f \in \mathcal{O}(\mathcal{E})$ are Hamiltonian. However,

LEMMA 3.2.3. *Every local action functional $f \in \mathcal{O}_{loc}(\mathcal{E})$ is Hamiltonian. Under the bijection between Hamiltonian functions up to constants and symplectic vector fields, local action functionals correspond to local symplectic vector fields.*

Let $f, g \in \mathcal{O}_{loc}(\mathcal{E})$. Then, we define the BV bracket
$$\{f, g\}$$
by
$$\{f, g\} = X_f g.$$
Note that this bracket is of cohomological degree 1. As usual, $\{f, g\}$ is symmetric and satisfies the graded Jacobi identity.

DEFINITION 3.2.4. *An even functional $S \in \mathcal{O}_{loc}(\mathcal{E})$ satisfies the classical master equation if*
$$\{S, S\} = 0.$$

3.3. Now let us turn to constructions of solutions of the classical master equation, analogous to the constructions we have already seen in the finite dimensional situation.

Let F be a vector bundle on M, whose space of global sections is denoted by $\mathcal{F}$.

DEFINITION 3.3.1. *A local Lie algebra structure on $\mathcal{F}$ is given by a continuous linear map*
$$\mathcal{F} \otimes \mathcal{F} \to \mathcal{F}$$
which is an anti-symmetric polydifferential operator, satisfying the Jacobi identity.

Let V be another vector bundle on M, whose space of global sections is denoted by $\mathcal{V}$. Let $\mathcal{F}$ be a local Lie algebra. *A* local Lie algebra action *of $\mathcal{F}$ on $\mathcal{V}$ is given by a Lie algebra map*
$$\mathcal{F} \to \mathrm{Der}(\mathcal{O}(\mathcal{V}))$$
such that the corresponding maps
$$\mathcal{F} \otimes \mathcal{V}^{\otimes n} \to \mathcal{V}$$
are polydifferential operators.

For example, the tangent bundle TM of M is a local Lie algebra. If $\mathfrak{g}$ is a finite dimensional Lie algebra, then $C^\infty(M) \otimes \mathfrak{g}$ is a local Lie algebra.

If the local Lie algebra $\mathcal{F}$ has local action on $\mathcal{V}$, then the infinite dimensional supermanifold $\mathcal{F}[1] \oplus \mathcal{V}$ has a Chevalley-Eilenberg differential, as usual. This is an element
$$X \in \mathrm{Der}\,\mathcal{O}(\mathcal{F}[1] \oplus \mathcal{V})$$

whose homogeneous components are of two types; firstly, we have the Lie bracket
$$\mathscr{F}^{\otimes 2} \to \mathscr{F}$$
and secondly, we have the action maps
$$\mathscr{F} \otimes \mathscr{V}^{\otimes n} \to \mathscr{V}.$$
Since all of these maps are polydifferential operators, the vector field X is local. As usual, the Jacobi identity, together with the axioms for a Lie algebra action, are equivalent to the identity
$$[X, X] = 0.$$

3.4. In this situation, we can apply the Batalin-Vilkovisky machinery as in the finite dimensional setting. We need to define the shifted cotangent bundle of $\mathscr{F}[1] \oplus \mathscr{V}$. For any vector bundle U on M, let
$$U^! = U^\vee \otimes \mathrm{Dens}(M).$$
Note that there is a natural pairing between $\Gamma(M, U^!)$ and $\Gamma(M, U)$. This pairing has no kernel.

Let E be the vector bundle on M defined by
$$E = F[1] \oplus V \oplus V^![-1] \oplus F^![-2].$$
The global sections of E will be denoted $\mathscr{E}$, which we write
$$\mathscr{E} = \mathscr{F}[1] \oplus \mathscr{V} \oplus \mathscr{V}^![-1] \oplus \mathscr{F}^![-2].$$
The graded pieces of the vector bundle E are the bundles of ghosts, fields, anti-fields and anti-ghosts, respectively.

The vector bundle E has a degree -1 symplectic pairing
$$E \otimes E \to \mathrm{Dens}(M),$$
arising from the natural pairings
$$F \otimes F^! \to \mathrm{Dens}(M)$$
$$V \otimes V^! \to \mathrm{Dens}(M).$$

3.5. We would like to extend the local vector field X on $\mathscr{F}[1] \oplus \mathscr{V}$ to a vector bundle on $\mathscr{E}$, which we view as the shifted cotangent bundle to $\mathscr{F}[1] \oplus \mathscr{V}$. Let us recall what happens in the finite dimensional situation. Let N be a finite dimensional supermanifold, and let X be a vector field on N. Then, by naturality, X extends to a vector field X' on the shifted cotangent bundle $T^*N[-1]$ of N. The Hamiltonian function of the vector field X' is linear on the fibres of the projection $T^*N[-1] \to N$; this property uniquely characterizes the lift X' of X.

Back in our infinite dimensional situation, we have the following.

3. THE CLASSICAL BV FORMALISM IN INFINITE DIMENSIONS

LEMMA 3.5.1. *There is a unique (up to the addition of a constant) local action functional*
$$S_{Gauge} \in \mathcal{O}_{loc}(\mathcal{E})$$
such that

(1) *S_{Gauge} is linear on the fibres of the projection map*
$$\mathcal{E} \to \mathscr{F}[1] \otimes \mathscr{V}.$$
That is,
$$S_{Gauge} \in \mathcal{O}(\mathscr{F}[1] \otimes \mathscr{V}) \otimes \operatorname{Hom}(\mathscr{V}^{!}[-1] \oplus \mathscr{F}^{!}[-2], \mathbb{R}) \subset \mathcal{O}(\mathcal{E}).$$

(2) *The Hamiltonian vector field X' associated to S_{Gauge} is a lift of the vector field X. This means that X' preserves the subalgebra*
$$\mathcal{O}(\mathscr{F}[1] \oplus \mathscr{V}) \subset \mathcal{O}(\mathcal{E})$$
and agrees with X on this sub-algebra.

Further, S_{Gauge} satisfies the classical master equation
$$\{S_{Gauge}, S_{Gauge}\} = 0.$$

PROOF. Uniqueness of S_{Gauge} is easy to see. To see existence, we will construct S_{Gauge} explicitly.

Let us $f \in \mathscr{F}$ be a ghost, $v \in \mathscr{V}$ a field, $v^{\vee} \in \mathscr{V}^{!}$ an anti-field, and $f^{\vee} \in \mathscr{F}^{!}$ an anti-ghost. Then let
$$S_{Gauge}(f, v, v^{\vee}, f^{\vee}) = \tfrac{1}{2} \langle [f,f], f^{\vee} \rangle + \langle f(v), v^{\vee} \rangle.$$

By $f(v)$ we mean the vector field $f \in \operatorname{Der}_{loc}(\mathcal{O}(\mathscr{V}))$ evaluated at the point $v \in \mathscr{V}$; the resulting element of $\mathscr{V} = T_v \mathscr{V}$ is paired with $v^{\vee}$.

The first term in S_{Gauge} is local because the Lie bracket map
$$\mathscr{F} \otimes \mathscr{F} \to \mathscr{F}$$
is a polydifferential operator.

The second term in S_{Gauge} is local because $f(v)$ arises from S_n-invariant polydifferential operators
$$\mathscr{F} \otimes \mathscr{V}^{\otimes n} \to \mathscr{V}.$$
for all n.

The associated local symplectic vector field $X' \in \operatorname{Der}_{loc}(\mathcal{O}(\mathcal{E}))$ preserves the sub-algebra
$$\mathcal{O}(\mathscr{F}[1] \oplus \mathscr{V}) \subset \mathcal{O}(\mathcal{E}).$$
It is easy to see that this vector field X' agrees with the Chevalley-Eilenberg vector field on the algebra $\mathcal{O}(\mathscr{F}[1] \oplus \mathscr{V})$.

The fact that the Chevalley-Eilenberg vector field X has square zero implies that
$$[X', X'] = 0.$$
It follows that the function S_{Gauge} satisfies $\{S_{Gauge}, S_{Gauge}\} = 0$. □

3.6. Let
$$S_{\mathscr{V}} \in \mathscr{O}_{loc}(\mathscr{V})$$
be a functional preserved by the action of the Lie algebra $\mathscr{F}$. We can pull back $S_{\mathscr{V}}$ to a function
$$S_{\mathscr{V}} \in \mathscr{O}_{loc}(\mathscr{E}).$$

The fact that this functional is invariant under the action of $\mathscr{F}$ implies that
$$X' S_{\mathscr{V}} = 0$$
where X' is the lift of the Chevalley-Eilenberg vector field X on $\mathscr{F}[1] \oplus \mathscr{V}$. This means that
$$\{S_{Gauge}, S_{\mathscr{V}}\} = 0.$$
Since $S_{\mathscr{V}}$ depends on only half of the variables of $\mathscr{E}$,
$$\{S_{\mathscr{V}}, S_{\mathscr{V}}\} = 0.$$
Thus, the BV action
$$S = S_{Gauge} + S_{\mathscr{V}}$$
satisfies the classical master equation.

4. Example: Chern-Simons theory

The main theory we will be concerned with is Yang-Mills theory. However, we will start by discussing a much simpler theory, Chern-Simons theory on an oriented three-manifold M. Let G be a compact Lie group, with Lie algebra $\mathfrak{g}$; let us fix an invariant pairing $\langle -, - \rangle_{\mathfrak{g}}$ on $\mathfrak{g}$.

4.1. The Chern-Simons field is a $\mathfrak{g}$-valued connection on M. For simplicity, we will assume we are perturbing around the trivial flat connection on the trivial $\mathfrak{g}$-bundle on M. Then, we can identify this space of fields with $\Omega^1(M) \otimes \mathfrak{g}$. The gauge group is the group
$$\mathscr{G} = \operatorname{Maps}(M, G)$$
of all smooth maps from M to G, a compact Lie group whose Lie algebra is $\mathfrak{g}$. We will mainly be interested in the Lie algebra of infinitesimal gauge symmetries, which we take to be
$$C^\infty(M) \otimes \mathfrak{g} = \Omega^0(M) \otimes \mathfrak{g}.$$
The action of this Lie algebra on the space of fields is the standard affine-linear action,
$$A \mapsto [X, A] + \mathrm{d} X$$
where $A \in \Omega^1(M) \otimes \mathfrak{g}$ and $X \in \Omega^0(M) \otimes \mathfrak{g}$.

Let $\langle -, - \rangle$ denote the pairing on $\Omega^*(M) \otimes \mathfrak{g}$ defined by
$$\langle \omega_1 \otimes E_1, \omega_2 \otimes E_2 \rangle = \int_M \omega_1 \wedge \omega_2 \langle E_1, E_2 \rangle_{\mathfrak{g}}$$
where $\langle -, - \rangle_{\mathfrak{g}}$ is our chosen invariant pairing on $\mathfrak{g}$.

4. EXAMPLE: CHERN-SIMONS THEORY

The Chern-Simons action S_{CS} on $\Omega^1(M) \otimes \mathfrak{g}$ is defined by

$$S_{CS}(A) = \tfrac{1}{2} \int_M \langle A, \mathrm{d}A \rangle + \tfrac{1}{6} \langle A, [A, A] \rangle$$

This is preserved by the Lie algebra $\Omega^0(M) \otimes \mathfrak{g}$ of infinitesimal gauge symmetries.

4.2. The Batalin-Vilkovisky space of fields associated for the Chern-Simons gauge theory is obtained by adding on ghosts, corresponding to elements of the Lie algebra of gauge symmetries; anti-fields, dual to the space of fields; and anti-ghosts, dual to the space of ghosts. The space we end up with is simply

$\Omega^0(M) \otimes \mathfrak{g}$	ghosts; degree -1
$\Omega^1(M) \otimes \mathfrak{g}$	fields; degree 0
$\Omega^2(M) \otimes \mathfrak{g}$	anti-fields; degree 1
$\Omega^3(M) \otimes \mathfrak{g}$	anti-ghosts; degree 2

In other words, the space of fields for Chern-Simons theory in the BV formalism is simply

$$\mathscr{E} = \Omega^*(M) \otimes \mathfrak{g}[1].$$

The degree -1 symplectic pairing on $\mathscr{E}$ arises from the degree -3 symmetric pairing on $\Omega^*(M) \otimes \mathfrak{g}$ described above.

4.3. The Batalin-Vilkovisky action on $\mathscr{E}$ is simply

$$S(e) = \tfrac{1}{2} \langle e, \mathrm{d}e \rangle + \tfrac{1}{6} \langle e, [e, e] \rangle$$

as before.

One can see this as follows. The BV action can be written

$$S = S_{Gauge} + S_{CS}$$

where

$$S_{CS} : \Omega^1(M) \otimes \mathfrak{g} \to \mathbb{R}$$

is the Chern-Simons action as above, and S_{Gauge} arises from the gauge action.

We will denote fields by A, ghosts by X, anti-fields by $A^\vee$ and anti-ghosts by $X^\vee$. Then, the Lie bracket on ghosts gives the term

$$\tfrac{1}{2} \langle [X, X], X^\vee \rangle.$$

The action of ghosts on fields gives the term

$$\langle \mathrm{d}X + [X, A], A^\vee \rangle.$$

4.4. In the BV formalism, we always integrate over an isotropic subspace of the space of fields $\mathscr{E}$. For Chern-Simons theory, the isotropic subspace is
$$\operatorname{Im} d^* \subset \mathscr{E}.$$
This, of course, requires the choice of a metric on M. The philosophy is that all constructions should be independent of the choice of gauge fixing condition. A precise statement along these lines is proved in (Cos07). See (CM08a) and (Iac08) for a more thorough discussion.

5. Example : Yang-Mills theory

The material in Section 3 leads to many examples of solutions of the classical master equation. The main example we will be concerned with in this book is associated to the Yang-Mills gauge theory in four dimensions.

5.1. Let M be an oriented 4-manifold equipped with a conformal class of metrics. Let $\mathfrak{g}$ be a semi-simple Lie algebra, and fix an invariant pairing $\langle -, - \rangle_\mathfrak{g}$ on $\mathfrak{g}$. The Yang-Mills field is a $\mathfrak{g}$-valued connection on M; since we are perturbing around a given flat connection, we can identify the space of fields with $\Omega^1(M) \otimes \mathfrak{g}$.

As with Chern-Simons theory, the gauge group is the group
$$\mathscr{G} = \operatorname{Maps}(M, G)$$
of all smooth maps from M to G, the compact Lie group associated to G. The Lie algebra of infinitesimal gauge symmetries is
$$C^\infty(M) \otimes \mathfrak{g} = \Omega^0(M) \otimes \mathfrak{g}.$$
The action of this Lie algebra on the space of fields is the standard affine-linear action,
$$A \mapsto [X, A] + dX$$
where $A \in \Omega^1(M) \otimes \mathfrak{g}$ and $X \in \Omega^0(M) \otimes \mathfrak{g}$.

Let $\langle -, - \rangle$ denote the pairing on $\Omega^*(M) \otimes \mathfrak{g}$ defined by
$$\langle \omega_1 \otimes E_1, \omega_2 \otimes E_2 \rangle = \int_M \omega_1 \wedge \omega_2 \langle E_1, E_2 \rangle_\mathfrak{g}$$
where $\langle -, - \rangle_\mathfrak{g}$ is our chosen invariant pairing on $\mathfrak{g}$.

5.2. The Yang-Mills action S_{YM} on $\Omega^1(M) \otimes \mathfrak{g}$ is defined by
$$S_{YM}(A) = \langle F(A)_+, F(A)_+ \rangle.$$
In this expression,
$$F(A)_+ = \tfrac{1}{2}(1 + *)F(A) \in \Omega^2_+(M) \otimes \mathfrak{g}$$

denotes the projection of the curvature onto the space of self-dual two-forms. This is not the standard way of writing the Yang-Mills action, but observe that

$$\begin{aligned}\langle F(A)_+, F(A)_+\rangle &= \langle \tfrac{1}{2}(1+*)F(A), \tfrac{1}{2}(1+*)F(A)\rangle \\ &= \tfrac{1}{2}\langle F(A), F(A)\rangle + \tfrac{1}{2}\langle *F(A), F(A)\rangle.\end{aligned}$$

The first term is topological, and so independent of the connection A. The second term is the usual way of writing the Yang-Mills action.

The Yang-Mills functional S_{YM} is invariant under the action of the group $\mathscr{G}$ of gauge symmetries, and under the Lie algebra $\Omega^0(M)\otimes\mathfrak{g}$ of infinitesimal gauge symmetries. The functional integral one is interested in making sense of is

$$\int_{A\in\Omega^1(M)\otimes\mathfrak{g}/\mathscr{G}} e^{S_{YM}(A)/\hbar}.$$

This is the integral over the space of all connections modulo gauge equivalence.

5.3. The Lie algebra $\Omega^0(M)\otimes\mathfrak{g}$ is local, as is the action of this Lie algebra on the space $\Omega^1(M)\otimes\mathfrak{g}$ of fields. Thus, we can apply the Batalin-Vilkovisky machinery, as described in Section 3. The resulting graded symplectic vector space of BV fields is

$$\mathscr{E} = \Omega^0(M)\otimes\mathfrak{g}[1] \oplus \Omega^1(M)\otimes\mathfrak{g} \oplus \Omega^3(M)\otimes\mathfrak{g}[-1] \oplus \Omega^4(M)\otimes\mathfrak{g}[-2].$$

As always, $V[-i]$ denotes the vector space V situated in degree i.

The degree -1 symplectic pairing on $\mathscr{E}$ is the obvious pairing between $\Omega^0(M)\otimes\mathfrak{g}$ and $\Omega^4(M)\otimes\mathfrak{g}$ and between $\Omega^1(M)\otimes\mathfrak{g}$ and $\Omega^3(M)\otimes\mathfrak{g}$.

5.4. The general BV procedure produces an action

$$S \in \mathscr{O}_{loc}(\mathscr{E})$$

which satisfies the classical master equation

$$\{S, S\} = 0.$$

Let us describe S explicitly. As in Section 3, S is a sum of two terms

$$S = S_{Gauge} + S_{YM}$$

where

$$S_{YM} : \Omega^1(M)\otimes\mathfrak{g} \to \mathbb{R}$$

is the Yang-Mills action as above.

The other term S_{Gauge} of S is constructed from the gauge action, as follows. Let us denote a ghost by X, a field by A, an anti-field by $A^\vee$ and an anti-ghost by $X^\vee$. Then, the Lie bracket on the space of ghosts contributes the term

$$\tfrac{1}{2}\langle [X, X], X^\vee\rangle.$$

The action of ghosts on fields contributes the term

$$\langle dX + [X, A], A^\vee\rangle.$$

5.5. The quadratic part of S induces a differential
$$Q : \mathcal{E} \to \mathcal{E}.$$
Because S satisfies the classical master equation,
$$Q^2 = 0.$$
The operator Q is explicitly described by the diagram
$$\Omega^0(M) \otimes \mathfrak{g} \xrightarrow{d} \Omega^1(M) \otimes \mathfrak{g} \xrightarrow{d*d} \Omega^3(M) \otimes \mathfrak{g} \xrightarrow{d} \Omega^4(M) \otimes \mathfrak{g}.$$

5.6. In the BV formalism, we need to integrate over an isotropic subspace in the space of fields $\mathcal{E}$. As with Chern-Simons theory, we would like to take our isotropic subspace to be of the form $\operatorname{Im} Q^{GF}$ for some differential operator
$$Q^{GF} : \mathcal{E} \to \mathcal{E}$$
which is of degree -1, square zero, and self-adjoint with respect to the symmetric pairing. These conditions are necessary in order to have a propagator of the kind that can be treated by the renormalization techniques described in this book.

The only obvious candidate is to take the operator Q^{GF} described by the diagram
$$\Omega^0(M) \otimes \mathfrak{g} \xleftarrow{d^*} \Omega^1(M) \otimes \mathfrak{g} \xleftarrow{d^**d^*} \Omega^3(M) \otimes \mathfrak{g} \xleftarrow{d^*} \Omega^4(M) \otimes \mathfrak{g}.$$
Such an operator requires the choice of a metric on M in the given conformal class.

Unfortunately, this operator doesn't satisfy certain conditions required to apply our renormalization techniques to the BV formalism. The problem is that Q^{GF} is of second order, so that the operator
$$D = [Q, Q^{GF}]$$
is of fourth order. The propagator is constructed from the heat kernel for D, and the heat kernel of a fourth order operator is not so well behaved.

This problem will be resolved in Chapter 6, by using the first order formulation of Yang-Mills theory.

6. *D*-modules and the classical BV formalism

In this section we will explain how to interpret some aspects of the classical BV formalism, as described in Section 3, using the theory of *D*-modules.

As we will see later, when constructing quantum theories in the BV formalism, obstructions to gauge symmetry can lie in the cohomology of the complex $\mathcal{O}_{loc}(\mathcal{E})$, with the differential given by bracketing with the (classical) BV action $\{S, -\}$. The classical master equation says that this differential is of square zero.

Thus, it is important to be able to compute the cohomology of the complex $\mathcal{O}_{loc}(\mathcal{E})$ with differential $\{S, -\}$. In this section we will show how to interpret this complex in terms of the Gel'fand-Fuchs cohomology of a certain D_M L_∞ algebra on M.

6.1. Let M be a C^∞ manifold. Let D_M denote the algebra of differential operators on M. A D_M-module is simply a sheaf of left modules over this sheaf of algebras on M. Any D_M-module is, in particular, a module over the sheaf C_M^∞ of smooth functions on M. Conversely, if R is a sheaf of C_M^∞-modules, the extra data which we need to make R into a D_M-module is a flat connection. This is, by definition, a map

$$\nabla : R \to R \otimes_{C_M^\infty} \Omega_M^1$$

satisfying the usual Liebniz rule, whose curvature $F(\nabla) \in R \otimes_{C_M^\infty} \Omega_M^2$ vanishes.

Thus, we can think of D_M-modules as (possibly infinite dimensional) vector bundles on M with a flat connection.

If E, F are vector bundles on M with flat connections, then the tensor product bundle $E \otimes F$ has an induced flat connection. In a similar way, if R, S are D_M-modules, then

$$R \otimes_{C_M^\infty} S$$

is a new D_M-module.

In this way, the category of D_M-modules becomes a symmetric monoidal category. Thus, one can talk about algebraic objects in the category of D_M-modules: for instance, commutative algebras, Lie algebras, L_∞ algebras, etc.

6.2. Any finite rank $\mathbb{Z}$ graded vector bundle E on M gives rise to a D_M-module $J(E)$, the D_M-module of jets of sections of E. The fibre of $J(E)$ at $x \in M$ is the space of formal germs at x of sections of E. The D_M-module structure on $J(E)$ is characterized by specifying the infinitesimal parallel transport, which identifies the fibre $J(E)_x$ with the fibre at an infinitesimally near point y. This parallel transport operation simply takes a formal germ of a section of E near x, and thinks of it as a formal germ of a section near y; this makes sense, as x and y are infinitesimally close.

Let us choose local coordinates $x_1, \ldots, x_n$ on an open neighbourhood U of M, and let us choose an isomorphism between E and a trivial graded vector bundle with fibre E_0, for some finite dimensional graded vector space E_0. Then, we can identify

$$\Gamma(U, J(E)) = C^\infty(U) \otimes_\mathbb{R} \mathbb{R}[[x_1, \ldots, x_n]] \otimes_\mathbb{R} E_0.$$

Thus, as a C_M^∞-module, $J(E)$ is an inverse limit of finite dimensional modules. We will endow the sheaf $J(E)$ with the topology of the inverse limit; all constructions we perform with D_M-modules like $J(E)$ must respect this topology.

The flat connection on $J(E)$ is given by the map
$$\nabla : C^\infty(U) \otimes_{\mathbb{R}} \mathbb{R}[[x_1, \ldots, x_n]] \otimes_{\mathbb{R}} E_0 \to \Omega^1(U) \otimes_{\mathbb{R}} \mathbb{R}[[x_1, \ldots, x_n]] \otimes_{\mathbb{R}} E_0$$
$$\nabla(f \otimes g \otimes v) = \mathrm{d}f \otimes g \otimes v + \sum f \mathrm{d}x_i \otimes \frac{\partial}{\partial x_i} g \otimes v.$$
This map is continuous for the topology on $J(E)$.

LEMMA 6.2.1. *Let E, F be vector bundles on M. Let $\mathscr{E}, \mathscr{F}$ denote the spaces of global sections of E and F respectively. Then,*
$$\mathrm{Diff}(\mathscr{E}, \mathscr{F}) = \mathrm{Hom}_{D_M}(J(E), J(F))$$
where Hom_{D_M} refers to the space of continuous D_M-module maps.

More generally, if $E_1, \ldots, E_n, F$ are vector bundles on M, then
$$\mathrm{PolyDiff}(\mathscr{E}_1 \otimes \cdots \otimes \mathscr{E}_n, \mathscr{F}) = \mathrm{Hom}_{D_M}\left(J(E_1) \otimes_{C_M^\infty} \cdots \otimes_{C_M^\infty} J(E_n), J(F)\right).$$
Here $J(E_1) \otimes_{C_M^\infty} \cdots \otimes_{C_M^\infty} J(E_n)$ refers to the completed tensor product of the pro-finitely generated C_M^∞-modules $J(E_i)$.

These identifications are compatible with composition.

It follows immediately that if E is a local Lie algebra on M, in the sense of Section 3, then $J(E)$ is a Lie algebra in the symmetric monoidal category of D_M-modules.

6.3. Let
$$J(E)^\vee = \mathrm{Hom}_{C_M^\infty}(J(E), C_M^\infty)$$
where $\mathrm{Hom}_{C_M^\infty}$ denotes the sheaf of continuous linear maps of C_M^∞-modules (in other words, $J(E)^\vee$ is the ind vector bundle dual to the pro vector bundle $J(E)$). Note that the usual formula for the flat connection on the dual of a vector bundle gives the C_M^∞-module $J(E)^\vee$ the structure of D_M-module.

For example, if E is the trivial vector bundle with fibre the vector space E_0, then
$$J(E)^\vee = D_M \otimes_{\mathbb{R}} E_0^\vee$$
as a D_M-module.

Let
$$\mathscr{O}(J(E)) = \widehat{\mathrm{Sym}}^* J(E)^\vee = \prod_{n \geq 0} \mathrm{Sym}^n J(E)^\vee$$
denote the completed symmetric algebra on $J(E)^\vee$. The symmetric product is taken over C_M^∞. Note that $\mathscr{O}(J(E))$ is a D_M-algebra.

6.4. As in Section 3, let
$$\mathrm{Der}_{loc}(\mathscr{O}(\mathscr{E})) = \prod_{n \geq 0} \mathrm{PolyDiff}(\mathscr{E}^{\otimes n}, \mathscr{E})_{S_n}$$
be the Lie algebra of local symmetries of $\mathscr{E}$. Lemma 6.2.1 immediately implies the following.

LEMMA 6.4.1. *The Lie algebra $\mathrm{Der}_{loc}(\mathscr{O}(\mathscr{E}))$ is canonically isomorphic to the Lie algebra of derivations of the D_M-algebra $\mathscr{O}(J(E))$.*

Let us suppose we have some

$$X \in \text{Der}_{loc}(\mathcal{O}(\mathcal{E}))$$

which is of cohomological degree 1, and which satisfies

$$[X, X] = 0.$$

Any classical theory in the BV formalism gives rise to a super vector bundle E with such an X, as explained in Section 3.

Thus, we can think of X as an element

$$X \in \text{Der}(\mathcal{O}(J(E)),$$

satisfying the equation $[X, X] = 0$.

There is a standard dictionary relating L_∞-algebras and square zero vector fields. Applied in our context, it shows the following. Let us consider the Taylor components of X as maps

$$X_n : (J(E)[-1])^{\otimes n} \to J(E)[-1]$$

of degree $2 - n$. Then, these give $J(E)[-1]$ the structure of D_M L_∞-algebra.

Thus, we have shown the following.

LEMMA 6.4.2. *Let E be a $\mathbb{Z}$-graded vector bundle, whose sections are the fields for a classical theory in the BV formalism. Then, the D_M-module $J(E)[-1]$ has a canonical structure of a D_M L_∞-algebra.*

We can consider $\mathcal{O}(J(E))$ as a cochain complex, with the differential $[X, -]$. This lemma shows that we can view $\mathcal{O}(J(E))$ as the Chevalley-Eilenberg cochain complex, taken in the symmetric monoidal category of D_M-modules, of the D_M L_∞-algebra $J(E)[-1]$. To be precise, we are taking the topological version of Chevalley-Eilenberg cochains, where we take the continuous dual of the completed exterior algebras of the topological D_M L_∞-algebra $J(E)[-1]$. This topological version of Chevalley-Eilenberg cochains is known as Gel'fand-Fuchs cohomology.

6.5. The L_∞-algebra $J(E)[-1]$ is something very familiar in the case of Chern-Simons theory. Let M be a three-dimensional manifold, and let $\mathfrak{g}$ be a Lie algebra with an invariant pairing. The vector bundle of fields for Chern-Simons theory on M with coefficients in $\mathfrak{g}$ is

$$E = \Omega_M^* \otimes \mathfrak{g}[1].$$

Thus, $E[-1] = \Omega_M^* \otimes \mathfrak{g}$ has the structure of a local differential graded Lie algebra. The differential is the de Rham differential, and the Lie bracket arises from wedge product of forms and the Lie bracket on $\mathfrak{g}$. This Lie algebra structure carries over to one on $J(E)[-1]$.

6.6. Let E and X be as above. The main object we would like to understand is the complex $\mathscr{O}_{loc}(\mathscr{E})$ of local action functionals on $\mathscr{E}$, with the differential arising from the local derivation X of $\mathscr{O}(\mathscr{E})$.

Let us consider $\mathscr{O}_{loc}(\mathscr{E})$ and $\mathscr{O}(J(E))$ both as cochain complexes, with differential arising from X. Recall that Dens_M, the line bundle of densities on M, is a right D_M-module in a natural way.

LEMMA 6.6.1. *There is an isomorphism of cochain complexes*
$$\mathscr{O}_{loc}(\mathscr{E}) = \mathrm{Dens}_M \otimes_{D_M} \mathscr{O}(J(E)).$$

PROOF. The vector bundle $\mathrm{Dens}_M \otimes_{C_M^\infty} \mathscr{O}(J(E))$ is, essentially by definition, the space of Lagrangians on the vector bundle E. Taking the tensor product over D_M amounts to quotienting by the space of Lagrangians which are a total derivative. A local action functional on $\mathscr{E}$ is defined to be something arising from the integral of a Lagrangian; thus, local action functionals are Lagrangians up to the addition of a total derivative.

It is straightforward to check that this isomorphism is compatible with differentials. □

I should emphasize that the tensor product appearing in the statement of this lemma is not derived. We have a slightly weaker result using the derived tensor product. Let $\mathbb{R} \subset \mathscr{O}_{loc}(\mathscr{E})$ denote the space of constant local action functionals; similarly, let $C_M^\infty \subset \mathscr{O}(J(E))$ denote the submodule spanned by the identity in $\mathscr{O}(J(E))$. Note that $\mathscr{O}(J(E))/C_M^\infty$ is a D_M-module, in fact, a non-unital D_M-algebra.

LEMMA 6.6.2. *There is a canonical quasi-isomorphism of cochain complexes*
$$\mathscr{O}_{loc}(\mathscr{E})/\mathbb{R} \simeq \mathrm{Dens}_M \otimes_{D_M}^{\mathbb{L}} (\mathscr{O}(J(E))/C_M^\infty)$$

PROOF. In light of the previous lemma, it suffices to check that
$$(\mathscr{O}(J(E))/C_M^\infty) \otimes_{D_M}^{\mathbb{L}} \mathrm{Dens}_M = (\mathscr{O}(J(E))/C_M^\infty) \otimes_{D_M} \mathrm{Dens}_M.$$
Note that
$$\mathscr{O}(J(E))/C_M^\infty = \prod_{n>0} \mathrm{Sym}^n J(E)^\vee.$$
Each D_M-module $\mathrm{Sym}^n J(E)^\vee$, for $n > 0$, is flat as a D_M-module. Indeed, $\mathrm{Sym}^n J(E)^\vee$ is a direct summand in $(J(E)^\vee)^{\otimes n}$. If we choose a local isomorphism between the vector bundle E on M and a trivial vector bundle with fibre E_0, we can identify
$$J(E)^\vee \cong E_0^\vee \otimes D_M$$
as D_M-modules. Thus, $J(E)^\vee$ is free, and therefore flat. More generally,
$$(J(E)^\vee)^{\otimes n} \cong (E_0^\vee)^{\otimes n} \otimes D_M^{\otimes n}$$
locally. Here, the tensor power of D_M is taken over C_M^∞. It is easy to see that $D_M^{\otimes n}$ is flat, and so $(J(E)^\vee)^{\otimes n}$ is. □

6.7. Let us suppose that $M = \mathbb{R}^n$. Further, let us assume that the vector bundle E on $\mathbb{R}^n$ is equipped with a flat connection; let E_0 be the fibre at $0 \in \mathbb{R}^n$. Suppose that the vector field X on $\mathcal{O}(\mathcal{E})$ is translation invariant, that is, preserved by the flat connection on E.

Translation gives an $\mathbb{R}^n$ action on $\mathcal{O}_{loc}(\mathcal{E})$, and the differential on $\mathcal{O}_{loc}(\mathcal{E})$ is preserved by this action.

We would like to be able to compute the cohomology of the $\mathbb{R}^n$-invariants of $\mathcal{O}_{loc}(\mathcal{E})$. The following lemma allows us to do this.

Recall that $\mathcal{O}(J(E))$, defined as $\prod_{n\geq 0} \operatorname{Sym}^n(J(E)^\vee)$, is a $D_{\mathbb{R}^n}$-algebra. Let $\mathcal{O}(J(E))_0$ denote its fibre at $0 \in \mathbb{R}^n$. The fact that $\mathcal{O}(J(E))$ is a $D_{\mathbb{R}^n}$-algebra implies that $\mathcal{O}(J(E))_0$ is acted on by the algebra $\mathbb{R}[\partial_1, \ldots, \partial_n]$. The generators ∂_i of this algebra act by derivations on $\mathcal{O}(J(E))_0$; that is, $\mathcal{O}(J(E))_0$ has an action of the Lie algebra $\mathbb{R}^n$ by derivations.

LEMMA 6.7.1. *There is a canonical quasi-isomorphism of complexes*
$$(\mathcal{O}_{loc}(\mathcal{E})/\mathbb{R})^{\mathbb{R}^n} \cong (\mathcal{O}(J(E))_0/\mathbb{R}) \otimes^{\mathbb{L}}_{\mathbb{R}[\partial_1,\ldots,\partial_n]} |\det|\,\mathbb{R}^n.$$

(Here $|\det|\,\mathbb{R}^n$ is the one dimensional $\mathbb{R}[\partial_i]$-module on which each ∂_i acts trivially, and on which GL_n acts by the absolute value of the determinant). This quasi-isomorphism is invariant under GL_n.

Note that this theorem can be restated as follows. Recall that $J(E)[-1]$ is a $D_{\mathbb{R}^n}$ L_∞-algebra. Thus, $J(E)_0[-1]$ is a topological L_∞-algebra. The differential graded algebra $\mathcal{O}(J(E))_0$ can be interpreted as the Gel'fand-Fuchs Lie algebra cochains $C^*(J(E)_0[-1])$. Thus, the lemma asserts that there is a quasi-isomorphism
$$(\mathcal{O}_{loc}(\mathcal{E})/\mathbb{R})^{\mathbb{R}^n} \simeq C^*_{red}(J(E)_0[-1]) \otimes_{\mathbb{R}[\partial_1,\ldots,\partial_4]} \mathbb{R}.$$

Here $C^*_{red}(J(E)_0[-1])$ refers to the reduced Gel'fand-Fuchs Lie algebra cohomology. We will use this form of the lemma when we consider Yang-Mills theory later.

PROOF. This is an almost immediate corollary of Lemma 6.6.1. Indeed, there is an isomorphism of complexes
$$\mathcal{O}_{loc}(\mathcal{E}) \cong \mathcal{O}(J(E)) \otimes_{D_{\mathbb{R}^n}} \operatorname{Dens}^n_{\mathbb{R}}.$$

We need to identify the translation invariant part of the right hand side.

Let us view $\mathcal{O}(J(E))$ as an infinite dimensional vector bundle on $\mathbb{R}^n$. The $\mathbb{R}^n$ action on E, and on $\mathbb{R}^n$, gives rise to an $\mathbb{R}^n$ action on $\mathcal{O}(J(E))$. Note that the flat connection on $\mathcal{O}(J(E))$ coming from this action does not coincide with the flat connection coming from the $D_{\mathbb{R}^n}$-module structure.

The $\mathbb{R}^n$ action on $\mathcal{O}(J(E))$ allows us to turn an element of $\mathcal{O}(J(E))_0$ into a translation invariant section of $\mathcal{O}(J(E))$. Thus,
$$\mathcal{O}(J(E))^{\mathbb{R}^n} = \mathcal{O}(J(E))_0.$$

It is easy to see that

$$(\mathcal{O}(J(E)) \otimes_{D(\mathbb{R}^n)} \mathrm{Dens}(\mathbb{R}^n))^{\mathbb{R}^n} = \mathcal{O}(J(E))^{\mathbb{R}^n} \otimes_{D(\mathbb{R}^n)^{\mathbb{R}^n}} (\mathrm{Dens}_{\mathbb{R}}^n)^{\mathbb{R}^n}$$
$$= \mathcal{O}(J(E))_0 \otimes_{\mathbb{R}[\partial_1,\ldots,\partial_n]} |\det|\mathbb{R}^n.$$

To complete the proof of the lemma, it suffices to observe that $\mathcal{O}(J(E))_0/\mathbb{R}$ is flat as a module for $\mathbb{R}[\partial_1,\ldots,\partial_n]$. □

7. BV theories on a compact manifold

In this section we will give a general definition of the notion of a quantum theory in the BV formalism. We will explain the set-up on a compact manifold first, as in that situation things are somewhat technically easier. Later, we will explain what needs to be changed when we work with translation invariant theories on $\mathbb{R}^n$, or when we work on a non-compact manifold.

DEFINITION 7.0.1. *A free theory on in the BV formalism is described by the following data:*

(1) A $\mathbb{Z}$-graded vector bundle E on M, over $\mathbb{R}$ or $\mathbb{C}$, whose space of global sections will be denoted $\mathscr{E}$. For simplicity, we will describe what happens over $\mathbb{C}$.

(2) E is equipped with anti-symmetric map of vector bundles on M, of cohomological degree -1,

$$\langle -, - \rangle_{loc} : E \otimes E \to \mathrm{Dens}(M).$$

This must be non-degenerate on each fibre.

This pairing induces an integration pairing on the space of global sections $\mathscr{E}$,

$$\langle -, - \rangle : \mathscr{E} \otimes \mathscr{E} \to \mathbb{C}$$
$$\langle e_1, e_2 \rangle = \int_M \langle e_1, e_2 \rangle_{loc}.$$

(3) A differential operator $Q : \mathscr{E} \to \mathscr{E}$ of cohomological degree one which is of square zero, and is skew self-adjoint for the pairing. Further, Q must make $\mathscr{E}$ into an elliptic complex.

Recall that the statement that $(\mathscr{E}, Q)$ is an elliptic complex means that the complex of vector bundles $(\sigma(Q), \pi^*E)$ on $T^*M \setminus M$, is exact.

Throughout, we will abuse notation and refer to the totality of data of a free BV theory by $\mathscr{E}$.

7.1. Recall that, as explained in Section 2, the differential graded manifold of fields that appear in the BV theory is the derived moduli space of solutions to the equations of motion. The equations of motion of a free classical field theory are linear equations. In addition, in a free gauge theory, the gauge group is Abelian and acts in an affine-linear way on the space of fields.

Thus, if we start with a free classical theory (which may be a gauge theory) then the derived moduli space of solutions to the equations of motion is linear, and thus described by a cochain complex. This is the cochain complex $(\mathscr{E}, Q)$.

We will discuss several examples of free BV theories, starting with a free massless scalar field theory on a manifold M. The equations of motion in this theory are that $\mathrm{D}\phi = 0$, where $\mathrm{D} : C^\infty(M) \to C^\infty(M)$ is the non-negative Laplacian on M.

Thus, the derived moduli space of solutions to the equations of motion is the two-term cochain complex

$$\mathscr{E} = C^\infty(M) \xrightarrow{\mathrm{D}} C^\infty(M)$$

concentrated in degrees 0 and 1. The degree -1 symplectic pairing on $\mathscr{E}$ is defined by

$$\langle \phi, \psi \rangle = \int_M \phi\psi \, \mathrm{dVol}_M$$

where ϕ is in the copy of $C^\infty(M)$ in degree 0, and ψ is in the copy in degree 1.

7.2. Next let us consider free gauge theory in this setting. Let $\mathfrak{g}$ be an Abelian Lie algebra with a non-degenerate invariant pairing. Let us consider the free Chern-Simons theory with gauge Lie algebra $\mathfrak{g}$. The fields are $\Omega^1(M) \otimes \mathfrak{g}$. The Lie algebra of gauge symmetries is the Abelian Lie algebra $\Omega^0(M) \otimes \mathfrak{g}$. The action of $\Omega^0(M) \otimes \mathfrak{g}$ on $\Omega^1(M) \otimes \mathfrak{g}$ is affine-linear, defined by

$$X(A) = A + \mathrm{d}X$$

for $A \in \Omega^1(M) \otimes \mathfrak{g}$, $X \in \Omega^0(M) \otimes \mathfrak{g}$.

Thus, the derived quotient of $\Omega^1(M) \otimes \mathfrak{g}$ by $\Omega^0(M) \otimes \mathfrak{g}$ is described by the two-term complex

$$\Omega^0(M) \otimes \mathfrak{g} \xrightarrow{\mathrm{d}} \Omega^0(M) \otimes \mathfrak{g},$$

concentrated in degrees 0 and -1.

The action on the space $\Omega^1(M) \otimes \mathfrak{g}$ of fields is $\int_M \langle A, \mathrm{d}A \rangle_{\mathfrak{g}}$, where $\langle -, - \rangle_{\mathfrak{g}}$ is the invariant pairing on $\mathfrak{g}$. Thus, the derived moduli space of solutions to the equations of motion is defined by the 4-term complex

$$\mathscr{E} = \Omega^0(M) \otimes \mathfrak{g} \to \Omega^1(M) \otimes \mathfrak{g} \to \Omega^2(M) \otimes \mathfrak{g} \to \Omega^3(M) \otimes \mathfrak{g}$$

concentrated in degrees $-1, 0, 1, 2$. The differential is the de Rham differential; thus, we can write

$$\mathscr{E} = \Omega^*(M) \otimes \mathfrak{g}[1].$$

The symplectic pairing on $\mathscr{E}$ is given by

$$\langle \omega_1 \otimes X_1, \omega_2 \otimes X_2 \rangle = \int_M \omega^1 \wedge \omega^2 \langle X_1, X_2 \rangle_{\mathfrak{g}},$$

where $\omega^i \in \Omega^i(M)$ and $X_i \in \mathfrak{g}[1]$.

7.3. Let us now consider the free Yang-Mills theory, where the gauge Lie algebra is again an Abelian Lie algebra $\mathfrak{g}$ with invariant pairing.

The space of fields is $\Omega^1(M) \otimes \mathfrak{g}$, and the gauge Lie algebra is $\Omega^0(M) \otimes \mathfrak{g}$. The action of $\Omega^0(M) \otimes \mathfrak{g}$ on $\Omega^1(M) \otimes \mathfrak{g}$ is, as with Chern-Simons theory, defined by
$$X(A) = A + \mathrm{d}X$$
for $A \in \Omega^1(M) \otimes \mathfrak{g}$ and $X \in \Omega^0(M) \otimes \mathfrak{g}$.

The action $S(A)$ is
$$S(A) = \int_M \langle \mathrm{d}A, *\mathrm{d}A \rangle_\mathfrak{g}$$
where d is the de Rham differential, and we are using the invariant pairing on $\mathfrak{g}$.

Thus, the derived moduli space of solutions to the equations of motion is the 4-term complex
$$\Omega^0(M) \otimes \mathfrak{g} \xrightarrow{\mathrm{d}} \Omega^1(M) \otimes \mathfrak{g} \xrightarrow{\mathrm{d}*\mathrm{d}} \Omega^3(M) \otimes \mathfrak{g} \xrightarrow{\mathrm{d}} \Omega^4(M) \otimes \mathfrak{g}.$$
concentrated in degrees $-1, 0, 1, 2$.

7.4. In the BV formalism, one always integrates over an isotropic subspace of the space $\mathscr{E}$ of fields. The subspaces we integrate over are always of the form $\operatorname{Im} Q^{GF}$ for a differential operator Q^{GF} on $\mathscr{E}$.

DEFINITION 7.4.1. *A gauge fixing operator on a free BV theory $\mathscr{E}$ is given by an operator*
$$Q^{GF} : \mathscr{E} \to \mathscr{E}$$
such that:

(1) *Q^{GF} is of cohomological degree -1, of square zero, and self-adjoint for the pairing $\langle -, - \rangle$.*
(2) *The commutator*
$$\mathrm{D} = [Q, Q^{GF}]$$
is required to be a generalized Laplacian, in the sense of (BGV92).

The scalar field theory has a very simple gauge fixing operator. If
$$\mathscr{E} = C^\infty(M) \oplus C^\infty(M)[-1]$$
is the graded vector space corresponding to the free scalar field theory on M, then
$$Q^{GF} : \mathscr{E}^1 = C^\infty(M) \to \mathscr{E}^0 = C^\infty(M)$$
is the identity. Since $Q : \mathscr{E}^0 \to \mathscr{E}^1$ is the Laplacian, we see that $[Q^{GF}, Q] : \mathscr{E}^i \to \mathscr{E}^i$ is again the Laplacian, and thus satisfies the required ellipticity properties.

In this case, the isotropic subspace $\operatorname{Im} Q^{GF} \subset \mathscr{E}$ which we are supposed to integrate over is simply $C^\infty(M)$. Thus, with this gauge fixing condition, functional integrals in the scalar field theory in the BV formalism coincide with the functional integrals considered in Chapter 2.

Chern-Simons theory on a 3-manifold M also has a simple gauge fixing operator satisfying these axioms, namely
$$Q^{GF} = d^* : \Omega^*(M) \otimes \mathfrak{g}[1] \to \Omega^*(M) \otimes \mathfrak{g}[1].$$
Again, since $[Q^{GF}, Q] = [d^*, d]$ is the Laplacian coupled to $\Omega^i(M) \otimes \mathfrak{g}$, it is clear that the required ellipticity conditions are satisfied.

Unfortunately, it seems that, in this formulation, Yang-Mills theory on M has no such gauge fixing operator. In Chapter 6, we will use a slightly different – but equivalent – formulation of classical Yang-Mills theory, called the first-order formulation. In the first-order formulation of Yang-Mills theory there is a natural gauge fixing operator.

8. Effective actions

8.1. A free theory in the BV formalism, together with a choice of gauge fixing condition, gives a free theory in the sense of Chapter 2, Section 13, as we will now see.

Recall that $\mathscr{E}^!$ denotes the space of sections of the bundle $E^\vee \otimes \mathrm{Dens}(M)$ on M. The pairing on E gives rise to an isomorphism
$$\mathscr{E}^! \to \mathscr{E}.$$
Thus, we can compose this isomorphism with the differential operator $Q^{GF} : \mathscr{E} \to \mathscr{E}$, to give an operator
$$D' : \mathscr{E}^! \to \mathscr{E}.$$
This differential operator, together with $D : \mathscr{E} \to \mathscr{E}$, define a free theory in the sense of Chapter 2, Section 13.

8.2. Given any $K \in \mathscr{E} \otimes \mathscr{E}$, let us define a convolution operator
$$K\star : \mathscr{E} \to \mathscr{E}$$
as follows.

If $e \in \mathscr{E}$, then $K \otimes e \in \mathscr{E} \otimes \mathscr{E} \otimes \mathscr{E}$. The pairing $\langle -, - \rangle$ on $\mathscr{E}$ gives a map
$$1 \otimes \langle -, - \rangle : \mathscr{E} \otimes \mathscr{E} \otimes \mathscr{E} \to \mathscr{E}$$
of cohomological degree -1, which is compatible with differentials.

We define
$$(K \star e) = (-1)^{|K|}(1 \otimes \langle -, - \rangle)(K \otimes e).$$
The reason for this choice of sign is the following.

LEMMA 8.2.1. *If $K \in \mathscr{E} \otimes \mathscr{E}$, then*
$$(QK) \star e = [Q, K\star]e.$$
In other words, the map
$$\star : \mathscr{E} \otimes \mathscr{E} \to \mathrm{End}(\mathscr{E})$$
is a cochain map.

(Here Q is the total differential on the complex $\mathscr{E} \otimes \mathscr{E}$. This differential is occasionally referred to as $Q \otimes 1 + 1 \otimes Q$).

8.3. There is a heat kernel
$$K_l \in \mathscr{E} \otimes \mathscr{E}$$
for the operator e^{-lD}, characterized by the equation
$$K_l \star e = e^{-lD} e.$$
This heat kernel is symmetric.

The propagator of our theory is given by
$$P(\varepsilon, L) = \int_{l=\varepsilon}^{L} (Q^{GF} \otimes 1) K_l \, dl.$$
This has the property that
$$P(\varepsilon, L) \star e = Q^{GF} \int_{\varepsilon}^{L} e^{-lD} e,$$
for any $e \in \mathscr{E}$.

8.4. As usual, let
$$\mathscr{O}(\mathscr{E}) = \widehat{\mathrm{Sym}}^* \mathscr{E}^\vee = \prod_{n \geq 0} \mathrm{Hom}(\mathscr{E}^{\otimes n}, \mathbb{C})_{S_n}$$
be the algebra of formal power series on $\mathscr{E}$.

We will let
$$\mathscr{O}_{loc}(\mathscr{E}) \subset \mathscr{O}(\mathscr{E})$$
denote the subspace of local action functionals.

DEFINITION 8.4.1. *A pre-theory is a collection of effective interactions $\{I[L]\}$ satisfying renormalization group flow and the locality axiom as defined in Chapter 2, Section 13.*

More precisely,

(1) *Each $I[L] \in \mathscr{O}^{+, ev}(\mathscr{E})[[\hbar]]$ is of degree 0 and at least cubic modulo $\hbar$.*
(2) *The renormalization group equation*
$$I[L] = W\left(P(\varepsilon, L), I[\varepsilon]\right)$$
is satisfied.
(3) *Each $I_{i,k}[L]$ has a small L asymptotic expansion in terms of local action functionals.*

We let $\widetilde{\mathscr{T}}^{(\infty)}(\mathscr{E})$ denote the set of pre-theories, and $\widetilde{\mathscr{T}}^{(n)}(\mathscr{E})$ denote the set of theories defined modulo $\hbar^{n+1}$.

We will reserve the word theory for collections of effective interactions $\{I[L]\}$ which satisfy the quantum master equation, as defined below.

The results of Chapter 2, Section 13 show that the space $\widetilde{\mathscr{T}}^{(n)}$ is a torsor over $\widetilde{\mathscr{T}}^{(n-1)}$ for the abelian group $\mathscr{O}_{loc}(\mathscr{E})$ of even local action functionals. The choice of renormalization scheme leads to a section of each torsor, and so to a bijection

$$\widetilde{\mathscr{T}}^{(\infty)} \cong \mathscr{O}_{loc}^{+}(\mathscr{E})[[\hbar]]$$

between the space of pre-theories and the space of local action functionals which are even at least cubic modulo $\hbar$.

9. The quantum master equation

9.1. As $\mathscr{E}$ has a degree -1 symplectic form, one can try to define the BV Laplacian on the space of functionals on $\mathscr{E}$. Unfortunately, this is ill-defined.

Let

$$\Delta_l = -\partial_{K_l} : \mathscr{O}(\mathscr{E}) \to \mathscr{O}(\mathscr{E})$$

denote the order two differential operator associated to the heat kernel K_l for $l > 0$.

Then, the BV Laplacian (if it were defined) would be the limit

$$\Delta_0 I = -\lim_{l \to 0} \partial_{K_l} I.$$

If $I \in \mathscr{O}_l(\mathscr{E})$ is local, then this expression has the same kind of singularities one finds in one-loop Feynman graphs. Thus, the naive quantum master equation

$$(Q + \hbar \Delta_0) e^{I/\hbar} = 0$$

is ill-defined.

9.2. The philosophy of this book is always that the fundamental object of a theory is not the local action functional $I \in \mathscr{O}_{loc}^{+}(\mathscr{E})[[\hbar]]$ but rather the effective interactions $\{I[L]\}$. The quantum master equation is an expression of gauge symmetry. Our philosophy suggests that the correct expression of gauge symmetry should be an equation imposed on the effective interactions $\{I[L]\}$.

If we make the unnatural choice of a renormalization scheme, then we find a bijection between local action functionals I and collections $\{I[L]\}$ of effective interactions. The quantum master equation we will impose on the effective interactions yields a renormalized version of the quantum master equation on the set of local action functionals.

DEFINITION 9.2.1. *A functional*

$$I \in \mathscr{O}^{+}(\mathscr{E})[[\hbar]]$$

satisfies the scale L quantum master equation if

$$(Q + \hbar \Delta_L) e^{I/\hbar} = 0.$$

176   5. GAUGE SYMMETRY AND THE BATALIN-VILKOVISKY FORMALISM

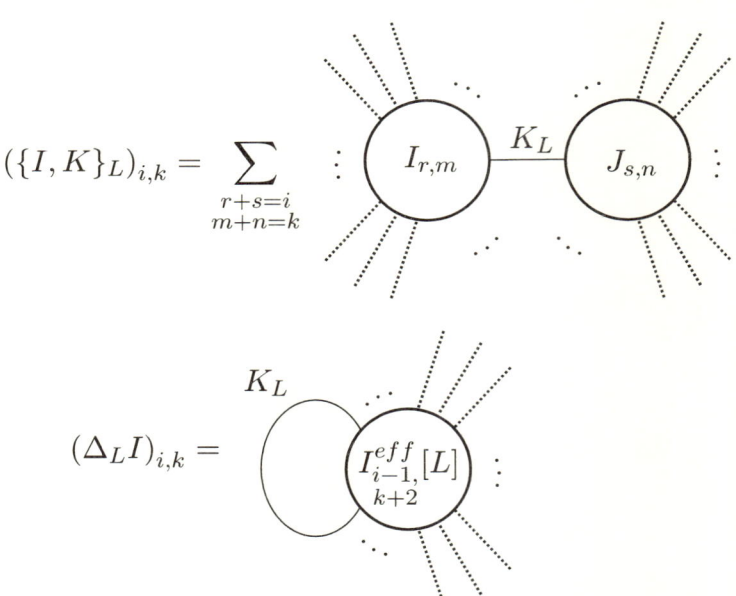

FIGURE 1. This diagram illustrates the bracket $\{I, J\}_L$ and the BV Laplacian Δ_L at scale L. As usual, the subscripts (i, k) indicate the part of the coefficient of $\hbar^i$ homogeneous of degree k in the field variable.

An equivalent expression of the scale L quantum master equation is as follows. There is a scale L BV bracket
$$\{-,-\}_L : \mathcal{O}(\mathcal{E}) \otimes \mathcal{O}(\mathcal{E}) \to \mathcal{O}(\mathcal{E})$$
defined by
$$\{I, J\}_L = \Delta_L(IJ) - \Delta_L(I)J - (-1)^{|I|}I\Delta_L(J).$$
The scale L quantum master equation can be rewritten as
$$QI + \tfrac{1}{2}\{I, I\}_L + \hbar \Delta_L I = 0.$$
A diagrammatic expression for $\{I, J\}_L$ and Δ_L is given in figure 1.

LEMMA 9.2.2. *A functional* $I \in \mathcal{O}^+(\mathcal{E})[[\hbar]]$ *satisfies the scale ε quantum master equation if and only if*
$$W(P(\varepsilon, L), I)$$
satisfies the scale L quantum master equation.

Thus, the renormalization group flow takes solutions of the scale ε QME to solutions of the scale L QME.

PROOF. Note that the propagator $P(\varepsilon, L)$ is the kernel for the operator
$$\int_\varepsilon^L Q^{GF} e^{-lD} dl.$$

This operator has a homological interpretation:

$$\left[Q, \int_\varepsilon^L Q^{GF} e^{-lD} dl\right] = \int_\varepsilon^L [Q, Q^{GF}] e^{-lD}$$
$$= \int_\varepsilon^L D e^{-lD}$$
$$= e^{-\varepsilon D} - e^{-LD}.$$

Thus, the operator $\int_\varepsilon^L Q^{GF} e^{-lD} dl$ is a chain homotopy between $e^{-\varepsilon D}$ and e^{-LD}.

It follows that

$$(Q \otimes 1 + 1 \otimes Q) P(\varepsilon, L) = K_\varepsilon - K_L$$

and that

(†) $$[\partial_{P(\varepsilon,L)}, Q] = \partial_{K_\varepsilon} - \partial_{K_L} = \Delta_L - \Delta_\varepsilon$$

Thus, the operator $\partial_{P(\varepsilon,L)}$ is a chain homotopy between the BV operator at scale ε and the BV operator at scale L.

Note that we can write

$$W(P(\varepsilon, L), I) = \hbar \log \left(e^{\hbar \partial_{P(\varepsilon,L)}} e^{I/\hbar} \right).$$

Equation (†) implies that

$$(Q + \hbar \Delta_L) \exp(W(P(\varepsilon, L), I)/\hbar) = (Q + \hbar \Delta_L) \left(e^{\hbar \partial_{P(\varepsilon,L)}} e^{I/\hbar} \right)$$
$$= e^{\hbar \partial_{P(\varepsilon,L)}} (Q + \hbar \Delta_\varepsilon) e^{I/\hbar}.$$

This shows that, if I satisfies the scale ε quantum master equation, then $W(P(\varepsilon, L), I)$ satisfies the scale L quantum master equation. The converse to this statement follows from the fact that the operator $W(P(\varepsilon, L), -)$ is invertible, with inverse $W(P(L, \varepsilon), -)$. □

9.3. Now we can define the notion of theory in the BV formalism.

DEFINITION 9.3.1. *A pre-theory, given by a collection of effective interactions $\{I[L]\}$, satisfies the quantum master equation if each $I[L]$ satisfies the scale L quantum master equation.*

A pre-theory which satisfies the QME will be called a theory. The set of theories will be denoted by $\mathscr{T}^{(\infty)}$, and the set of theories defined modulo $\hbar^{n+1}$ will be denoted by $\mathscr{T}^{(n)}$.

If we choose a renormalization scheme, then we can identify the set of pre-theories $\{I[L]\}$ with the set of local action functionals $I \in \mathscr{O}_{loc}^+(\mathscr{E})[[\hbar]]$. Under this correspondence, the statement that $\{I[L]\}$ satisfies the quantum master equation imposes an equation on I, which we call the *renormalized quantum master equation*.

However, we will always regard the choice of renormalization scheme as unnatural. The effective interactions are always considered to be the fundamental objects.

9.4. The classical master equation. As we have seen in Section 3, the scale zero bracket
$$\{I, J\} = \lim_{L \to 0} \{I, J\}_L$$
is well defined. The bracket only involves tree level Feynman diagrams, and so does not have any singularities.

LEMMA 9.4.1. *Let $I^{(0)} \in \mathcal{O}_{loc}(\mathcal{E})$ be a local action functional which is at least cubic. Let*
$$I^{(0)}[L] = W(P(0, L), I) \text{ modulo } \hbar \in \mathcal{O}(\mathcal{E})$$
be the corresponding tree-level effective interaction. Then $I^{(0)}[L]$ satisfies the classical master equation
$$QI^{(0)}[L] + \frac{1}{2}\{I^{(0)}[L], I^{(0)}[L]\}_L = 0$$
if and only if the local action functional $I^{(0)}$ satisfies the classical master equation
$$QI^{(0)} + \frac{1}{2}\{I^{(0)}, I^{(0)}\} = 0.$$

10. Homotopies between theories

10.1. We have already talked about homotopies between solutions of the quantum master equation in the BV formalism in finite dimensions (see Section 2.7). Let us briefly recall the picture. Let V be a finite dimensional graded vector space equipped with a symplectic pairing of degree -1, and with a differential Q preserving the symplectic pairing. A functional $I \in \mathcal{O}(V)$ satisfies the quantum master equation if
$$(Q + \hbar\Delta)e^{I/\hbar} = 0.$$
A homotopy between two solutions I_0, I_1 of the quantum master equation is an element
$$I(t, \mathrm{d}t) \in \mathcal{O}(V) \otimes \Omega^*([0, 1])$$
of cohomological degree zero, such that
$$(Q + \mathrm{d}_{DR} + \hbar\Delta)e^{I(t,\mathrm{d}t)/\hbar} = 0$$
and $I(0, 0) = I_0$, $I(1, 0) = I_1$.

If we write
$$I(t, \mathrm{d}t) = I(t) + J(t)\mathrm{d}t$$
then the quantum master equation for $I(t, \mathrm{d}t)$ says that, for all values of t, $I(t) \in \mathcal{O}(V)$ satisfies the quantum master equation, and

(†) $$\frac{\mathrm{d}}{\mathrm{d}t}I(t) + QJ(t) + \{I(t), J(t)\} + \hbar\Delta J(t) = 0.$$

Let
$$X(t) = -\{J(t), -\}$$
be the one-parameter family of Hamiltonian vector fields associated to the functionals $-J(t)$. There is a corresponding one-parameter family of symplectic diffeomorphisms
$$\phi(t) : V \to V$$
such that
$$\phi(t+\varepsilon) = (1 + \varepsilon X(t))\phi(t) \text{ modulo } \varepsilon^2.$$
Equation (†) says that
$$e^{\langle v, Qv\rangle/\hbar + I(t)/\hbar} dV = \phi(t)^* \left(e^{\langle v, Qv\rangle/\hbar + I_0/\hbar} dV \right)$$
where dV is the Lebesgue measure on V.

Thus, we see that

> Two solutions I_0, I_1 of the quantum master equation are homotopic if and only if there is a symplectic diffeomorphism $\phi = \phi(1)$, in the connected component of the identity, such that
> $$e^{\langle v, Qv\rangle/\hbar + I_1/\hbar} dV = \phi^* \left(e^{\langle v, Qv\rangle/\hbar + I_0/\hbar} dV \right).$$

Thus, the notion of homotopy for solutions of the quantum master equation is enough to tell us about changes of coordinates on the space of fields V.

10.2. We can also think in terms of homotopies to describe the dependence of the BV integral on the choice of a gauge fixing condition.

Recall that, in the finite dimensional setting, a gauge fixing condition is given by choosing an isotropic subspace $L \subset V$ with the property that the pairing $\langle v, Qv \rangle$ is non-degenerate on L. Let us choose such a subspace. Then, $L \oplus QL$ is a symplectic subspace of V, whose orthogonal complement we will denote by H. H again has a symplectic form of degree -1.

Let us further assume that $Q = 0$ on H. This implies that H is the cohomology $H^*(E, Q)$ of E.

Then, given any interaction functional I satisfying the quantum master equation, we find a solution of the quantum master on H by the formula

$$(\star) \qquad I_H(h) = \hbar \log \int_{l \in L} \exp\left(\frac{1}{\hbar} \langle l, Ql \rangle + \frac{1}{\hbar} I(l+h) \right).$$

If we vary the isotropic subspace $L \subset E$, then the solution I_H to the quantum master equation on H changes by a homotopy. Similarly, if we vary the solution I to the quantum master equation on V by a homotopy, the corresponding solution I_H on H varies by a homotopy.

One way to phrase this is to say that the set of gauge fixing conditions has an enrichment to a simplicial set, where the n-simplices are just smooth families of isotropic subspaces $L \subset E$ parameterized by the n-simplex, satisfying the transversality condition given above. Let us denote this simplicial set by $\mathscr{GF}(E, Q)$. The set of solutions to the quantum master equation also has an enrichment to a simplicial set, whose 1-simplices are homotopies, and 2-simplices are homotopies between homotopies, etc. Let us denote this set by $\mathcal{QME}(E, Q)$.

Then, the functional integral $(\star)$ defines a map of simplicial sets

$$\mathcal{GF}(E, Q) \times \mathcal{QME}(E, Q) \to \mathcal{QME}(H, 0).$$

10.3. We would like to extend this picture to the infinite dimensional space of fields $\mathcal{E}$. In finite dimensions, there are two ways to say that solutions to the quantum master equation are equivalent: either they are homotopic, or they are related by some non-linear change of coordinates. These two definitions are equivalent.

In infinite dimensions, it is difficult to make sense of non-linear changes of coordinates, as the change in the non-existent Lebesgue measure in $\mathcal{E}$ should be taken into account. Thus, in infinite dimensions, only the homotopy picture can be used. In finite dimensions, two solutions of the QME are homotopic if and only if they are related by a symplectic change of coordinates. Thus, two homotopic theories on $\mathcal{E}$ should be viewed as being related by a renormalized version of a local symplectic change of coordinates on $\mathcal{E}$.

In finite dimensions, we constructed a simplicial set of gauge fixing conditions, and a simplicial set of solutions to the quantum master equation. The functional integral was then a map to the simplicial set of solutions to the quantum master equation on cohomology.

This picture translates directly to infinite dimensions, with some extra subtleties. In infinite dimensions, the notion of theory – and so the notion of solution to the quantum master equation – depends on the choice of a gauge fixing condition.

We will introduce a simplicial set $\mathcal{GF}$ of gauge fixing conditions which, in most examples, is contractible. The simplicial set of solutions to the quantum master equation, which we denote by $\mathcal{QME}$, is a simplicial set fibred over $\mathcal{GF}$.

This shows that the concept of theory is independent, up to homotopy, of the choice of gauge fixing conditions.

Like in finite dimensions, the functional integral is a map of simplicial sets from $\mathcal{QME}$ to a simplicial set of solutions to the QME on cohomology.

10.4. Let us fix a free BV theory $(\mathcal{E}, Q, \langle -, - \rangle)$ on M. We will be interested in families of theories over the algebra of forms on the n-simplex. Our gauge fixing conditions will also come in families.

DEFINITION 10.4.1. *A family of gauge fixing conditions for $\mathcal{E}$, over $\Omega^*(\Delta^n)$, is an $\Omega^*(\Delta^n)$ linear differential operator*

$$Q^{GF} : \mathcal{E} \otimes \Omega^*(\Delta^n) \to \mathcal{E} \otimes \Omega^*(\Delta^n)$$

of cohomological degree -1, with the following properties.

(1) Q^{GF} is self-adjoint for the $\Omega^*(\Delta^n)$ linear pairing $\mathscr{E}\otimes\mathscr{E}\otimes\Omega^*(\Delta^n) \to \Omega^*(\Delta^n)$.
(2) $Q^{GF2} = 0$.
(3) The operator
$$\mathrm{D} = [Q + \mathrm{d}_{DR}, Q^{GF}]$$
is a generalized Laplacian. This means that the symbol of D is a smooth family of Riemannian metrics on the bundle T^*M, parameterized by Δ^n.

EXAMPLE. Let us consider Chern-Simons theory on a three-manifold M, with coefficients in a Lie algebra $\mathfrak{g}$ with an invariant pairing. Then, the space of fields is $\mathscr{E} = \Omega^*(M) \otimes \mathfrak{g}[1]$.

The choice of a smooth family of metrics on M parameterized by Δ^n leads to an operator
$$\mathrm{d}^* : \mathscr{E} \to \mathscr{E} \otimes C^\infty(\Delta^n).$$
We define Q^{GF} to be the $\Omega^*(\Delta^n)$ linear extension of this to
$$\mathscr{E} \otimes \Omega^*(\Delta^n) \to \mathscr{E} \otimes \Omega^*(\Delta^n).$$

The choice of such a family of gauge fixing conditions over $\Omega^*(\Delta^n)$ leads to a heat kernel and a propagator, as usual. The heat kernel is the element
$$K_l \in \mathscr{E} \otimes \mathscr{E} \otimes \Omega^*(\Delta^n)$$
which is the kernel for the $\Omega^*(\Delta^n)$ linear operator $e^{-l\mathrm{D}}$. The propagator is the element
$$P(\varepsilon, L) = \int_\varepsilon^L (Q^{GF} \otimes 1) K_l \mathrm{d}l \in \mathscr{E} \otimes \mathscr{E} \otimes \Omega^*(\Delta^n).$$

As before, the propagator $P(\varepsilon, L)$ is a chain homotopy between the heat kernels K_ε and K_L:
$$(Q + \mathrm{d}_{dR})P(\varepsilon, L) = K_\varepsilon - K_L$$
where $Q + \mathrm{d}_{dR}$ refers to the total differential on the cochain complex $\mathscr{E} \otimes \mathscr{E} \otimes \Omega^*(\Delta^n)$.

If
$$\Delta^m \to \Delta^n$$
is a face or degeneracy map, one can pull a family of gauge fixing operators over $\Omega^*(\Delta^n)$ to one over $\Omega^*(\Delta^m)$. In this way, gauge fixing operators for $(\mathscr{E}, Q)$ form a simplicial set, which we will denote $\mathscr{GF}(\mathscr{E}, Q)$.

10.5. Let us fix a family of gauge fixing conditions over $\Omega^*(\Delta^n)$. This allows us to consider families of pre-theories over $\Omega^*(\Delta^n)$. Such a family is given by a collection of effective interactions $\{I[L]\}$, where each $I[L]$ is a functional on $\mathscr{E}$ with values in $\Omega^*(\Delta^n)$:
$$I[L] \in \mathscr{O}^+(\mathscr{E}, \Omega^*(\Delta^n))[[\hbar]].$$

Each $I[L]$ must be of cohomological degree zero, where the cohomological degree incorporates the cohomological grading on $\mathscr{E}$ as well as that on $\Omega^*(\Delta^n)$. As usual, the collection $\{I[L]\}$ of effective interactions must satisfy the renormalization group equation

$$I[L] = W\left(P(\varepsilon, L), I[\varepsilon]\right)$$

as well as the locality axiom, which states that $I[L]$ has a small L asymptotic expansion in terms of local action functionals.

As with gauge fixing operators, one can pull back pre-theories along face and degeneracy maps of simplices. In this way, we find a simplicial set $\widetilde{\mathscr{T}}^{(\infty)}(\mathscr{E}, Q)$, whose n-simplices are families of gauge fixing conditions over $\Omega^*(\Delta^n)$, together with a family of pre-theories for this family of gauge fixing conditions.

Thus, there is a natural map of simplicial sets

$$\widetilde{\mathscr{T}}^{(\infty)}(\mathscr{E}, Q) \to \mathscr{GF}.$$

The renormalization results from Chapter 2 apply in this situation, and allow us to describe the simplicial set $\widetilde{\mathscr{T}}^{(\infty)}(\mathscr{E}, Q)$ explicitly. Let $\mathscr{O}_{loc}^{+,0}(\mathscr{E})[[\hbar]]$ denote the simplicial set whose n-simplices are cohomological degree zero elements of $\mathscr{O}_{loc}^+(\mathscr{E}, \Omega^*(\Delta^n))[[\hbar]]$.

Then, if we choose a renormalization scheme, we find an isomorphism of simplicial sets

$$\widetilde{\mathscr{T}}^{(\infty)}(\mathscr{E}, Q) \cong \mathscr{GF} \times \mathscr{O}_{loc}^+(\mathscr{E})[[\hbar]].$$

We can reformulate this statement in a renormalization-scheme independent way, as follows. Let $\widetilde{\mathscr{T}}^{(n)}(\mathscr{E}, Q)$ denote the simplicial set of pre-theories defined modulo $\hbar^{n+1}$. Then, the map

$$\widetilde{\mathscr{T}}^{(n+1)}(\mathscr{E}, Q) \to \widetilde{\mathscr{T}}^{(n)}(\mathscr{E}, Q)$$

makes $\widetilde{\mathscr{T}}^{(n+1)}(\mathscr{E}, Q)$ into a principal bundle for the simplicial abelian group whose n-simplices are closed degree zero elements of $\mathscr{O}_{loc}(\mathscr{E}, \Omega^*(\Delta^n))$.

10.6. There is an analog of the quantum master equation in this situation.

DEFINITION-LEMMA 10.6.1. *A functional*

$$I \in \mathscr{O}^+(\mathscr{E}, \Omega^*(\Delta^k))[[\hbar]]$$

satisfies the scale L quantum master equation if

$$(Q + \mathrm{d}_{DR} + \hbar \Delta_L) e^{I/\hbar} = 0.$$

Here d_{DR} is the de Rham differential on $\Omega^*(\Delta_k)$, and Δ_L is the operator of contraction with the heat kernel $K_L \in \mathscr{E} \otimes \mathscr{E} \otimes \Omega^*(\Delta^n)$. Everything, as always, is linear over $\Omega^*(\Delta^n)$.

As before, a functional I satisfies the scale ε quantum master equation if and only if $W(P(\varepsilon, L), I)$ satisfies the scale L quantum master equation.

A theory is a pre-theory $\{I[L]\}$ where each $I[L]$ satisfies the scale L quantum master equation.

The simplicial set of theories will be denoted by $\mathscr{T}^{(\infty)}(\mathscr{E}, Q)$, as before.

Thus, a homotopy between two theories is a 1-simplex of $\mathscr{T}^{(\infty)}$ connecting the two 0-simplices. Two theories which are homotopic should be viewed as being equivalent: the philosophy is that they differ by a change of coordinates on the space $\mathscr{E}$.

Later we will see that the map $\mathscr{T}^{(\infty)}(\mathscr{E}, Q) \to \mathscr{GF}(\mathscr{E}, Q)$ of simplicial sets is a fibration. In all examples, $\mathscr{GF}(\mathscr{E}, Q)$ is a contractible simplicial set. This means that the concept of theory is independent of the choice of gauge fixing condition: we can take the fibre of $\mathscr{T}^{(\infty)}(\mathscr{E}, Q)$ over any 0-simplex of $\mathscr{GF}(\mathscr{E}, Q)$, and this fibre is independent of the choice of point of $\mathscr{GF}(\mathscr{E}, Q)$.

10.7. In finite dimensions, we have seen that the functional integral yields a map of simplicial sets from the simplicial set $\mathscr{QME}$ of solutions to the quantum master equation to the simplicial set of solutions to the quantum master equation on cohomology.

A similar result holds in infinite dimensions, as long as we impose some extra constraints on our gauge fixing conditions.

DEFINITION 10.7.1. *A family of gauge fixing conditions parameterized by Δ^n, given by a $C^\infty(\Delta^n)$ linear map*

$$Q^{GF} : \mathscr{E} \otimes C^\infty(\Delta^n) \to \mathscr{E} \otimes C^\infty(\Delta^n),$$

is non-negative *if the operator*

$$\mathrm{D} = [Q, Q^{GF}]$$

is symmetric for some Hermitian metric on the vector bundle E, with non-negative eigenvalues, and if we have a direct sum decomposition

$$\mathscr{E} = \operatorname{Ker} \mathrm{D} \oplus \operatorname{Im} Q \oplus \operatorname{Im} Q^{GF}.$$

Here, $\operatorname{Ker} \mathrm{D}$, $\operatorname{Im} Q^{GF}$, and $\operatorname{Im} Q$ are both regarded as sub $C^\infty(\Delta^n)$-modules of $\mathscr{E} \otimes C^\infty(\Delta^n)$.

We will let

$$\mathscr{H} = \operatorname{Ker} \mathrm{D} \subset \mathscr{E} \otimes C^\infty(\Delta^n)$$

denote the sub $C^\infty(\Delta^n)$-module of harmonic elements of $\mathscr{E}$.

In this situation, $\operatorname{Ker} \mathrm{D}$ is canonically isomorphic to the Q-cohomology of $\mathscr{E} \otimes C^\infty(\Delta^n)$.

PROPOSITION 10.7.2. *Let $\mathscr{GF}(\mathscr{E}, Q)^+$ denote the simplicial set of positive gauge fixing conditions, and let $\mathscr{T}^{(\infty)}(\mathscr{E}, Q)^+$ denote the fibre product of $\mathscr{T}^{(\infty)}(\mathscr{E}, Q)$ with $\mathscr{GF}(\mathscr{E}, Q)^+$, over $\mathscr{GF}(\mathscr{E}, Q)$.*

Let $\mathscr{QME}(H^*(\mathscr{E}, Q))$ denote the simplicial set of solutions to the quantum master equation on the graded symplectic vector space $H^*(\mathscr{E}, Q)$.

There is a canonical map of simplicial sets

$$\mathscr{T}^{(\infty)}(\mathscr{E}, Q)^+ \to \mathscr{QME}(H^*(\mathscr{E}, Q))$$

which, on 0-*simplices, takes an element* $\{I[L]\}$ *of* $\mathscr{T}^{(\infty)}(\mathscr{E},Q)^+$ *to the restriction of* $I[\infty]$ *to the space of harmonic elements of* $\mathscr{E}$.

PROOF. Suppose that
$$Q^{GF} : \mathscr{E} \otimes C^\infty(\Delta^n) \to \mathscr{E} \otimes C^\infty(\Delta^n)$$
is a family of non-negative gauge fixing conditions.

Suppose that
$$\{I[L] \in \mathscr{O}^+(\mathscr{E}, \Omega^*(\Delta^n))[[\hbar]]\}$$
is a family of theories for this family of gauge fixing conditions.

We need to show how to construct a function in
$$\mathscr{O}^+(H^*(\mathscr{E},Q), \Omega^*(\Delta^n))[[\hbar]]$$
which satisfies the quantum master equation. Further, this construction is required to be compatible with face and degeneracy maps of simplices.

The first step in the construction is to embed $H^*(\mathscr{E},Q) \otimes \Omega^*(\Delta^n)$ as a sub $\Omega^*(\Delta^n)$-module of $\mathscr{E} \otimes \Omega^*(\Delta^n)$.

The positivity conditions we imposed on the operator
$$[Q, Q^{GF}] : \mathscr{E} \otimes C^\infty(\Delta^n) \to \mathscr{E} \otimes C^\infty(\Delta^n)$$
imply that there is a canonical isomorphism
$$\mathrm{Ker}[Q, Q^{GF}] \cong H^*(\mathscr{E},Q) \otimes C^\infty(\Delta^n).$$
This gives a map
$$\Phi : H^*(\mathscr{E},Q) \to \mathscr{E} \otimes C^\infty(\Delta^n) \subset \mathscr{E} \otimes \Omega^*(\Delta^n)$$
sending $\alpha \in H^*(\mathscr{E},Q)$ to the family $\Phi(\alpha)(t_1,\ldots,n)$ of harmonic representatives of α for the family of Laplacians $[Q, Q^{GF}]$. Here, $t_1 < \cdots < t_n$ refer to the coordinates on the n-simplex.

However, the resulting map
$$H^*(\mathscr{E},Q) \to \mathscr{E} \otimes \Omega^*(\Delta^n)$$
is not a cochain map. Although $Q\Phi(\alpha) = 0$, $\mathrm{d}_{dR}\Phi(\alpha)$ is not necessarily zero. We need to modify it so that it becomes a cochain map.

Let
$$A = [Q, Q^{GF}]^{-1} Q^{GF} : \mathscr{E} \otimes \Omega^*(\Delta^n) \to \mathscr{E} \otimes \Omega^*(\Delta^n).$$
The positivity assumptions we impose on $[Q, Q^{GF}]$ implies that this operator exists.

The modification is defined by
$$\alpha \mapsto \Phi(\alpha) - A \cdot \mathrm{d}_{dR}\Phi(\alpha) + (A \cdot \mathrm{d}_{dR})^2 \Phi(\alpha) - \cdots$$
$$= \sum_{l \geq 0} (-A \cdot \mathrm{d}_{dR})^l \Phi(\alpha).$$

Note that $(A \cdot \mathrm{d}_{dR})^{n+1} = 0$, so that this series converges.

Since
$$[Q, A \cdot \mathrm{d}_{dR}] = \mathrm{d}_{dR}$$
$$[Q, \mathrm{d}_{dR}] = 0$$
$$Q\Phi(\alpha) = 0$$

we find that
$$(Q + \mathrm{d}_{dR}) \sum_{l \geq 0} (-A \cdot \mathrm{d}_{dR})^n \Phi(\alpha)$$
$$= \sum -(-A \cdot \mathrm{d}_{dR})^k [Q, A \cdot \mathrm{d}_{dR}](-A \cdot \mathrm{d}_{dR})^m \Phi(\alpha)$$
$$+ \mathrm{d}_{dR} \sum (-A \cdot \mathrm{d}_{dR})^l \Phi(\alpha)$$
$$= \sum -(-A \cdot \mathrm{d}_{dR})^k \mathrm{d}_{dR} (-A \cdot \mathrm{d}_{dR})^m \Phi(\alpha)$$
$$+ \mathrm{d}_{dR} \sum (-A \cdot \mathrm{d}_{dR})^l \Phi(\alpha)$$
$$= \sum -\mathrm{d}_{dR} (-A \cdot \mathrm{d}_{dR})^l \Phi(\alpha)$$
$$+ \mathrm{d}_{dR} \sum (-A \cdot \mathrm{d}_{dR})^l \Phi(\alpha)$$
$$= 0.$$

Thus, we have constructed a cochain map
$$H^*(\mathscr{E}, Q) \to \mathscr{E} \otimes \Omega^*(\Delta^n).$$

The image of the resulting map
$$H^*(\mathscr{E}, Q) \otimes \Omega^*(\Delta^n) \to \mathscr{E} \otimes \Omega^*(\Delta^n)$$
is the kernel of $[Q + \mathrm{d}_{dR}, Q^{GF}]$.

The positivity assumptions we have imposed on the operator $[Q, Q^{GF}]$ imply that, if K_t is the heat kernel for the operator $[Q + \mathrm{d}_{dR}, Q^{GF}]$, then K_∞ exists. Further, the propagator
$$P(\varepsilon, \infty) = \int_\varepsilon^\infty Q^{GF} K_t \mathrm{d}t$$
exists and is smooth for all $\varepsilon > 0$.

It follows that if
$$\{I[L] \in \mathscr{O}^+(\mathscr{E}, \Omega^*(\Delta^n))[[\hbar]]\}$$
satisfies the renormalization group equation, and the quantum master equation, then $I[\infty]$ exists.

The kernel K_∞ is the kernel for the operator of projection onto
$$\mathrm{Ker}[Q + \mathrm{d}_{dR}, Q^{GF}] = H^*(\mathscr{E}, Q) \otimes \Omega^*(\Delta^n).$$

It follows that
$$K_\infty \in H^*(\mathscr{E}, Q) \otimes H^*(\mathscr{E}, Q) \subset \mathscr{E} \otimes \mathscr{E} \otimes \Omega^*(\Delta^n)$$
is the inverse to the natural degree -1 symplectic pairing on $H^*(\mathscr{E}, Q)$.

Thus, if $I[L] \in \mathscr{O}^+(\mathscr{E}, \Omega^*(\Delta^n))[[\hbar]]$ solves the renormalization group equation and the quantum master equation, then
$$I[\infty]\,|_{H^*(\mathscr{E},Q) \otimes \Omega^*(\Delta^n)}$$
solves the quantum master equation on $H^*(\mathscr{E}, Q) \otimes \Omega^*(\Delta^n)$, as desired. □

11. Obstruction theory

We have seen that the simplicial set $\widetilde{\mathscr{T}}^{(n+1)}(\mathscr{E}, Q)$ of pre-theories defined modulo $\hbar^{n+2}$ is a torsor over the simplicial set $\widetilde{\mathscr{T}}^{(n)}(\mathscr{E}, Q)$ for the simplicial abelian group of local action functionals. We would like to have a similar result for the simplicial set $\mathscr{T}^{(n+1)}(\mathscr{E}, Q)$ of theories.

What we find is that there is a cohomological obstruction to extending a point of $\mathscr{T}^{(n)}(\mathscr{E}, Q)$ to a point in $\mathscr{T}^{(n+1)}(\mathscr{E}, Q)$. This obstruction is a local action functional. The set of possible lifts is the set of ways of making the obstruction cochain exact; thus, the set of possible lifts is a torsor for the abelian group of closed elements of the space of local action functionals.

11.1. Let us fix a free BV theory $(\mathscr{E}, Q)$ on a compact manifold M, and let $\{I[L]\} \in \mathscr{T}^{(n)}(\mathscr{E}, Q)[k]$ be a k-simplex in the space of theories defined modulo $\hbar^{n+1}$.

Let us lift, arbitrarily, $\{I[L]\}$ to an element $\{\widetilde{I}[L]\}$ of $\widetilde{\mathscr{T}}^{(n+1)}(\mathscr{E}, Q)[k]$, that is, a pre-theory defined modulo $\hbar^{n+2}$.

Let us define the scale L obstruction by
$$O_{n+1}[L] = \hbar^{-n-1}\left(Q\widetilde{I}[L] + \tfrac{1}{2}\{\widetilde{I}[L], \widetilde{I}[L]\}_L + \hbar\Delta_L \widetilde{I}[L]\right).$$

This expression is independent of $\hbar$, because of the $\hbar^{-n-1}$ and the fact that $\widetilde{I}[L]$ satisfies the quantum master equation modulo $\hbar^{n+1}$.

LEMMA 11.1.1. *Let ε be a parameter of square zero and cohomological degree -1. Let $I_0[L]$ be $I[L]$ modulo $\hbar$. Then,*
$$I_0[L] + \varepsilon O_{n+1}[L]$$
satisfies both the scale L classical master equation and the classical renormalization group equation. Thus, $I_0[L] + \varepsilon O_{n+1}[L]$ defines a classical theory in the BV formalism.

The set of lifts of $\{I[L]\}$ to a k-simplex of $\mathscr{T}^{(n+1)}(\mathscr{E}, Q)$ is the set of degree 0 elements $J[L] \in \mathscr{O}_{loc}(\mathscr{E}, \Omega^*(\Delta^k))$ such that $I_0[L] + \delta J[L]$ satisfies the classical renormalization group equation and locality axiom modulo δ^2, and such that
$$QJ[L] + \{I_0[L], J[L]\} = O_{n+1}[L].$$

PROOF. We will prove this using a simple lemma about differential graded Lie algebras. Let $\mathfrak{g}$ be a differential graded Lie algebra, and let $X \in \mathfrak{g}$ be an odd element. Let
$$O(X) = d_{\mathfrak{g}} X + \tfrac{1}{2}[X, X].$$

Then,
$$d_{\mathfrak{g}} O(X) + [X, O(X)] = 0.$$

Let us make
$$\mathscr{O}(\mathscr{E}, \Omega^*(\Delta^k))[[\hbar]]/\hbar^{n+2}[1]$$
into a differential graded Lie algebra, with bracket $\{-,-\}_L$, and with differential $Q + \hbar \Delta_L$. (Here, Q incorporates the de Rham differential on $\Omega^*(\Delta^k)$). Then,
$$\hbar^{n+1} O_{n+1}[L] = (Q + \hbar \Delta_L) I[L] + \tfrac{1}{2}\{I[L], I[L]\}_L$$
from which it follows that
$$Q O_{n+1}[L] + \{I_0[L], O_{n+1}[L]\}_L = 0$$
as desired.

Next, we need to show that that $I_0[L] + \varepsilon O_{n+1}[L]$ satisfies the classical renormalization group equation. In fact, the classical renormalization group equation can be expressed as a classical master equation. Let us consider the algebra
$$\mathscr{O}(\mathscr{E}, \Omega^*(\Delta^k) \otimes \Omega^*((0,\infty)_L)).$$
Let us give this algebra a BV structure, where the differential is the total differential coming from $Q : \mathscr{E} \to \mathscr{E}$ and the de Rham differential on $\Omega^*(\Delta^k)$ and $\Omega^*((0,\infty)_L)$. The BV operator defined by the formula
$$\Delta_L + \mathrm{d}L \partial_{(Q^{GF} \otimes 1) K_L}.$$

An element
$$J[L] \in \mathscr{O}(\mathscr{E}, \Omega^*(\Delta^k) \otimes C^\infty((0,\infty)_L))[[\hbar]]$$
satisfies the renormalization group equation and the quantum master equation if and only if, when we view it as an element of
$$\mathscr{O}(\mathscr{E}, \Omega^*(\Delta^k) \otimes \Omega^*((0,\infty)_L)),$$
it satisfies the quantum master equation.

This formulation is just a rephrasing of the infinitesimal form of the renormalization group equation.

Note that $\widetilde{I}[L]$ satisfies the renormalization group equation. It follows that, if we view $\widetilde{I}[L]$ as an element
$$\widetilde{I}[L] \in \mathscr{O}(\mathscr{E}, \Omega^*(\Delta^k) \otimes \Omega^*((0,\infty)_L)),$$
that $O_{n+1}[L]$ is the obstruction to $\widetilde{I}[L]$ satisfying the renormalization group equation. Thus, the previous argument applies to show that $I_0[L] + \varepsilon O_{n+1}[L]$ satisfies the classical RGE as well as the classical master equation.

For the final clause, observe that any other lift of $I[L]$ to a k-simplex of $\widetilde{\mathscr{T}}^{(n+1)}(\mathscr{E}, Q)$ is of the form $\widetilde{I}[L] + \hbar^{n+1} J[L]$ where $I_0[L] + \delta J[L]$ satisfies the classical RGE modulo δ^2. The obstruction vanishes for such a lift if and only if
$$QJ[L] + \{I_0[L], J[L]\} = O_{n+1}[L].$$

□

Since $I_0[L] + \varepsilon O_{n+1}[L]$ satisfies the classical master equation, it follows that the $L \to 0$ limit of $O_{n+1}[L]$ exists, and is an element $O_{n+1} \in \mathscr{O}_{loc}(\mathscr{E}, \Omega^*(\Delta^k))$. Further, if $I_0 \in \mathscr{O}_{loc}(\mathscr{E}, \Omega^*(\Delta^k))$ refers to the classical action at scale 0, then

$$QO_{n+1} + \{I_0, O_{n+1}\} = 0$$

where the bracket $\{-,-\}$ is now taken at scale zero.

Thus, we have shown the following.

COROLLARY 11.1.2. *Let $\{I[L]\} \in \mathscr{T}^{(n)}(\mathscr{E}, Q)[k]$. Then, there is an obstruction*

$$O_{n+1} \in \mathscr{O}_{loc}(\mathscr{E}, \Omega^*(\Delta^k))$$

which is a closed, degree 1 element, where the cochain complex has the differential $Q + \{I_0, -\}$.

The set of lifts of $\{I[L]\}$ to a k-simplex of $\mathscr{T}^{(n+1)}(\mathscr{E}, Q)$ is the set of degree 0 elements $J \in \mathscr{O}_{loc}(\mathscr{E}, \Omega^*(\Delta^k))$ making O_{n+1} exact.

11.2. Let us consider $\mathscr{O}_{loc}(\mathscr{E})$ as a cochain complex, with differential $Q + \{I_0, -\}$, where $I_0 = \lim_{L \to 0} I_0[L]$ is the classical interaction. There is a simplicial abelian group associated to this cochain complex, by the Dold-Kan correspondence; the n-simplices are elements $\mathscr{O}_{loc}(\mathscr{E}, \Omega^*(\Delta^k))$ which are closed and of degree 0. We will refer to this simplicial abelian group by the same notation, $\mathscr{O}_{loc}(\mathscr{E})$.

We are also interested in the simplicial abelian group $\mathscr{O}_{loc}(\mathscr{E})[1]$, whose n-simplices are closed, degree 1 elements of $\mathscr{O}_{loc}(\mathscr{E}, \Omega^*(\Delta^k))$.

Now let us choose, arbitrarily, a section $\widetilde{\mathscr{T}}^{(n)}(\mathscr{E}, Q) \to \widetilde{\mathscr{T}}^{(n+1)}(\mathscr{E}, Q)$. This could come, for example, from the choice of a renormalization scheme.

The obstruction O_{n+1} described above gives a map of simplicial sets

$$O_{n+1} : \mathscr{T}^{(n)}(\mathscr{E}, Q) \to \mathscr{O}_{loc}(\mathscr{E})[1].$$

We can rephrase the obstruction theory results above in more homotopy-theoretic terms as follows.

COROLLARY 11.2.1. *There is a homotopy fibre diagram of simplicial sets*

$$\begin{array}{ccc} \mathscr{T}^{(n+1)}(\mathscr{E}, Q) & \longrightarrow & 0 \\ \downarrow & & \downarrow \\ \mathscr{T}^{(n)}(\mathscr{E}, Q) & \xrightarrow{O_{n+1}} & \mathscr{O}_{loc}(\mathscr{E}, Q)[1] \end{array}$$

This corollary just says that a point of $\mathscr{T}^{(n+1)}(\mathscr{E}, Q)$ is the same as a point of $\mathscr{T}^{(n)}(\mathscr{E}, Q)$ together with a homotopy between the obstruction and zero.

PROOF. Let us define a simplicial set $P\mathscr{O}_{loc}(\mathscr{E}, Q)[1]$ whose k-simplices are pairs

$$\alpha, \beta \in \mathscr{O}_{loc}(\mathscr{E}, \Omega^*(\Delta^k))$$

which are of degrees 0 and 1 respectively, such that β is the coboundary of α,
$$(Q + \mathrm{d}_{dr})\alpha + \{I_0, \alpha\} = \beta.$$
There is a map
$$P\mathcal{O}_{loc}(\mathcal{E}, Q)[1] \to \mathcal{O}_{loc}(\mathcal{E}, Q)[1]$$
sending $(\alpha, \beta) \to \beta$. This map is a fibration, and $P\mathcal{O}_{loc}(\mathcal{E}, Q)[1]$ is contractible.

The obstruction theory results above show that
$$\mathcal{T}^{(n+1)}(\mathcal{E}, Q) = \mathcal{T}^{(n)}(\mathcal{E}, Q) \times_{\mathcal{O}_{loc}(\mathcal{E}, Q)[1]} P\mathcal{O}_{loc}(\mathcal{E}, Q)[1].$$

It follows from the fact that the map $P\mathcal{O}_{loc}(\mathcal{E}, Q)[1] \to \mathcal{O}_{loc}(\mathcal{E}, Q)[1]$ is a fibration that this fibre product is in fact a homotopy fibre product. Since $P\mathcal{O}_{loc}(\mathcal{E}, Q)[1]$ is contractible, we find that $\mathcal{T}^{(n+1)}(\mathcal{E}, Q)$ is the homotopy fibre product
$$\mathcal{T}^{(n+1)}(\mathcal{E}, Q) \simeq \mathcal{T}^{(n)}(\mathcal{E}, Q) \times^{\mathbb{L}}_{\mathcal{O}_{loc}(\mathcal{E}, Q)[1]} \{0\}$$
as desired. □

COROLLARY 11.2.2. *The map*
$$\mathcal{T}^{(n+1)}(\mathcal{E}, Q) \to \mathcal{T}^{(n)}(\mathcal{E}, Q)$$
is a fibration of simplicial sets, as is the map
$$\mathcal{T}^{(\infty)}(\mathcal{E}, Q) \to \mathcal{GF}(\mathcal{E}, Q).$$

PROOF. This follows from the previous corollary, and the fact that $\mathcal{T}^{(0)}(\mathcal{E}, Q)$ is a product of $\mathcal{GF}(\mathcal{E}, Q)$ with the simplicial set whose k-simplices are elements of $\mathcal{O}_{loc}(\mathcal{E}, \Omega^*(\Delta^k))$ satisfying the scale 0 classical master equation. □

12. BV theories on $\mathbb{R}^n$

Now we will see how to define theories in the BV formalism on $\mathbb{R}^n$. The main difference from the discussion on a compact manifold is that on $\mathbb{R}^n$ we can talk about renormalizability. This requires us to assume that our bundle of fields has an extra grading, by dimension; this will allow us to apply the renormalization group flow.

DEFINITION 12.0.1. *A free theory on $\mathbb{R}^n$ in the BV formalism is described by the following data:*

(1) *A bigraded vector space E. The first grading is called the cohomological grading, and the second grading is the dimension grading.*
   We will think of E as a trivial vector bundle on $\mathbb{R}^n$, and let
$$\mathcal{E} = E \otimes \mathscr{S}(\mathbb{R}^n)$$
be the space of Schwartz sections.

The grading by dimension on $\mathscr{E}$ induces an $\mathbb{R}_{>0}$ action on $\mathscr{E} = \mathscr{S}(\mathbb{R}^n) \otimes E$, by

$$R_l(f(x)e) = f(\lambda x)\lambda^i e$$

where $f(x) \in \mathscr{S}(\mathbb{R}^n)$ and e is of dimension i.

(2) *E is equipped with a non-degenerate degree -1 anti-symmetric pairing*

$$\langle -, - \rangle : E \otimes E \to \det(\mathbb{R}^n)$$

which respects dimension, so that E_i pairs with E_{n-i}.
The pairing on E induces an integration pairing

$$\langle -, - \rangle : \mathscr{E} \otimes \mathscr{E} \to \mathbb{C}$$

$$\langle f_1 e_1, f_2 e_2 \rangle = \int_{\mathbb{R}^n} f_1 f_2 \langle e_1, e_2 \rangle$$

where $e_i \in E$ and $f_i \in \mathscr{S}(\mathbb{R}^n)$. This pairing is of dimension zero.

(3) *A differential operator $Q : \mathscr{E} \to \mathscr{E}$ which is translation invariant, of cohomological degree one, preserves dimension, is of square zero, and is skew self adjoint for the pairing.*

Throughout, we will abuse notation and refer to the totality of data of a free BV theory by $\mathscr{E}$.

12.1. As on a compact manifold, we need to choose a gauge fixing operator $Q^{GF} : \mathscr{E} \to \mathscr{E}$ in order to discuss the renormalization group flow. The set of possible gauge fixing conditions will form a simplicial set $\mathscr{GF}$; the following definition describes the set of m-simplices.

DEFINITION 12.1.1. *Let $(\mathscr{E}, Q)$ be a free BV theory. A family of gauge fixing operators on $\mathscr{E}$, parameterized by $\Omega^*(\Delta^m)$, is an $\Omega^*(\Delta^m)$ linear differential operator*

$$Q^{GF} : \mathscr{E} \otimes \Omega^*(\Delta^m) \to \mathscr{E} \otimes \Omega^*(\Delta^m)$$

such that:

(1) *Q^{GF} is of cohomological degree -1, translation-invariant, of square zero, and self-adjoint for the pairing $\langle -, - \rangle$.*
(2) *Q^{GF} is of dimension -2.*
(3) *The commutator*

$$\mathrm{D} = [Q + \mathrm{d}_{dR}, Q^{GF}]$$

is a sum of two terms

$$\mathrm{D} = \mathrm{D}' + \mathrm{D}'',$$

where:

(a) *D' is the tensor product of the Laplacian on $\mathbb{R}^n$ with the identity on E.*
(b) *D'' is a nilpotent operator commuting with D'.*

12. BV THEORIES ON $\mathbb{R}^n$

12.2. We will use the notation of Chapter 3, Section 3. Thus,

$$\mathcal{O}(\mathcal{E}, \Omega^*(\Delta^m)) = \prod_{k>0} \operatorname{Hom}(\mathcal{D}_g(\mathbb{R}^{nk}, \Omega^*(\Delta^m)) \otimes E^{\otimes k})_{S_k}$$

where $\mathcal{D}_g(\mathbb{R}^{nk}, \Omega^*(\Delta^m))$ refers to the space $\Omega^*(\Delta^m)$-valued distributions on $\mathbb{R}^{nk}$, which are invariant under the action of $\mathbb{R}^n$ by translation, and of rapid decay away from the diagonal, as detailed in Chapter 3, Section 3.

We will let

$$\mathcal{O}_{loc}(\mathcal{E}, \Omega^*(\Delta^m)) \subset \mathcal{O}(\mathcal{E}, \Omega^*(\Delta^m))$$

denote the subspace of local action functionals, which are translation invariant, as always.

12.3. There is a heat kernel

$$K_l \in C^\infty(\mathbb{R}^n \times \mathbb{R}^n) \otimes E \otimes E \otimes \Omega^*(\Delta^m)$$

for the operator $e^{-l\mathrm{D}}$. If we pick a basis e_i for E, then

$$K_l = \sum \Phi_{i,j}(x, y, l) e_i \otimes e_j$$

where, for all basis elements e_k and functions $g(x) \in \mathscr{S}(\mathbb{R}^n)$,

$$(-1)^{|e_k|} \sum_{i,j} \int_{y \in \mathbb{R}^n} \Phi_{i,j}(x,y,l) g(y) e_i \langle e_j, e_k \rangle = e^{-lD} g(x) e_k$$

Here $|e_k|$ denotes the parity of the basis element e_k.

The functions $\Phi_{i,j}(x, y, l)$ appearing in the heat kernel are of the form

$$\Phi_{i,j}(x, y, l) = \Psi(l^{1/2}, l^{-1/2}, x - y) e^{-\|x-y\|^2/l}$$

where

$$\Psi \in \Omega^*(\Delta^m)[l^{1/2}, l^{-1/2}, x - y]$$

is a polynomial in the variables $l^{\pm 1/2}$ and $x - y$, with coefficients in $\Omega^*(\Delta^m)$.

The propagator of our theory is given by

$$P(\varepsilon, L) = \int_{l=\varepsilon}^{L} (Q^{GF} \otimes 1) K_l \, \mathrm{d}l.$$

This propagator satisfies the conditions of Chapter 3, Section 3, so that the bijection between collections of effective interactions and local action functionals holds in this situation.

Further, the propagator $P(\varepsilon, L)$ scales as

$$R_l P(\varepsilon, L) = P(l^{-2}\varepsilon, l^{-2}L).$$

Thus, the conditions of Chapter 4, Section 6 are satisfied.

DEFINITION 12.3.1. *As on a compact manifold, a* pre-theory *is a collection of effective interactions $\{I[L]\}$ satisfying renormalization group flow and the locality axiom as defined in Chapter 3, Section 3.*

More formally, we define a simplicial set $\widetilde{\mathscr{T}}^{(\infty)}(\mathcal{E}, Q)$ of pre-theories. An m-simplex of $\widetilde{\mathscr{T}}^{(\infty)}(\mathcal{E}, Q)$ consists of

(1) An m-simplex Q^{GF} of the simplicial set of gauge fixing operators.
(2) A collection
$$I[L] \in \mathscr{O}^{+,0}(\mathscr{E}, \Omega^*(\Delta^m))[[\hbar]]$$
of effective interactions, which are of cohomological degree 0 and at least cubic modulo $\hbar$.

The following axioms must be satisfied:
(1) The renormalization group equation
$$I[L] = W\left(P(\varepsilon, L), I[\varepsilon]\right)$$
is satisfied.
(2) The locality axiom of Chapter 3, Section 3 holds. If we consider $I_{i,k}[L]$ as an element
$$I_{i,k}[L] \in \mathcal{D}_g(\mathbb{R}^{nk}, \Omega^*(\Delta^m)) \otimes (\mathscr{E}^\vee)^{\otimes k},$$
then, if $e \in \mathscr{S}(\mathbb{R}^{nk}) \otimes E^{\otimes k}$ has compact support away from the small diagonal,
$$I_{i,k}[L](e) \to 0 \text{ as } L \to 0.$$

We let $\widetilde{\mathscr{T}}^{(\infty)}(\mathscr{E}, Q)$ denote the simplicial set of theories, and $\widetilde{\mathscr{T}}^{(n)}(\mathscr{E}, Q)$ denote the simplicial set of theories defined modulo $\hbar^{n+1}$.

We will reserve the word theory for collections of effective interactions $\{I[L]\}$ which satisfy the quantum master equation, as defined below.

12.4. As in Chapter 4, there is a natural action of $\mathbb{R}_{>0}$ on the set of pre-theories. This action is called the local renormalization group flow.

The local renormalization group flow also acts on families of theories over the n-simplex, and so gives an action of $\mathbb{R}_{>0}$ on the simplicial set $\widetilde{\mathscr{T}}^{(\infty)}(\mathscr{E}, Q)$.

As before, if $\{I[L]\} \in \widetilde{\mathscr{T}}^{(\infty)}\mathscr{E}, Q[m]$ is an m-simplex in the space of pre-theories, then
$$\mathcal{RG}_l(I[L]) \in \mathscr{O}^+(\mathscr{E}, \Omega^*(\Delta^m))[[\hbar]] \otimes \mathbb{C}[l, l^{-1}, \log l].$$

We let
$$\widetilde{\mathscr{R}}^{(\infty)}(\mathscr{E}, Q) \subset \widetilde{\mathscr{T}}^{(\infty)}(\mathscr{E}, Q)$$
$$\widetilde{\mathscr{M}}^{(\infty)}(\mathscr{E}, Q) \subset \widetilde{\mathscr{T}}^{(\infty)}(\mathscr{E}, Q)$$
be the sub simplicial sets of relevant and marginal pre-theories, respectively; where, as before, an m-simplex $\{I[L]\}$ in the space of pre-theories is relevant if
$$\mathcal{RG}_l(I[L]) \in \mathscr{O}^+(\mathscr{E}, \Omega^*(\Delta^m))[[\hbar]] \otimes \mathbb{C}[l, \log l].$$
and marginal if
$$\mathcal{RG}_l(I[L]) \in \mathscr{O}^+(\mathscr{E}, \Omega^*(\Delta^m))[[\hbar]] \otimes \mathbb{C}[\log l].$$

(Note that all theories we are considering are translation invariant)

12.5. Let us define a simplicial abelian group whose m-simplices are degree 0 elements of $\mathscr{O}_{loc}(\mathscr{E}, \Omega^*(\Delta^m))$. We will denote this simplicial abelian group by $\mathscr{O}_{loc}(\mathscr{E})$. We will let
$$\mathscr{O}_{loc}^i(\mathscr{E}) \subset \mathscr{O}_{loc}(\mathscr{E})$$
denote the sub simplicial abelian group whose m simplices are degree 0 and dimension i elements of $\mathscr{O}_{loc}(\mathscr{E}, \Omega^*(\Delta^m))$. These elements are not required to be closed, as we are not currently dealing with the quantum master equation. Similarly, we will denote by $\mathscr{O}_{loc}^{\geq 0}(\mathscr{E})$ the simplicial abelian group of local action functionals of dimension ≥ 0.

The results of Chapter 4 show the following:

THEOREM 12.5.1. (1) The simplicial set $\widetilde{\mathscr{T}}^{(n+1)}(\mathscr{E}, Q)$ is a torsor over $\widetilde{\mathscr{T}}^{(n)}(\mathscr{E}, Q)$ for the simplicial abelian group $\mathscr{O}_{loc}(\mathscr{E})$.
  (2) The simplicial set $\widetilde{\mathscr{R}}^{(n+1)}(\mathscr{E}, Q)$ is a torsor over $\widetilde{\mathscr{R}}^{(n)}(\mathscr{E}, Q)$ for the simplicial abelian group $\mathscr{O}_{loc}^{\geq 0}$.
  (3) The simplicial set $\widetilde{\mathscr{M}}^{(n+1)}(\mathscr{E}, Q)$ is a torsor over $\widetilde{\mathscr{M}}^{(n)}(\mathscr{E}, Q)$ for the simplicial abelian group $\mathscr{O}_{loc}^{\geq 0}$.

12.6. Next we will define a simplicial set of theories, which are pre-theories satisfying the quantum master equation. As on a compact manifold, we can define a scale L BV operator Δ_L and a scale L bracket $\{-,-\}_L$ on the space
$$\mathscr{O}^+(\mathscr{E}, \Omega^*(\Delta^m))[[\hbar]].$$
We say that an element $I \in \mathscr{O}^+(\mathscr{E}, \Omega^*(\Delta^m))[[\hbar]]$ satisfies the scale L quantum master equation if
$$(Q + \mathrm{d}_{dR})I[L] + \hbar \Delta_L I[L] + \{I[L], I[L]\}_L$$
where d_{dR} is the de Rham differential on $\Omega^*(\Delta^m)$.

As on a compact manifold, we have the following.

LEMMA 12.6.1. *A functional*
$$I \in \mathscr{O}^+(\mathscr{E}, \Omega^*(\Delta^m))[[\hbar]]$$
satisfies the scale ε quantum master equation if and only if
$$W(P(\varepsilon, L), I)$$
satisfies the scale L quantum master equation.

We can now define the notion of theory in the BV formalism.

DEFINITION 12.6.2. *Define a sub-simplicial set*
$$\mathscr{T}^{(\infty)}(\mathscr{E}, Q) \subset \widetilde{\mathscr{T}}^{(\infty)}(\mathscr{E}, Q)$$
of theories, inside the simplicial set of pre-theories, as follows.

An m-simplex in $\mathscr{T}^{(\infty)}(\mathscr{E}, Q)$ is an m-simplex in $\widetilde{\mathscr{T}}^{(\infty)}(\mathscr{E}, Q)$, described by a collection of effective interactions
$$\{I[L] \in \mathscr{O}^+(\mathscr{E}, \Omega^*(\Delta^m))[[\hbar]]\}$$

such that each $I[L]$ satisfies the scale L quantum master equation.

Similarly, let us define simplicial sets $\mathscr{R}^{(\infty)}(\mathscr{E}, Q)$ and $\mathscr{M}^{(\infty)}(\mathscr{E}, Q)$ of relevant and marginal theories, respectively, by

$$\mathscr{R}^{(\infty)}(\mathscr{E}, Q) = \widetilde{\mathscr{R}}^{(\infty)}(\mathscr{E}, Q) \cap \mathscr{T}^{(\infty)}(\mathscr{E}, Q)$$

$$\mathscr{M}^{(\infty)}(\mathscr{E}, Q) = \widetilde{\mathscr{M}}^{(\infty)}(\mathscr{E}, Q) \cap \mathscr{T}^{(\infty)}(\mathscr{E}, Q).$$

12.7. Suppose we have an m-simplex $\{I[L]\}$ in the space $\mathscr{T}^{(n)}(\mathscr{E}, Q)$ of theories defined modulo $\hbar^{n+1}$. Then, as in Section 11, there is an obstruction

$$O_{n+1}(\{I[L]\}) \in \mathscr{O}_{loc}(\mathscr{E}, \Omega^*(\Delta^m))$$

to lifting $\{I[L]\}$ to an m-simplex of $\mathscr{T}^{(n+1)}(\mathscr{E}, Q)$. This obstruction is closed,

$$QO_{n-1}(\{I[L]\}) + \mathrm{d}_{dR}O_{n+1}(\{I[L]\}) + \{I_0, O_{n+1}(\{I[L]\})\} = 0,$$

and of cohomological degree 1. Here I_0 is the scale zero classical action. Lifts of $\{I[L]\}$ to $\mathscr{T}^{(n+1)}(\mathscr{E}, Q)$ are the same as degree zero cochains of $\mathscr{O}_{loc}(\mathscr{E}, \Omega^*(\Delta^m))$ which bound $O_{n+1}(\{I[L]\})$.

12.8. Recall that the obstruction $O_{n+1}(\{I[L]\})$ depends on a lift of $\{I[L]\}$ to a pre-theory defined modulo $\hbar^{n+2}$. (The cohomology class of the obstruction, however, is independent of such a lift).

If $\{I[L]\} \in \mathscr{R}^{(n)}(\mathscr{E}, Q)$ is a relevant theory defined modulo $\hbar^{n+1}$, then we can lift it to a relevant pre-theory $\widetilde{I}[L]$, defined modulo $\hbar^{n+2}$.

LEMMA 12.8.1. *In this situation, the obstruction $O_{n+1}(\{I[L]\}$ (defined using a relevant lift $\widetilde{I}[L]$) is of dimension ≥ 0.*

PROOF. This is a simple calculation. Recall that there is a scale L obstruction defined by

$$\hbar^{n+1}O_{n+1}[L] = (Q + \mathrm{d}_{dR})\widetilde{I}[L] + \hbar\Delta_L\widetilde{I}[L] + \{\widetilde{I}[L], \widetilde{I}[L]\}_L.$$

The scale zero obstruction we have been using is the $L \to 0$ limit of $O_{n+1}[L]$.

Let R_l^* denote the operation induced from the rescaling map $R_l : \mathscr{E} \to \mathscr{E}$, on all spaces of functionals on $\mathscr{E}$.

We need to show that

$$R_l^*O_{n+1}[l^2 L] \in \mathscr{O}(\mathscr{E}, \Omega^*(\Delta^m)) \otimes \mathbb{C}[l].$$

In other words, we need to check that only positive powers of l appear.

Because $O_{n+1}[L]$ satisfies a version of the classical renormalization group equation, it is automatic that no powers of $\log l$ can appear in this expression.

The fact that $\widetilde{I}[L]$ is a relevant theory means that

$$R_l^*\widetilde{I}[l^2 L] \in \mathscr{O}(\mathscr{E}, \Omega^*(\Delta^m)) \otimes \mathbb{C}[l, \log l].$$

In addition, the bracket $\{-, -\}_L$ and the BV operator Δ_L have a compatibility with the operator R_l^*, namely

$$R_l^*\{I, J\}_{l^2 L} = \{R_l^*I, R_l^*J\}_L$$
$$R_l^*(\Delta_{l^2 L}I) = \Delta_L R_l^*I.$$

Putting these identities together yields the proof. □

In a similar way, if $\{I[L]\}$ is an n-simplex in the space of marginal theories, we can arrange it so that the obstruction $O_{n+1}(\{I[L]\})$ is a dimension zero local action functional.

From this lemma, we see that to lift an element $\{I[L]\} \in \mathscr{R}^{(n)}(\mathscr{E}, Q)$ to an element of $\mathscr{R}^{(n+1)}(\mathscr{E}, Q)$ is the same as giving a local action functional

$$J \in \mathscr{O}_{loc}(\mathscr{E}, \Omega^*(\Delta^m)),$$

of cohomological degree 0 and dimension ≥ 0, which kills the obstruction O_{n+1}:

$$(Q + \mathrm{d}_{dR})J + \{I_0, J\} = O_{n+1}.$$

A similar statement holds for marginal theories.

12.9. Let us denote by $\mathscr{O}_{loc}(\mathscr{E}, Q)$ the simplicial abelian group whose m-simplices are closed, degree 0 elements of $\mathscr{O}_{loc}(\mathscr{E}, \Omega^*(\Delta^m))$. Similarly, let $\mathscr{O}_{loc}^{\geq 0}(\mathscr{E}, Q)$ denote the simplicial abelian groups whose n-simplices are closed degree 0 elements of $\mathscr{O}_{loc}^{\geq 0}(\mathscr{E}, \Omega^*(\Delta^m))$, and let $\mathscr{O}_{loc}^{0}(\mathscr{E}, Q)$ be the simplicial abelian group whose n-simplices are closed degree 0 elements of $\mathscr{O}_{loc}^{\geq 0}(\mathscr{E}, \Omega^*(\Delta^m))$. Recall that the superscripts ≥ 0 and 0 indicate that we are looking at elements of dimension ≥ 0 or 0 respectively.

The obstructions discussed above are maps of simplicial sets

$$O_{n+1} : \mathscr{T}^{(n)}(\mathscr{E}, Q) \to \mathscr{O}_{loc}(\mathscr{E}, Q)[1]$$
$$O_{n+1} : \mathscr{R}^{(n)}(\mathscr{E}, Q) \to \mathscr{O}_{loc}^{\geq 0}(\mathscr{E}, Q)[1]$$
$$O_{n+1} : \mathscr{M}^{(n)}(\mathscr{E}, Q) \to \mathscr{O}_{loc}^{0}(\mathscr{E}, Q)[1].$$

Just like on a compact manifold, we have the following theorem.

THEOREM 12.9.1. *There are homotopy Cartesian diagrams*

$$\begin{array}{ccc} \mathscr{T}^{(n+1)}(\mathscr{E}, Q) & \longrightarrow & 0 \\ \downarrow & & \downarrow \\ \mathscr{T}^{(n)}(\mathscr{E}, Q) & \xrightarrow{O_{n+1}} & \mathscr{O}_{loc}(\mathscr{E}, Q)[1] \end{array}$$

for all theories,

$$\begin{array}{ccc} \mathscr{R}^{(n+1)}(\mathscr{E}, Q) & \longrightarrow & 0 \\ \downarrow & & \downarrow \\ \mathscr{R}^{(n)}(\mathscr{E}, Q) & \xrightarrow{O_{n+1}} & \mathscr{O}_{loc}^{\geq 0}(\mathscr{E}, Q)[1] \end{array}$$

for relevant theories, and

$$\begin{array}{ccc} \mathscr{M}^{(n+1)}(\mathscr{E},Q) & \longrightarrow & 0 \\ \downarrow & & \downarrow \\ \mathscr{M}^{(n)}(\mathscr{E},Q) & \xrightarrow{O_{n+1}} & \mathscr{O}_{loc}^{0}(\mathscr{E},Q)[1] \end{array}$$

for marginal theories.

13. The sheaf of BV theories on a manifold

In this section, we will describe the BV formalism in the most general set-up we will need. We will show that there is a simplicial sheaf of theories, on any manifold, compact or not. On a compact manifold, global sections will be theories in the sense we described earlier.

13.1. We will set things up in somewhat greater generality than we have used before. Everything in this chapter will depend on some auxiliary differential graded ring.

DEFINITION 13.1.1. *A differential graded manifold is a manifold X, possibly with corners, equipped with a sheaf of commutative graded algebras A over the sheaf C_X^∞ of smooth functions on X, together with a differential operator*

$$\mathrm{d}_{\mathscr{A}} : \mathscr{A} \to \mathscr{A}$$

where $\mathscr{A} = \Gamma(X, A)$, with the following properties.

(1) *As a C_X^∞-module, A is locally trivial of finite rank, so that A arises as the sheaf of sections of some graded vector bundle.*
(2) $\mathrm{d}_{\mathscr{A}}$ *is a derivation of degree 1 and square zero.*
(3) *There is a sheaf of ideals $m \subset A$ which is nilpotent (that is, $m^k = 0$ for some $k \gg 0$), such that $A/m = C_X^\infty$.*

A differential graded vector bundle on (X, A) is a sheaf E of locally free, finite rank A-modules, whose space of global sections will be denoted $\mathscr{E}$; together with a differential operator $\mathrm{d}_{\mathscr{E}} : \mathscr{E} \to \mathscr{E}$ which makes $\mathscr{E}$ into a differential graded $\mathscr{A}$-module.

If $(X, A), (Y, B)$ are dg manifolds, a map $(X, A) \to (Y, B)$ is a smooth map $f : X \to Y$, together with a map of sheaves of graded C_X^∞-algebras $f^* B \to A$, compatible with differentials.

If (X, A) and (Y, B) are dg manifolds, we can take their product; this is the dg manifold $(X \times Y, A \boxtimes B)$. Any ordinary manifold M is a dg manifold, where $A = C_M^\infty$.

DEFINITION 13.1.2. *Let M be a manifold. A family of free BV theories on M, parameterized by a dg manifold (X, A), is the following data.*

(1) *A dg bundle E on $(M \times X, A)$, whose differential will be denoted $Q : \mathscr{E} \to \mathscr{E}$.*

(2) *A map*
$$E \otimes_A E \to \mathrm{Dens}_M \otimes A$$
which is of degree -1, *anti-symmetric, and leads to an isomorphism*
$$\mathrm{Hom}_A(E, A) \otimes \mathrm{Dens}_M \to E$$
of sheaves of A-modules on $M \times X$.
  This pairing leads to a degree -1 *anti-symmetric pairing*
$$\langle -, - \rangle : \mathscr{E} \otimes_{\mathscr{A}} \mathscr{E} \to \mathscr{A}.$$

(3) *If $e, e' \in \mathscr{E}$,*
$$\mathrm{d}_{\mathscr{A}} \langle e, e' \rangle = \langle Qe, e' \rangle + (-1)^{|e|} \langle e, Qe' \rangle.$$

DEFINITION 13.1.3. *Let $(E, Q, \langle -, - \rangle)$ be a family of free BV theories on M parameterized by $\mathscr{A}$. A gauge fixing condition on $\mathscr{E}$ is an $\mathscr{A}$-linear differential operator*
$$Q^{GF} : \mathscr{A} \to \mathscr{A}$$
such that
$$\mathrm{D} = [Q, Q^{GF}] : \mathscr{E} \to \mathscr{E}$$
is a generalized Laplacian in the sense of (BGV92). This means that the symbol $\sigma(\mathrm{D})$ of D, is a smooth family of metrics on T^*M, parameterized by X, times the identity on E.

Throughout this section, we will fix a family of free theories on M, parameterized by $\mathscr{A}$. We will take $\mathscr{A}$ to be our base ring throughout; thus, tensor products, etc. will be taken over $\mathscr{A}$.

If $U \subset M$ is an open subset, then a family of free theories $(\mathscr{E}, Q)$ on M, parameterized by $\mathscr{A}$, restricts to one $(\mathscr{E}|_U, Q)$ on U. Similarly, a gauge fixing condition for $(\mathscr{E}, Q)$ restricts to a gauge fixing condition for $(\mathscr{E}|_U, Q)$. Thus, we have a sheaf $\mathscr{GF}_{sh}(\mathscr{E}, Q)$ of gauge fixing conditions for $(\mathscr{E}, Q)$ on M.

All of our constructions will be functorial with respect to change of the base ring $\mathscr{A}$. Note that the forms on the n-simplex $\Omega^*(\Delta^n)$ provide a cosimplicial dg manifold. Thus, the sheaf of gauge fixing conditions on $(\mathscr{E}, Q)$ over $\mathscr{A}$ has a natural enrichment to a simplicial presheaf, whose n-simplices are gauge fixing conditions on $\mathscr{E} \otimes \Omega^*(\Delta^n)$ over $\mathscr{A} \otimes \Omega^*(\Delta^n)$. This simplicial presheaf will be denote $\mathscr{GF}_{sh,\Delta}(\mathscr{E}, Q)$.

A similar remark holds for theories over $\mathscr{A}$, which we will define shortly.

13.2. If we were working on a compact manifold, one could construct the propagator for the theory using the heat kernel for D. Heat kernels on non-compact manifolds are more difficult to deal with. In Chapter 2, Section 14, we dealt with this problem by using a "fake heat kernel" to define the propagator.

A fake heat kernel is defined in Chapter 2, Section 14 to be an element
$$K_l \in \mathscr{E} \otimes_{\mathscr{A}} \mathscr{E} \otimes C^\infty((0, \infty)_l)$$

satisfying the same small l asymptotics one expects for the actual heat kernel.

Let us choose such a fake heat kernel. This allows us to define the fake propagator

$$P(\varepsilon, L) = \int_\varepsilon^L (Q^{GF} \otimes 1) K_l \mathrm{d}l \in \mathscr{E} \otimes_{\mathscr{A}} \mathscr{E}.$$

We will assume that our fake heat kernel K_l is such that the fake propagator $P(\varepsilon, L)$ is symmetric.

13.3. Let $\mathscr{E}_c$ denote the space of sections of the bundle E on $M \times X$ which are supported on a subset of M which is proper over X. Recall that in Chapter 2, Section 14 we defined a subspace

$$\operatorname{Hom}_p(\mathscr{E}_c^{\otimes_{\mathscr{A}} n}, \mathbb{R}) \subset \operatorname{Hom}(\mathscr{E}_c^{\otimes_{\mathscr{A}} n}, \mathbb{R})$$

of properly supported $E^\vee$-valued distributions on $M^n \times X$. An element of $\operatorname{Hom}(\mathscr{E}_c^{\otimes_{\mathscr{A}} n}, \mathbb{R})$ is in $\operatorname{Hom}_p(\mathscr{E}_c^{\otimes_{\mathscr{A}} n}, \mathbb{R})$ if its support $C \subset M^n \times X$ has the property that all of the projection maps $C \to M \times X$ are proper.

We will let

$$\mathscr{O}_p(\mathscr{E}_c) = \prod_{n>0} \operatorname{Hom}_p(\mathscr{E}_c^{\otimes_{\mathscr{A}} n}, \mathbb{R})_{S_n}.$$

DEFINITION 13.3.1. *A pre-theory for $(\mathscr{E}, Q)$ on M consists of a gauge fixing condition $Q^{GF} : \mathscr{E} \to \mathscr{E}$, and a collection of effective interactions*

$$I[L] \in \mathscr{O}_p^+(\mathscr{E}_c)[[\hbar]]$$

satisfying the axioms of Definition 14.6.1. That is, the renormalization group equation

$$I[L] = W\left(P(\varepsilon, L), I[\varepsilon]\right).$$

must hold, and $I[L]$ must admit a small L asymptotic expansion in terms of local action functionals.

We will let $\widetilde{\mathscr{T}}^{(n)}(\mathscr{E}, Q)$ denote the set of pre-theories defined modulo $\hbar^n$.

In Chapter 2, Section 14 we showed the following.

(1) The notion of pre-theory is independent of the choice of fake heat kernel.
(2) If $I[L]$ defines a pre-theory on M, then we can restrict $I[L]$ in a canonical way to get a pre-theory on any open subset $U \subset M$. In this way, the set of pre-theories forms a sheaf on the manifold M, which we will denote by $\widetilde{\mathscr{T}}_{sh}^{(n)}(\mathscr{E}, Q)$.
(3) The map $\widetilde{\mathscr{T}}_{sh}^{(n)}(\mathscr{E}, Q) \to \widetilde{\mathscr{T}}_{sh}^{(n-1)}(\mathscr{E}, Q)$ makes the sheaf $\widetilde{\mathscr{T}}_{sh}^{(n)}(\mathscr{E}, Q)$ into a torsor for the sheaf $\mathscr{O}_{loc}(\mathscr{E})$ of cohomological degree zero local action functionals on M.
(4) The choice of renormalization scheme gives rise to an isomorphism of sheaves

$$\widetilde{\mathscr{T}}_{sh}^{(\infty)}(\mathscr{E}) \cong \mathscr{O}_{loc}^+(\mathscr{E})[[\hbar]]$$

between the sheaf of pre-theories and the sheaf of local action functionals which are at least cubic modulo $\hbar$.

13.4. Next, we will define a subsheaf of the sheaf of pre-theories consisting of those pre-theories which satisfy the quantum master equation, i.e., which are theories.

Let us fix, as above, a fake heat kernel K_l. Unfortunately, as the fake heat kernel does not satisfy the heat equation (except in an asymptotic sense as $l \to 0$), we do not have the identity

$$(Q \otimes 1 + 1 \otimes Q)P(\varepsilon, L) = K_\varepsilon - K_L.$$

This makes it a little more tricky to define the quantum master equation than on a compact manifold. It turns out that the correct BV operator to use is not that defined by K_ε, but something slightly different.

LEMMA 13.4.1. *The limit*

$$\lim_{\varepsilon \to 0} \left(K_\varepsilon - (Q \otimes 1 + 1 \otimes Q)P(\varepsilon, L) \right)$$

exists, and is an element of $\mathscr{E} \otimes_\mathscr{A} \mathscr{E}$ (that is, it is a smooth section of $E \boxtimes_A E$ on $M^2 \times X$).

Let us denote this limit by K'_L. Note that

$$(Q \otimes 1 + 1 \otimes Q)P(\varepsilon, L) = K'_\varepsilon - K'_L.$$

Let

$$\Delta_L : \mathscr{O}(\mathscr{E}_p) \to \mathscr{O}(\mathscr{E}_p)$$

be the operator of contraction with $-K'_L$. As in Section 7, one can also define a bracket $\{-,-\}_L$ associated to the kernel K_L.

DEFINITION 13.4.2. *A functional $I \in \mathscr{O}_p(\mathscr{E})^+[[\hbar]]$ satisfies the scale L quantum master equation if*

$$QI[L] + \tfrac{1}{2}\{I[L], I[L]\} + \hbar \Delta_L I[L] = 0.$$

A pre-theory $\{I[L]\}$ is a theory if each $I[L]$ satisfies the scale L quantum master equation. As before, the renormalization group equation implies that if $I[L]$ satisfies the scale L QME for one value of L, then it does for all others.

We will let $\mathscr{T}^{(n)}(\mathscr{E}, Q)$ denote the space of theories on M.

13.5. Everything, so far, has depended on the choice of a fake heat kernel. However, we have seen in Chapter 2, Section 14 that the notion of pre-theory is independent of the choice of fake heat kernel.

Indeed, suppose $K_l, \widetilde{K}_l$ are two fake heat kernels, with associated fake propagators $P(\varepsilon, L), \widetilde{P}(\varepsilon, L)$. Suppose that $\{I[L]\}$ defines a theory for the fake propagator $P(\varepsilon, L)$. Then,

$$\widetilde{I}[L] = W\left(\widetilde{P}(0, L) - P(0, L), I[L]\right)$$

defines a theory for the fake propagator $\widetilde{P}(0,L)$. This expression is well-defined because $\widetilde{P}(0,L) - P(0,L)$ is smooth, that is, an element of $\mathscr{E} \otimes \mathscr{E}$.

LEMMA 13.5.1. *Suppose that the theory $\{I[L]\}$ satisfies the quantum master equation defined using the fake heat kernel K_l. Then $\{\widetilde{I}[L]\}$ satisfies the quantum master equation defined using the fake heat kernel $\widetilde{K}_l$.*

PROOF. Recall that we defined
$$K'_l = \lim_{\varepsilon \to 0} \left(K_\varepsilon - (Q \otimes 1 + 1 \otimes Q) P(\varepsilon, L) \right).$$
Let us define $\widetilde{K}'_l$ in the same way, using $\widetilde{K}_l$ and $\widetilde{P}(\varepsilon, L)$ in place of K_l and $P(\varepsilon, L)$.

Then,
$$(Q \otimes 1 + 1 \otimes Q)(P(0,L) - \widetilde{P}(0,L)) = \widetilde{K}'_L - K'_L.$$
Let Δ_L, $\widetilde{\Delta}_L$ denote the two BV operators. It follows from this identity that
$$[Q, \partial_{P(0,L) - \widetilde{P}(0,L)}] = \widetilde{\Delta}_L - \Delta_L$$
where, as usual, $\partial_{P(0,L) - \widetilde{P}(0,L)}$ refers to the operator on $\mathscr{O}_p(\mathscr{E}_c)$ defined by contracting with $P(0,L) - \widetilde{P}(0,L)$.

This identity says that $\partial_{P(0,L) - \widetilde{P}(0,L)}$ is a chain homotopy between the two BV operators. As in Lemma 9.2.2, it follows that if $I[L]$ satisfies the Δ_L quantum master equation, then
$$W\left(\widetilde{P}(0,L) - P(0,L), I[L]\right)$$
satisfies the $\widetilde{\Delta}_L$ quantum master equation.
□

13.6. So far, we have seen that the set of pre-theories $\widetilde{\mathscr{T}}^{(n)}(\mathscr{E}, Q)$ arises as the set of global sections of a sheaf $\widetilde{\mathscr{T}}^{(n)}_{sh}(\mathscr{E}, Q)$ on M. We have not yet shown that the set of theories forms a subsheaf. We need to show that the quantum master equation is local in nature, and thus compatible with restriction.

This we will do using the obstruction techniques of 11.

Let $\{I[L]\}$ be a theory on M, defined modulo $\hbar^{n+1}$. In Section 11, we showed how to construct an obstruction
$$O_{n+1}(\{I[L]\}) \in \mathscr{O}_{loc}(\mathscr{E})$$
to the theory $\{I[L]\}$ satisfying the quantum master equation modulo $\hbar^{n+2}$. This obstruction is of degree 1, and satisfies the equation
$$QO_{n+1}(\{I[L]\}) + \{I_0, O_{n+1}(\{I[L]\})\} = 0.$$

The formula for the obstruction given there applies in this more general setting, also, and we find, as in Section 11, that lifts of $\{I[L]\}$ to a theory

defined modulo $\hbar^{n+2}$ are the same as local action functionals $J \in \mathscr{O}_{loc}(\mathscr{E})$ such that
$$QJ + \{I_0, J\} = O_{n+1}(\{I[L]\}).$$

The cochain representative of the obstruction depends on a lift of $I[L]$ to a pre-theory $\widetilde{I}[L]$ defined modulo $\hbar^{n+2}$, even though the cohomology class of the obstruction is independent of such a lift. However, the choice of a renormalization scheme gives us a section of the map $\widetilde{\mathscr{T}}^{(n+1)} \to \widetilde{\mathscr{T}}^{(n)}$, thus giving a lift of any theory defined modulo $\hbar^{n+1}$ to a pre-theory defined modulo $\hbar^{n+2}$. When we speak about the obstruction, we will mean the obstruction associated to the choice of a renormalization scheme in this way.

The following lemma shows that the obstruction does not depend on the choice of a fake heat kernel, and is local in nature, so that the set of theories forms a subsheaf of the sheaf of pre-theories.

LEMMA 13.6.1. (1) Let $\{I[L]\}$ be a theory on M, defined modulo $\hbar^{n+1}$. Then, the restriction $\{I_U[L]\}$ of $\{I[L]\}$ to any open subset of U is a theory, that is, satisfies the quantum master equation.
(2) The restriction of the obstruction $O_{n+1}(\{I[L]\})$ to U is the obstruction to the restriction $\{I_U[L]\}$, that is,
$$O_{n+1}(\{I_U[L]\}) = O_{n+1}(\{I[L]\})|_U \in \mathscr{O}_{loc}(\mathscr{E}|_U).$$
(3) The obstruction $O_{n+1}(\{I[L]\})$ is independent of the choice of a fake heat kernel.

PROOF. We will assume, by induction, that the first statement holds: that is, that the restriction of a theory $\{I[L]\}$ defined modulo $\hbar^{n+1}$ to U is a theory on U.

Let $\widetilde{I}[L]$ denote the lift of $I[L]$ to a pre-theory defined modulo $\hbar^{n+2}$, which arises from the choice of a renormalization scheme.

Recall that we have a scale L obstruction defined by
$$O_{n+1}[L] = \hbar^{-n-1}\left(Q\widetilde{I}[L] + \{\widetilde{I}[L], \widetilde{I}[L]\}_L + \hbar\Delta_L \widetilde{I}[L]\right).$$

This scale L obstruction is such that $I_0[L] + \varepsilon O_{n+1}[L]$ satisfies the classical renormalization group equation and the classical master equation, where ε is a parameter of degree -1. The $L \to 0$ limit of this scale L obstruction exists, and is the local obstruction $O_{n+1}(\{I[L]\}) \in \mathscr{O}_{loc}(\mathscr{E})$.

Let us suppose that the pre-theory $\widetilde{I}[L]$ corresponds to a local action functional $I \in \mathscr{O}_{loc}(\mathscr{E})[\hbar]/\hbar^{n+2}$, under the bijection between pre-theories and local action functionals arising from the choice of a renormalization scheme. It follows that $\widetilde{I}[L]$ restricted to U corresponds to I restricted to U.

It is immediate from the graphical expression of $O_{n+1}[L]$ that the small L asymptotics of $O_{n+1}[L]$, when restricted to U, only depends on the small L asymptotics of the heat kernel K_L and on the $I_{i,k}$ restricted to U.

This proves the second and third statements of the lemma, and we can continue by induction. □

Thus, the set $\mathscr{T}^{(n)}(\mathscr{E}, Q)$ of theories arises as the global sections of a sheaf of theories on M, which we will denote by $\mathscr{T}_{sh}^{(n)}(\mathscr{E}, Q)$. The set of sections of this sheaf over an open subset $U \subset M$ is the set $\mathscr{T}^{(n)}(\mathscr{E}|_U, Q)$ of theories for the restriction of the free theory $(\mathscr{E}, Q)$ to U.

13.7. So far we have been fixing a base dg manifold (X, A). If $f : (Y, B) \to (X, A)$ is a map of dg manifolds, then we can pull back $\mathscr{E}$ to a dg vector bundle on $M \times (X, A)$. Global sections of this dg vector bundle are

$$f^*\mathscr{E} = \mathscr{E} \otimes_{\mathscr{A}} \mathscr{B}.$$

The sheaves of theories and pre-theories are natural with respect to these pull-backs; there are maps

$$f^* : \mathscr{T}_{sh}^{(n)}(\mathscr{E}, Q) \to \mathscr{T}_{sh}^{(n)}(f^*(\mathscr{E}, Q)).$$

This allows us to enrich our sheaf $\mathscr{T}_{sh}^{(n)}(\mathscr{E}, Q)$ into a simplicial presheaf, which we will denote by $\mathscr{T}_{sh,\Delta}^{(n)}(\mathscr{E}, Q)$. The set $\mathscr{T}_{sh,\Delta}^{(n)}(\mathscr{E}, Q)(U)[k]$ of k-simplices of this simplicial set of sections over U will be defined by

$$\mathscr{T}_{sh,\Delta}^{(n)}(\mathscr{E}, Q)(U)[k] = \mathscr{T}^{(n)}(\mathscr{E}|_U \otimes \Omega^*(\Delta^k), Q + \mathrm{d}_{dR}).$$

Here, $\mathscr{E}|_U \otimes \Omega^*(\Delta^k)$ is viewed as a module for $\mathscr{A} \otimes \Omega^*(\Delta^k)$, and we work relative to the base ring $\mathscr{A} \otimes \Omega^*(\Delta^k)$.

13.8. Let us view $\mathscr{O}_{loc}(\mathscr{E}, Q)$ as a simplicial presheaf on M, whose k-simplices, on an open subset $U \subset M$, are closed degree 0 elements of $\mathscr{O}_{loc}(\mathscr{E}|_U, \Omega^*(\Delta^k))$.

Then, the obstruction map

$$O_{n+1} : \mathscr{T}_{sh,\Delta}^{(n)}(\mathscr{E}, Q) \to \mathscr{O}_{loc}(\mathscr{E}, Q)[1]$$

is a map of simplicial presheaves.

When we talked about the simplicial set of theories on a compact manifold, we saw that the obstruction map allows us to represent $\mathscr{T}^{(n+1)}(\mathscr{E}, Q)$ as a homotopy fibre product of simplicial sets. We would like to have a similar statement in this context.

Instead of delving into the homotopy theory of simplicial presheaves – which can be somewhat technical – I will simply explain how to write $\mathscr{T}_{sh,\Delta}^{(n+1)}(\mathscr{E}, Q)$ as an actual fibre product.

Let us define a simplicial $P\mathscr{O}_{loc}(\mathscr{E}, Q)[1]$ by saying that the sections of the presheaf of k-simplices, on an open subset U, are pairs of elements

$$\alpha, \beta \in \mathscr{O}_{loc}(\mathscr{E}|_U, \Omega^*(\Delta^k))$$

such that α is of degree 0, β is of degree 1, and

$$(Q + \mathrm{d}_{dR})\alpha + \{I_0, \alpha\} = \beta.$$

There is a map
$$P\mathcal{O}_{loc}(\mathcal{E},Q)[1] \to \mathcal{O}_{loc}(\mathcal{E},Q)[1]$$
$$(\alpha,\beta) \mapsto \beta.$$

Further, $P\mathcal{O}_{loc}(\mathcal{E},Q)[1]$ is contractible.

THEOREM 13.8.1. *There is a fibre diagram of simplicial presheaves*
$$\begin{array}{ccc} \mathcal{T}^{(n+1)}_{sh,\Delta}(\mathcal{E},Q) & \longrightarrow & P\mathcal{O}_{loc}(\mathcal{E},Q)[1] \\ \downarrow & & \downarrow \\ \mathcal{T}^{(n)}(\mathcal{E},Q) & \longrightarrow & \mathcal{O}_{loc}(\mathcal{E},Q)[1] \end{array}$$

PROOF. This is an immediate corollary of the obstruction theoretic results above. □

Note that since $P\mathcal{O}_{loc}(\mathcal{E},Q)[1]$ is contractible, to show that we have a homotopy fibre diagram
$$\begin{array}{ccc} \mathcal{T}^{(n+1)}_{sh,\Delta}(\mathcal{E},Q) & \longrightarrow & 0 \\ \downarrow & & \downarrow \\ \mathcal{T}^{(n)}(\mathcal{E},Q) & \longrightarrow & \mathcal{O}_{loc}(\mathcal{E},Q)[1] \end{array}$$
it suffices to show that the map
$$P\mathcal{O}_{loc}(\mathcal{E},Q)[1] \to \mathcal{O}_{loc}(\mathcal{E},Q)[1]$$
is a fibration in the appropriate model structure.

14. Quantizing Chern-Simons theory

In this section, we will give a very simple obstruction theoretic proof of the existence of a quantization of Chern-Simons theory on any 3-manifold, modulo constants. For a more direct approach, including stronger results, see (Cos07; CM08a; Iac08). This section is written as an example in the obstruction-theoretic techniques developed in this book; I do not prove any new results about Chern-Simons theory.

14.1. Let M be a 3-manifold and let $\mathfrak{g}$ be a Lie algebra with an invariant pairing. Let us assume that $H^i(\mathfrak{g}) = 0$ if i is $1, 2$ or 4. (This is satisfied, for example, if $\mathfrak{g}$ is semi-simple).

The space of fields for Chern-Simons theory with values in $\mathfrak{g}$ is
$$\mathcal{E}_{CS}(M) = \Omega^*(M) \otimes \mathfrak{g}[1].$$
The operator $Q : \mathcal{E}_{CS}(M) \to \mathcal{E}_{CS}(M)$ arising from the quadratic part of the Chern-Simons operator is simply the de Rham differential, and the cubic

part of the action is, as usual,

$$I_{CS} = \tfrac{1}{6} \int_M \langle \alpha, [\alpha, \alpha] \rangle_{\mathfrak{g}}$$

where $\langle -, - \rangle_{\mathfrak{g}}$ refers to the chosen invariant pairing on $\mathfrak{g}$.

14.2. Recall that if we choose a Riemannian metric on M, we get an associated gauge fixing condition for Chern-Simons theory on M, where the gauge fixing operator is $Q^{GF} = d^*$.

In Section 10 we showed how to construct a simplicial set of gauge fixing conditions, and a simplicial set of theories living over the simplicial set of gauge fixing conditions. We will apply this construction to Chern-Simons theory.

Let $\text{Met}(M)$ denote the simplicial set of Riemannian metrics on M. Thus, the n-simplices are smooth families of Riemannian metrics parameterized by $\triangle^n$. Let $\mathscr{T}^{(\infty)}(\mathscr{E}_{CS}(M)) \to \text{Met}(M)$ denote the simplicial set, constructed in Section 10, of quantizations of our classical Chern-Simons theory. Thus, a 0-simplex of $\mathscr{T}^{(\infty)}(\mathscr{E}_{CS}(M))$ is a metric on M, and a quantization of Chern-Simons theory with the corresponding gauge fixing condition. An n-simplex of $\mathscr{T}^{(\infty)}(\mathscr{E}_{CS}(M))$ is a smooth family of metrics on M and a family of quantizations for this family of metrics.

In Section 11 we showed that the map

$$\mathscr{T}^{(\infty)}(\mathscr{E}_{CS}(M)) \to \text{Met}(M)$$

is a fibration of simplicial sets.

Let

$$\mathcal{CS}^{\infty}(M) = \Gamma(\text{Met}(M), \mathscr{T}^{(\infty)}(\mathscr{E}_{CS}(M)))$$

denote the simplicial set of sections of $\mathscr{T}^{(\infty)}(\mathscr{E}_{CS}(M))$.

A point in $\mathcal{CS}^{\infty}(M)$ is a "universal" quantization of Chern-Simons theory on M; that is, a quantization of Chern-Simons theory for every metric on M, which varies in a homotopically trivial fashion when the metric varies.

Let $\mathcal{CS}^n(M)$ be defined in the same way as $\mathcal{CS}^{\infty}(M)$, except that quantizations are only considered modulo $\hbar^{n+1}$.

In this section, we will prove the following result.

THEOREM 14.2.1. *For any connected 3 manifold M, there is an isomorphism of simplicial sets*

$$\mathcal{CS}^{\infty}(M) \cong \hbar H^3(\mathfrak{g})[[\hbar]]$$

where the right hand side is regarded as a constant simplicial set.

Note that $H^3(\mathfrak{g})$ is the natural home of the level of Chern-Simons theory. Indeed, if $\mathfrak{g}$ is a semi-simple Lie algebra with no Abelian factors, then $H^3(\mathfrak{g})$ is canonically isomorphic to the tangent space to the space of non-degenerate invariant bilinear forms on $\mathfrak{g}$. Thus, this result shows that specifying a quantization of Chern-Simons theory is equivalent to specifying an $\mathbb{R}[[\hbar]]$-valued invariant form on $\mathfrak{g}$, which modulo $\hbar$ is a given form; or equivalently, to specify a $\hbar$-dependent level.

This is certainly not the strongest result in this direction; see (Cos07; CM08a; Iac08) for stronger results. However, the result proved here is probably the strongest result one can prove which does not require a detailed analysis of the Feynman graphs appearing in Chern-Simons theory. The proof given below is a simple application of the obstruction theoretic techniques we developed earlier.

14.3. Let us now turn to the proof of Theorem 14.2.1. First, we will state some simple results regarding the group of obstructions to the quantization of Chern-Simons theory.

The deformation-obstruction complex for quantizing Chern-Simons theory is, as we have seen in 11, the complex $\mathscr{O}_{loc}(\mathscr{E}_{CS}(M))/\mathbb{R}$ with differential $Q + \{I_{CS}, -\}$. (Here we are working modulo constants).

In Section 6, we showed how to compute the deformation-obstruction complex for quantizing any theory using an auxiliary D_M Lie algebra, or more generally, a D_M L_∞-algebra. For Chern-Simons theory, the D_M L_∞-algebra is $J(\Omega^*(M) \otimes \mathfrak{g}))$, the D_M L_∞-algebra of jets of forms with coefficients in the Lie algebra $\mathfrak{g}$. This is quasi-isomorphic to the D_M L_∞-algebra $\mathfrak{g} \otimes C_M^\infty$, which arises from the locally constant sheaf of Lie algebras $\mathfrak{g}$.

Thus, the results of Section 11 show the following.

LEMMA 14.3.1. *There is a quasi-isomorphism*
$$(\mathscr{O}_{loc}(\mathscr{E}_{CS}(M)), Q + \{I_{CS}, -\}) \simeq \omega_M \otimes_{D_M} (C^*_{red}(\mathfrak{g}) \otimes C_M^\infty).$$

It follows from this and from our assumptions about the cohomology of $\mathfrak{g}$ that the space of obstructions to quantizing $\alpha \in \mathcal{CS}^n(M)$ to order $n+1$ is $H^1(M, H^3(\mathfrak{g}))$, and that the space of deformations of any such quantization is $H^0(M, H^3(\mathfrak{g}))$.

14.4. Now we are ready to prove 14.2.1. We will prove the result by induction. Thus, let us assume, by induction, that for all 3-manifolds M, every
$$\alpha_{k-1} \in \mathcal{CS}^{k-1}(M)$$
extends to an element
$$\alpha_k \in \mathcal{CS}^k(M).$$
To prove the theorem, we need to show that for every $\alpha_k \in \mathcal{CS}^k(M)$ the obstruction
$$O(\alpha_k) \in H^1(M, H^3(\mathfrak{g}))$$
to extending α_k to the next order vanishes.

Recall that, if $U \subset M$ is an open subset, there is a restriction map
$$\mathcal{CS}^k(M) \to \mathcal{CS}^k(U).$$

Further, this restriction map is compatible with the formation of the obstruction class. That is, the following diagram commutes:

$$\begin{array}{ccc} \mathcal{CS}^k(M) & \longrightarrow & \mathcal{CS}^k(U) \\ \downarrow & & \downarrow \\ H^1(M, H^3(\mathfrak{g})) & \longrightarrow & H^1(U, H^3(\mathfrak{g})) \end{array}$$

where the vertical arrows are the obstruction maps.

We will use this fact to deduce that the obstruction $O(\alpha_k)$ is zero.

Let us choose a basis a_i of $H_1(M, \mathbb{R})$ where each a_i is the fundamental class of an embedded circle in M. Let U_i be an embedded torus in M which supports the homology class a_i. Let $\alpha_k \mid_{U_i}$ denote the restriction of α_k to U_i. We need to show that

$$O(\alpha_k \mid_{U_i}) = 0 \in H^1(U_i, H^3(\mathfrak{g})).$$

Let us choose an embedding of U_i into $\mathbb{R}^3$. There is a commutative square

$$\begin{array}{ccc} \mathcal{CS}^k(\mathbb{R}^3) & \longrightarrow & \mathcal{CS}^k(U_i) \\ \downarrow & & \downarrow \\ \mathcal{CS}^{k-1}(\mathbb{R}^3) & \longrightarrow & \mathcal{CS}^{k-1}(U_i). \end{array}$$

The vertical maps are torsors for the same abelian group, $H^0(U_i, H^3(\mathfrak{g})) = H^0(\mathbb{R}^3, H^3(\mathfrak{g}))$. It follows that this is a fibre square. This allows us to deduce, by induction, that the map

$$\mathcal{CS}^k(\mathbb{R}^3) \to \mathcal{CS}^k(U_i)$$

is a bijection.

From this, we see that the obstruction $O(\alpha_k \mid_{U_i}) \in H^1(U_i, H^3(\mathfrak{g}))$ is the restriction of a class in $H^1(\mathbb{R}^3, H^3(\mathfrak{g}))$, and is thus zero, as desired.

This completes the proof of Theorem 14.2.1.

CHAPTER 6

Renormalizability of Yang-Mills theory

1. Introduction

In this chapter we prove the following theorem.

THEOREM 1.0.1. *Pure Yang-Mills theory on $\mathbb{R}^4$, or on any four-manifold with a flat metric, with coefficients in any semi-simple Lie algebra $\mathfrak{g}$, is perturbatively renormalizable.*

The proof is more conceptual than existing proofs in the physics literature (for instance, we don't rely on any graph combinatorics). The proof is an application of the results developed in Chapter 5, in particular of Theorem 12.9.1. To apply this result to Yang-Mills theory, we need to verify that certain cohomology groups of the complex of classical local action functionals vanish. This turns out to be a computation in Gel'fand-Fuchs cohomology. This obstruction theoretic argument is in some ways unsatisfactory; it relies on apparently fortuitous vanishing of certain Lie algebra cohomology groups (the fact that $H^5(\mathfrak{su}(3))^{\text{Out}(\mathfrak{su}(3))} = 0$ plays a crucial role). However, I don't know of any more direct argument.

If our Lie algebra $\mathfrak{g}$ is a direct sum of k simple Lie algebras, we find that there are k coupling constants. This means that the space of renormalizable quantizations of pure Yang-Mills theory we construct depends on k parameters, term-by-term in $\hbar$; the space of such quantizations is the space of series $\hbar \mathbb{R}[[\hbar]]^{\oplus k}$.

In addition, we show that the renormalizable quantizations of pure Yang-Mills theory constructed here are *universal* in the Wilsonian sense described in Chapter 4: any deformation is equivalent, in the low energy limit, to a renormalizable deformation. However, the usefulness of Wilsonian universality in this situation is counteracted by the phenomenon of asymptotic freedom, which implies that at low energy the perturbation expansion for the theory is not well-behaved.

We only need to prove the result on $\mathbb{R}^4$, as any theory on $\mathbb{R}^4$ which is invariant under Euclidean symmetries gives a theory on any four-manifold with a flat metric.

2. First-order Yang-Mills theory

We have seen in Chapter 5 how to put the usual formulation of Yang-Mills theory into the Batalin-Vilkovisky formalism. Further, we have seen that the lack of a suitable gauge fixing condition prevents us from applying our renormalization techniques to this theory. In this section we will introduce the first-order formulation of Yang-Mills theory, which resolves this problem.

I should emphasize that the first-order formulation of Yang-Mills theory is *completely equivalent* to the usual second-order formulation. In the course of this chapter we will give a careful proof of the equivalence of classical Yang-Mills theory in the first and second order formulations. The first-order formulation is a familiar device in the physics literature, where the equivalence of first and second order formulations is taken as established (see, for example, (CCRF+98)).

Let $\mathfrak{g}$ be a Lie algebra equipped with an invariant pairing. Throughout this chapter, $\Omega^i(\mathbb{R}^n)$ will denote the space of Schwartz i-forms on $\mathbb{R}^n$.

The first-order formulation of Yang-Mills theory has two fields: a connection $A \in \Omega^1(\mathbb{R}^4) \otimes \mathfrak{g}$, as before, as well as a self-dual two form $B \in \Omega^2_+(\mathbb{R}^4) \otimes \mathfrak{g}$. The Lie algebra of infinitesimal gauge symmetries is $\Omega^0(\mathbb{R}^4) \otimes \mathfrak{g}$. If $X \in \Omega^0(\mathbb{R}^4) \otimes \mathfrak{g}$, the fields A and B transform as

$$A \mapsto [X, A] + \mathrm{d}X$$
$$B \mapsto [X, B].$$

In other words, A transforms as a connection, and B transforms as a form.

The action, with coupling constant c, is

$$S(A, B) = \langle F(A), B \rangle + c \langle B, B \rangle = \langle F(A)_+, B \rangle + c \langle B, B \rangle$$

where $\langle -, - \rangle$ denotes the inner product on the space $\Omega^*(\mathbb{R}^4) \otimes \mathfrak{g}$ given by

$$\langle \omega_1 \otimes X_1, \omega_2 \otimes X_2 \rangle = \int_{\mathbb{R}^4} \omega_1 \wedge \omega_2 \langle X_1, X_2 \rangle_\mathfrak{g}.$$

The functional integral for this first-order treatment of Yang-Mills is thus

$$\int_{(A,B) \in \left(\Omega^1(\mathbb{R}^4) \otimes \mathfrak{g} \oplus \Omega^2_+(\mathbb{R}^4) \otimes \mathfrak{g}\right)/\mathscr{G}} e^{S(A,B)/\hbar}$$

where we integrate over the quotient of the space $\Omega^1(\mathbb{R}^4) \otimes \mathfrak{g} \oplus \Omega^2_+(\mathbb{R}^4) \otimes \mathfrak{g}$ of fields by the action of the gauge group $\mathscr{G} = \mathrm{Maps}(\mathbb{R}^4, G)$.

The quadratic part of the action $S(A, B)$ is

$$\langle \mathrm{d}A, B \rangle + c \langle B, B \rangle.$$

2.1. Let us now apply the Batalin-Vilkovisky machine to first-order Yang-Mills theory. The extended space of fields we end up with is described

2. FIRST-ORDER YANG-MILLS THEORY

in the following diagram:

$$\Omega^0(\mathbb{R}^4) \otimes \mathfrak{g} \qquad \text{ghosts, degree } -1$$
$$\Omega^1(\mathbb{R}^4) \otimes \mathfrak{g} \oplus \Omega^2_+(\mathbb{R}^4) \otimes \mathfrak{g} \qquad \text{fields, degree } 0$$
$$\Omega^2_+(\mathbb{R}^4) \otimes \mathfrak{g} \oplus \Omega^3(\mathbb{R}^4) \otimes \mathfrak{g} \qquad \text{anti-fields, degree } 1$$
$$\Omega^4(\mathbb{R}^4) \otimes \mathfrak{g} \qquad \text{anti-ghosts, degree } 2$$

Let us denote this extended space of fields by $\mathscr{E}$.

The graded vector space $\mathscr{E}$ is naturally an odd symplectic vector space; the ghosts pair with the anti-ghosts and the fields with the anti-fields.

To describe the action, we need names for variables living in the various direct summands of $\mathscr{E}$. Let us denote by X a ghost variable, that is, an element of $\Omega^0(\mathbb{R}^4) \otimes \mathfrak{g}$; the field variables will be denoted by $A \in \Omega^1(\mathbb{R}^4) \otimes \mathfrak{g}$ and $B \in \Omega^2_+(\mathbb{R}^4) \otimes \mathfrak{g}$; the anti-field variables are $A^\vee \in \Omega^2_+(\mathbb{R}^4) \otimes \mathfrak{g}$ and $B^\vee \in \Omega^2_+(\mathbb{R}^4) \otimes \mathfrak{g}$; and finally the anti-ghost variable $X^\vee \in \Omega^2_+(\mathbb{R}^4) \otimes \mathfrak{g}$.

Then the action, with coupling constant c, can be written

$$S_{FO}(c) = \tfrac{1}{2} \langle [X,X], X^\vee \rangle + \langle [X,B], B^\vee \rangle + \langle \mathrm{d}X, A^\vee \rangle$$
$$+ \langle [X,A], A^\vee \rangle + \langle F(A), B \rangle + c \langle B, B \rangle.$$

The subscript FO indicates "first-order". The first term in this expression gives the Lie bracket on the space of ghosts; the next three terms account for the action of this Lie algebra on the space of fields; and the final two terms give the first-order Yang-Mills action on the space of fields.

One convenient way to describe this BV action is to introduce an auxiliary differential graded algebra, $\mathscr{Y}$. We define $\mathscr{Y}$ and its differential Q by the following diagram:

$$\begin{array}{cccc} \mathscr{Y}^0 & \mathscr{Y}^1 & \mathscr{Y}^2 & \mathscr{Y}^3 \\ \| & \| & \| & \| \end{array}$$

$$\Omega^0(\mathbb{R}^4) \xrightarrow{\mathrm{d}} \Omega^1(\mathbb{R}^4) \xrightarrow{\mathrm{d}} \Omega^2_+(\mathbb{R}^4)$$
$$\oplus \quad \xrightarrow{2c\,\mathrm{Id}} \quad \oplus$$
$$\Omega^2_+(\mathbb{R}^4) \xrightarrow{\mathrm{d}} \Omega^3(\mathbb{R}^4) \xrightarrow{\mathrm{d}} \Omega^4(\mathbb{R}^4)$$

In this diagram, and throughout this chapter, we abuse notation and write

$$\mathrm{d} : \Omega^1(\mathbb{R}^4) \to \Omega^2_+(\mathbb{R}^4)$$

for the de Rham differential composed with the projection onto the summand $\Omega^2_+(\mathbb{R}^4)$ of $\Omega^2(\mathbb{R}^4)$.

The algebra structure on $\mathscr{Y}$ is as follows. The middle row of the diagram forms an algebra in a natural way: if $\alpha, \beta \in \Omega^1(\mathbb{R}^4)$, their product is

$$\pi(\alpha \wedge \beta) \in \Omega^2_+(\mathbb{R}^4)$$

where π is the projection from $\Omega^2(\mathbb{R}^4)$ onto $\Omega^2_+(\mathbb{R}^4)$. Thus, the middle row of the diagram is the quotient of the ordinary de Rham complex of $\mathbb{R}^4$ by the differential ideal consisting of $\Omega^2_-(\mathbb{R}^4)$, $\Omega^3(\mathbb{R}^4)$ and $\Omega^4(\mathbb{R}^4)$.

The bottom row of this diagram is a module over the middle row, in an evident way. This defines the commutative algebra structure.

The algebra $\mathscr{Y}$ has a trace

$$\mathrm{Tr} : \mathscr{Y} \to \mathbb{R}$$

of degree -3, defined by

$$\mathrm{Tr}(a) = \int_{\mathbb{R}^4} a$$

if $a \in \mathscr{Y}^3 = \Omega^4(\mathbb{R}^4)$, and $\mathrm{Tr}\, a = 0$ otherwise.

We can identify the space $\mathscr{Y} \otimes \mathfrak{g}[1]$ with the Batalin-Vilkovisky space of fields for the first-order formulation of Yang-Mills theory. Note that $\mathscr{Y}$ has an odd symmetric pairing defined by $\mathrm{Tr}(ab)$. Thus, $\mathscr{Y} \otimes \mathfrak{g}$ has an odd symmetric pairing defined by

$$\langle a \otimes E, a' \otimes E' \rangle = \mathrm{Tr}(aa') \langle E, E' \rangle_{\mathfrak{g}}.$$

Thus, $\mathscr{Y} \otimes \mathfrak{g}[1]$ has an odd symplectic pairing; this is the same as the Batalin-Vilkovisky odd symplectic pairing.

The general BV procedure produces an action on $\mathscr{Y} \otimes \mathfrak{g}[1]$.

LEMMA 2.1.1. *The Batalin-Vilkovisky action for first-order Yang-Mills theory is the Chern-Simons type action*

$$S_{FO}(a \otimes E) = \tfrac{1}{2} \langle a \otimes E, Q a \otimes E \rangle + \tfrac{1}{6} \langle a \otimes E, [a \otimes E, a \otimes E] \rangle.$$

This is a simple direct computation.

3. Equivalence of first-order and second-order formulations

3.1. The idea behind the equivalence between the first- and second-order formulations of Yang-Mills is very simple. In the first-order formulation, the fields in cohomological degree zero are a connection A and a self-dual two form B. If we perform the gauge invariant change of variables

$$B \mapsto B - \tfrac{1}{2c} F(A)_+$$

then the first-order action

$$\langle F(A)_+, B \rangle + c \langle B, B \rangle$$

changes to

$$-\tfrac{1}{4c} \langle F(A)_+, F(A)_+ \rangle + c \langle B, B \rangle.$$

The first term is the usual Yang-Mills action. Since this change of coordinates is upper triangular, and therefore formally preserves the measure, we can formally conclude that the theory given by the first-order action is equivalent to the theory given by the new action $-\tfrac{1}{4c} \langle F(A)_+, F(A)_+ \rangle + c \langle B, B \rangle$. The field B is non-interacting in this new action; thus, we can integrate it out to leave the field A with the usual Yang-Mills action.

3. EQUIVALENCE OF FIRST-ORDER AND SECOND-ORDER FORMULATIONS

3.2. In this section we will show how to make this argument more precise in the context of the BV formalism. Thus, let us return to considering the space $\mathscr{Y} \otimes \mathfrak{g}[1]$ of BV fields for the first-order Yang-Mills theory. As before, let us denote a ghost by $X \in \Omega^0(\mathbb{R}^4) \otimes \mathfrak{g}$, an anti-ghost by $X^\vee \in \Omega^4(\mathbb{R}^4) \otimes \mathfrak{g}$, an anti-field for B by $B^\vee \in \Omega^2_+(\mathbb{R}^4) \otimes \mathfrak{g}$, and an anti-field for A by $A^\vee \in \Omega^3(\mathbb{R}^4) \otimes \mathfrak{g}$.

Recall that we can write the second-order Yang-Mills action, in the BV formalism, with coupling constant g, as

$$S_{YM}(g) = \tfrac{1}{2} \langle [X,X], X^\vee \rangle + \langle [X,A], X^\vee \rangle + \langle \mathrm{d}X, A^\vee \rangle - \frac{1}{g} \langle F(A)_+, F(A)_+ \rangle.$$

Let

$$H = -\tfrac{1}{2c} \langle F(A)_+, B^\vee \rangle$$

In the classical BV formalism, the first-order Yang-Mills action $S_{FO}(c)$ is equivalent to the action

$$S_{FO}(c) + \varepsilon\{H, S_{FO}(c)\}.$$

Let us consider the one-parameter family of equivalent actions

$$S(t) \in \mathscr{O}_{loc}(\mathscr{Y} \otimes \mathfrak{g}[1])$$

where

$$S_0 = S_{FO}(c)$$
$$\frac{\mathrm{d}}{\mathrm{d}t} S(t) = \{H, S(t)\}.$$

LEMMA 3.2.1. *The action S_1 is*

$$S_1(X, A, B, A^\vee, B^\vee, X^\vee) = S_{YM}(4c) + c \langle B, B \rangle + \langle [X, B], B^\vee \rangle.$$

where $S_{YM}(4c)$ is the standard second-order Yang-Mills action in the BV formalism with coupling constant $4c$, as above.

PROOF. Recall that the action S can be written as

$$S_{FO}(c) = S_{Gauge} + \langle F(A)_+, B \rangle + c \langle B, B \rangle$$

where S_{Gauge} encodes the Lie bracket on the space of ghosts, as well as the action of this Lie algebra on the space of fields A and B.

Because the functional

$$H = -\tfrac{1}{2c} \langle F(A)_+, B^\vee \rangle$$

is gauge invariant,

$$\{S_{Gauge}, H\} = 0.$$

Note also that

$$\{H, \langle F(A)_+, B \rangle\} = -\tfrac{1}{2c} \langle F(A)_+, F(A)_+ \rangle$$
$$\{H, \langle B, B \rangle\} = -\tfrac{1}{c} \langle F(A)_+, B \rangle$$

Thus,

$$S_t = S_{Gauge} + a(t) \langle F(A)_+, F(A)_+ \rangle + b(t) \langle F(A)_+, B \rangle + c \langle B, B \rangle$$

where
$$\frac{d}{dt}a(t) = -\tfrac{1}{2c}b(t)$$
$$\frac{d}{dt}b(t) = -1$$
$$b(0) = 1$$
$$a(0) = 0.$$

Thus,
$$b(t) = 1 - t$$
$$a(t) = -\tfrac{1}{2c}t + \tfrac{1}{4c}t^2$$

so that
$$S_1 = S_{Gauge} - \tfrac{1}{4c}\langle F(A)_+, F(A)_+\rangle + c\langle B, B\rangle$$
$$= S_{YM}(4c) + \langle [X,B], B^\vee\rangle + c\langle B, B\rangle$$

as desired. □

3.3. Thus, we see that at the classical level, the action $S_{FO}(c)$ for the first-order formulation of Yang-Mills is equivalent, in the BV formalism, to the action $S_{YM}(4c) + \langle [X,B], B^\vee\rangle + c\langle B, B\rangle$.

People often argue, heuristically, that this equation is true at the quantum level as well, by saying that the change of coordinates we have performed is upper triangular and thus does not affect the (non-existent) "Lebesgue measure".

For us, these considerations are of no importance. The main theorem in this chapter is a classification of all possible renormalizable quantizations of classical Yang-Mills theory. The proof of this classification is of a purely cohomological nature, and works for any version of classical Yang-Mills theory. The only difficulty with the most familiar version is the lack of a suitable gauge fixing condition.

3.4. The BV formalism allows us to "integrate out" some of the fields in a theory. If we split our odd symplectic vector space as a symplectic direct sum of two such spaces, and then integrate over a Lagrangian subspace of one of them, we are left with an action functional on the other symplectic vector space.

Let us apply this procedure to integrate out the field B (and its anti-field $B^\vee$). We will see that we are left with the usual second-order formulation of Yang-Mills theory, in the BV formalism. This argument is somewhat heuristic, as we are dealing with infinite dimensional integrals. However, it is rather straightforward to see that this integrating out procedure works at the classical level, and yields an equivalence between classical Yang-Mills in its usual formulation and the classical first-order Yang-Mills we are using.

Let us write the space of fields $\mathscr{Y} \otimes \mathfrak{g}[1]$ as a direct sum

$$\mathscr{Y} \otimes \mathfrak{g}[1] = \mathscr{E}_1 \oplus \mathscr{E}_2$$

where

$$\mathscr{E}_1 = \left(\Omega^0(\mathbb{R}^4) \otimes \mathfrak{g}[1]\right) \oplus \left(\Omega^1(\mathbb{R}^4) \otimes \mathfrak{g}\right) \oplus \left(\Omega^3(\mathbb{R}^4) \otimes \mathfrak{g}[-1]\right) \oplus \left(\Omega^4(\mathbb{R}^4) \otimes \mathfrak{g}[-2]\right)$$

and

$$\mathscr{E}_2 = \left(\Omega^2_+(\mathbb{R}^4) \otimes \mathfrak{g}\right) \oplus \left(\Omega^2_+(\mathbb{R}^4) \otimes \mathfrak{g}[-1]\right).$$

This is a symplectic direct sum, and $\mathscr{E}_1$ is the space of BV fields for second-order Yang-Mills theory.

Let

$$L \subset \mathscr{E}_2$$

be the obvious Lagrangian subspace,

$$L = \Omega^2_+(\mathbb{R}^4) \otimes \mathfrak{g}.$$

Recall that the action on the space $\mathscr{Y} \otimes \mathfrak{g}[1]$ we are using is

$$S_{YM}(4c) + \langle [X, B], B^\vee \rangle + c \langle B, B \rangle.$$

Integrating out the fields in $\mathscr{E}_2$ leaves the functional on $\mathscr{E}_1$ defined by

$$(3.4.1) \qquad \hbar \log \left(\int_{B \in L} e^{S_{YM}(4c)/\hbar + c \langle B, B \rangle/\hbar} \right) = S_{YM}(4c) + C$$

where C is a constant.

Thus, we see that integrating out the field B leaves the second-order version of Yang-Mills theory in the BV formalism.

4. Gauge fixing

As we have seen, the main difficult with working with second-order Yang-Mills theory is that there seems to be no gauge fixing condition suitable for our renormalization techniques. In this section we will construct such a gauge fixing condition for first-order Yang-Mills theory.

Let

$$Q^{GF} : \mathscr{Y} \to \mathscr{Y}$$

be the differential operator defined by the diagram

$$\begin{array}{ccccc} \Omega^0(\mathbb{R}^4) & \xleftarrow{d^*} & \Omega^1(\mathbb{R}^4) & \xleftarrow{2d^*} & \Omega^2_+(\mathbb{R}^4) \\ & & \oplus & & \oplus \\ & & \Omega^2_+(\mathbb{R}^4) & \xleftarrow{2d^*} & \Omega^3(\mathbb{R}^4) & \xleftarrow{d^*} & \Omega^4(\mathbb{R}^4) \end{array}.$$

In the map $\Omega^3(\mathbb{R}^4) \to \Omega^2_+(\mathbb{R}^4)$, we are composing $2d^*$ with the projection onto $\Omega^2_+(\mathbb{R}^4)$ but, as usual, we omit this from hte notation.

Then, the Laplacian-type operator

$$\mathrm{D} = [Q, Q^{GF}]$$

can be decomposed as
$$D = D' + 4cD''$$
where both operators D' and D'' are independent of the coupling constant c. An elementary calculation shows that D' is the usual Laplacian on the spaces of forms, and that D'' is given by the diagram

degree -1	degree 0	degree 1	degree 2
$\Omega^0(\mathbb{R}^4)$	$\Omega^1(\mathbb{R}^4)$	$\Omega^2_+(\mathbb{R}^4)$	
	$d^* \uparrow$	$d^* \uparrow$	
	$\Omega^2_+(\mathbb{R}^4)$	$\Omega^3(\mathbb{R}^4)$	$\Omega^4(\mathbb{R}^4)$

Note that
$$[D', D''] = 0$$
$$(D'')^2 = 0.$$
This implies that the gauge fixing condition satisfies the technical requirements of Chapter 5, and so that there is a propagator given by the integral of a heat kernel satisfying the various axioms we need.

5. Renormalizability

We have already defined pure Yang-Mills theory in the first-order formulation at the classical level. In this section we will state the main theorem of this chapter, which allows us to construct this theory at the quantum level.

Let us split up the first-order action S_{FO} as a sum
$$S_{FO}(a \otimes E) = \langle a \otimes E, Qa \otimes E \rangle + I^{(0)}(a \otimes E)$$
where, as before, Q is described by the diagram

$$\Omega^0(\mathbb{R}^4) \xrightarrow{d} \Omega^1(\mathbb{R}^4) \xrightarrow{d} \Omega^2_+(\mathbb{R}^4)$$
$$\oplus \quad \nearrow^{2c\,\mathrm{Id}} \quad \oplus$$
$$\Omega^2_+(\mathbb{R}^4) \xrightarrow{d} \Omega^3(\mathbb{R}^4) \xrightarrow{d} \Omega^4(\mathbb{R}^4).$$

The functional
$$I^{(0)}(a \otimes E) \in \mathscr{O}_{loc}(\mathscr{Y} \otimes \mathfrak{g}[1])$$
is the interacting part of the action. $I^{(0)}$ is cubic, of cohomological degree zero, and satisfies
$$QI^{(0)} + \tfrac{1}{2}\{I^{(0)}, I^{(0)}\} = 0.$$
We would like to turn this into a quantum theory. At the classical level, Yang-Mills theory is conformally invariant. Thus, $I^{(0)} \in \mathscr{M}^{(0)}$. We would like to classify all lifts of $I^{(0)}$ to elements of $\mathscr{R}^{(\infty)}$, that is, to a relevant quantum theory.

Let
$$H^{i,j}(\mathcal{O}_{loc}(\mathscr{Y} \otimes \mathfrak{g}[1])^{\mathbb{R}^4})$$
denote the cohomology of the complex $\mathcal{O}_{loc}(\mathscr{Y} \otimes \mathfrak{g}[1])^{\mathbb{R}^4}$ of translation invariant local action functionals, with differential $Q + \{I^{(0)}, -\}$, in cohomological degree i and scaling dimension j.

Suppose we have an element $X^{(n)} \in \mathscr{R}^{(n)}$ which is a lift of the classical theory to a theory defined modulo $\hbar^{n+1}$. Theorem 12.9.1 implies that the obstruction to lifting $X^{(n)}$ to $\mathscr{R}^{(n+1)}$ is an element of $H^{1,\geq 0}(\mathcal{O}_{loc}(\mathscr{Y} \otimes \mathfrak{g}[1])^{\mathbb{R}^4})$. If the obstruction vanishes, the moduli space of lifts up to equivalence is a quotient of $H^{0,\geq 0}(\mathcal{O}_{loc}(\mathscr{Y} \otimes \mathfrak{g}[1])^{\mathbb{R}^4})$ by some action of the space $H^{-1,\geq 0}(\mathcal{O}_{loc}(\mathscr{Y} \otimes \mathfrak{g}[1])^{\mathbb{R}^4})$. If $H^{-1,\geq 0}(\mathcal{O}_{loc}(\mathscr{Y} \otimes \mathfrak{g}[1])^{\mathbb{R}^4})$ also vanishes, then the space of lifts up to equivalence is isomorphic to $H^{0,\geq 0}(\mathcal{O}_{loc}(\mathscr{Y} \otimes \mathfrak{g}[1])^{\mathbb{R}^4})$.

Thus, the key question in understanding renormalizability of Yang-Mills is the computation of the cohomology groups $H^{i,j}(\mathcal{O}_{loc}(\mathscr{Y} \otimes \mathfrak{g}[1])^{\mathbb{R}^4})$ when $j \geq 0$ and $i = -1, 0, 1$. Since, at the classical level, the theory is invariant under $SO(4)$, we can restrict ourselves to considering quantizations which are also $SO(4)$-invariant. Thus, we need only compute the groups $H^{i,j}\left(\mathcal{O}_{loc}(\mathscr{Y} \otimes \mathfrak{g}[1])^{\mathbb{R}^4 \ltimes SO(4)}\right)$ when $j \geq 0$ and $i = -1, 0, 1$.

THEOREM 5.0.1. *Let $\mathfrak{g}$ be a semi-simple Lie algebra. For any non-zero value of the coupling constant c, there are natural isomorphisms*

$$H^{i,0}\left(\mathcal{O}_{loc}(\mathscr{Y} \otimes \mathfrak{g}[1])^{\mathbb{R}^4 \ltimes SO(4)}\right) = \begin{cases} 0 & \text{if } i < 0 \\ H^0(\mathfrak{g}, \operatorname{Sym}^2 \mathfrak{g}) & \text{if } i = 0 \\ H^5(\mathfrak{g}) & \text{if } i = 1 \end{cases}$$

Further,

$$H^{i,j}\left(\mathcal{O}_{loc}(\mathscr{Y} \otimes \mathfrak{g}[1])^{\mathbb{R}^4 \ltimes SO(4)}\right) = 0 \text{ if } j > 0 \text{ and } i \leq 2.$$

The proof of this theorem will occupy most of this chapter.

I should remark that the same result holds if we replace $\mathscr{Y} \otimes \mathfrak{g}[1]$ by the space of BV fields for second-order Yang-Mills theory. This is because the theorem is a calculation of the cohomology of the deformation complex for the classical theory, and first-order and second-order Yang-Mills are equivalent at the classical level and so have quasi-isomorphic deformation complexes.

5.1. Since the cohomology groups $H^{0,j}\left(\mathcal{O}_{loc}(\mathscr{Y} \otimes \mathfrak{g}[1])^{\mathbb{R}^4 \ltimes SO(4)}\right)$ vanish if $j > 0$, we see that any relevant lift $X^{(n)} \in \mathscr{R}^{(n)}$ of classical Yang-Mills is equivalent to a marginal lift $Y^{(n)} \in \mathscr{M}^{(n)}$. Hence we will consider only marginal lifts.

However, there are potential obstructions to constructing a marginal lift, lying in the Lie algebra cohomology group

$$H^5(\mathfrak{g}) = H^{1,0}\left(\mathscr{O}_{loc}(\mathscr{Y} \otimes \mathfrak{g}[1])^{\mathbb{R}^4 \rtimes SO(4)}\right).$$

Unfortunately this group is non-zero if the semi-simple Lie algebra $\mathfrak{g}$ contains a factor of $\mathfrak{su}(n)$ where $n \geq 3$.

We can ensure that such obstructions must vanish by asking that our quantizations respect an additional symmetry. Let

$$H \subset \operatorname{Out} \mathfrak{g}$$

be the group of outer automorphisms of $\mathfrak{g}$ which preserve the decomposition

$$\mathfrak{g} = \mathfrak{g}_1 \oplus \mathfrak{g}_2 \oplus \cdots \oplus \mathfrak{g}_k$$

of $\mathfrak{g}$ into simple factors. The classical Yang-Mills action $I^{(0)}$ is invariant under the action of H. Thus, we can ask for quantizations which are invariant under H; if the quantization $X^{(n)} \in \mathscr{M}^{(n)}$ is H-invariant, then the obstruction

$$O_{n+1}(X^{(n)}) \in H^{i,0}\left(\mathscr{O}_{loc}(\mathscr{Y} \otimes \mathfrak{g}[1])^{\mathbb{R}^4 \rtimes SO(4)}\right) = H^5(\mathfrak{g})$$

will also be H-invariant.

LEMMA 5.1.1. *For any semi-simple Lie algebra* $\mathfrak{g}$,

$$H^5(\mathfrak{g})^H = 0.$$

Thus, if $X^{(n)}$ is H-invariant, the obstruction $O_{n+1}(X^{(n)})$ vanishes.

PROOF OF LEMMA. If $\mathfrak{g} = \mathfrak{g}_1 \oplus \mathfrak{g}_2 \oplus \cdots \oplus \mathfrak{g}_k$ is the decomposition of $\mathfrak{g}$ into simple factors, then

$$H^5(\mathfrak{g}) = \oplus_i H^5(\mathfrak{g}_i).$$

Thus, it suffices to prove that for any simple Lie algebra $\mathfrak{g}$,

$$H^5(\mathfrak{g})^{\operatorname{Out}(\mathfrak{g})} = 0.$$

Recall that if G is the compact Lie group associated to $\mathfrak{g}$,

$$H^*(\mathfrak{g}) = H^*(G, \mathbb{R}).$$

If $\mathfrak{g} \neq \mathfrak{su}(n)$, then standard results on the cohomology of compact Lie groups imply that $H^5(\mathfrak{g}) = 0$. The only problems arise when $\mathfrak{g} = \mathfrak{su}(n)$ when $n \geq 3$.

If $n > 3$, the map $SU(3) \to SU(n)$ induces an isomorphism on H^5. Further,

$$\operatorname{Out}(SU(n)) = \operatorname{Out}(\mathfrak{su}(n)) = \mathbb{Z}/2,$$

and this group acts by taking a unitary matrix $A \in SU(n)$ to its complex conjugate $\overline{A}$. Therefore the map $SU(3) \to SU(n)$ is equivariant for the action of $\mathbb{Z}/2$.

Thus, what we need to show is that

$$H^5(SU(3))^{\mathbb{Z}/2} = 0.$$

The fundamental class in $H^8(SU(3))$ is the cup product of a generator of $H^3(SU(3))$ with a generator of $H^5(SU(3))$. The $\mathbb{Z}/2$ action on $H^3(SU(3))$ is trivial; thus, to show that $H^5(SU(3))^{\mathbb{Z}/2} = 0$, it suffices to show that $H^8(SU(3))^{\mathbb{Z}/2} = 0$; or, in other words, that the non-trivial element of $\mathbb{Z}/2$ acts on $SU(3)$ in an orientation reversing way.

Thus, it suffices to show that the map
$$\mathfrak{su}(3) \to \mathfrak{su}(3)$$
$$A \mapsto \overline{A}$$
is orientation reversing. This is a simple computation. $\square$

As a corollary of this lemma and of Theorem 5.0.1, we find the following. Let
$$\mathscr{M}_{YM}^{(\infty)} \subset \mathscr{M}^{(\infty)}$$
$$\mathscr{R}_{YM}^{(\infty)} \subset \mathscr{R}^{(\infty)}$$
denote the sub-simplicial set of marginal (respectively, relevant) theories which coincide at the classical level with Yang-Mills theory.

COROLLARY 5.1.2. *The inclusion*
$$\mathscr{M}_{YM}^{(\infty)} \to \mathscr{R}_{YM}^{(\infty)}$$
is an isomorphism on π_0.

There is a (non-canonical) bijection
$$\pi_0\left(\mathscr{M}_{YM}^{(\infty)}\right) \cong H^0(\mathfrak{g}, \operatorname{Sym}^2 \mathfrak{g}) \otimes \hbar \mathbb{R}[[\hbar]].$$

Thus, the set of renormalizable quantizations of pure Yang-Mills theory is the set of deformations of the chosen pairing on $\mathfrak{g}$ to a symmetric invariant pairing
$$\mathfrak{g} \otimes \mathfrak{g} \to \mathbb{R}[[\hbar]]$$
which, modulo $\hbar$, is the original pairing.

6. Universality

The quantizations of Yang-Mills constructed above are universal: any other quantization of Yang-Mills is equivalent, in the low-energy limit, to one of the quantizations constructed in Theorem 5.1.2.

The precise statement is the following.

THEOREM 6.0.1. *Let*
$$I \in \left(\mathscr{T}_{YM}^{(\infty)}\right)^{\mathbb{R}^4 \rtimes SO(4)}$$

be any quantization of first-order Yang-Mills theory which is invariant under the Euclidean symmetries of $\mathbb{R}^4$. Then there is a theory I' which is equivalent to I – that is, in the same connected component of the simplicial set $\left(\mathscr{T}_{YM}^{(\infty)}\right)^{\mathbb{R}^4 \rtimes SO(4)}$) – and which satisfies the following.

(1) I' has only marginal and irrelevant terms, that is, for all L,
$$\mathcal{RG}_l(I'[L]) \in \mathscr{O}(\mathscr{Y} \otimes \mathfrak{g}[1])[[\hbar]] \otimes \mathbb{R}[l^{-1}, \log l]$$
as a function of l.

(2) There is a marginal theory $J \in \mathscr{M}_{YM}^{(\infty)}$ which is in the same universality class as I', that is,
$$\mathcal{RG}_l(I'[L]) - \mathcal{RG}_l(I[L]) \in \mathscr{O}(\mathscr{Y} \otimes \mathfrak{g}[1])[[\hbar]] \otimes l^{-1}\mathbb{R}[l^{-1}, \log l],$$
so that $\mathcal{RG}_l(I'[L]) - \mathcal{RG}_l(I[L])$ tends to zero as $l \to \infty$.

Further, the marginal theory J is uniquely determined (up to contractible choice) by these properties.

PROOF. The first statement follows immediately from the fact that the cohomology of $\mathscr{O}_{loc}(\mathscr{Y} \otimes \mathfrak{g}[1])^{\mathbb{R}^4 \rtimes SO(4)}$ vanishes in positive scaling dimension.

The second statement is also straightforward. We have
$$\mathcal{RG}_l(I'[L]) \in \mathscr{O}(\mathscr{Y} \otimes \mathfrak{g}[1])[[\hbar]] \otimes \mathbb{R}[l^{-1}, \log l].$$
Define $J_l[L]$ by discarding those terms of $\mathcal{RG}_l(I'[L])$ which have negative powers of l. Then,
$$J_l[L] \in \mathscr{O}(\mathscr{Y} \otimes \mathfrak{g}[1])[[\hbar]] \otimes \mathbb{R}[l^{-1}, \log l].$$
It is straightforward to check that $J_l[L]$ satisfies the renormalization group equations and the quantum master equation:
$$W(P(\varepsilon, L), J_l[\varepsilon]) = J_l[L]$$
$$\mathcal{RG}_m(J_l[L]) = J_{lm}[L]$$
$$(Q + \hbar \Delta_L) e^{J_l[L]/\hbar} = 0.$$
These three equations are simple consequences of the corresponding equations for $\mathcal{RG}_*(I'[L])$.

Thus, we let
$$J[L] = J_1[L].$$
This set of effective interactions defines a marginal theory in the same universality class as I', as desired. □

7. Cohomology calculations

In this section we will prove Theorem 5.0.1.
Let
$$Y = (\mathscr{Y})^{\mathbb{R}^4}$$
so that
$$\mathscr{Y} = Y \otimes \mathscr{S}(\mathbb{R}^4).$$

7. COHOMOLOGY CALCULATIONS

Explicitly,
$$Y = \begin{cases} \Omega^0 & \text{degree } 0 \\ \Omega^1 \oplus \Omega^2_+ & \text{degree } 1 \\ \Omega^2_+ \oplus \Omega^3 & \text{degree } 2 \\ \Omega^4 & \text{degree } 3 \end{cases}$$

where Ω^i refers to the space of translation invariant forms on $\mathbb{R}^4$.

The space
$$\mathscr{O}_{loc}(\mathscr{Y} \otimes \mathfrak{g}[1])^{\mathbb{R}^4}$$
of translation invariant functionals on $\mathscr{Y} \otimes \mathfrak{g}[1]$ is an odd Lie algebra under the Batalin-Vilkovisky bracket. Let
$$Q : \mathscr{O}_{loc}(\mathscr{Y} \otimes \mathfrak{g}[1])^{\mathbb{R}^4} \to \mathscr{O}_{loc}(\mathscr{Y} \otimes \mathfrak{g}[1])^{\mathbb{R}^4}$$
be the differential coming from the differential on $\mathscr{Y}$, and let
$$I^{(0)} \in \mathscr{O}_{loc}(\mathscr{Y} \otimes \mathfrak{g}[1])^{\mathbb{R}^4}$$
be the interaction. This satisfies the classical master equation
$$QI^{(0)} + \frac{1}{2}\{I^{(0)}, I^{(0)}\} = 0.$$

Thus, $Q + \{I^{(0)}, -\}$ defines a differential on $\mathscr{O}_{loc}(\mathscr{Y} \otimes \mathfrak{g}[1])^{\mathbb{R}^4}$.

We are interested in the complex $\mathscr{O}_{loc}(\mathscr{Y} \otimes \mathfrak{g}[1])^{\mathbb{R}^4}$ with respect to the differential $Q + \{I^{(0)}, -\}$. This complex $\mathscr{O}_{loc}(\mathscr{Y} \otimes \mathfrak{g}[1])^{\mathbb{R}^4}$ is bigraded; the first grading is the usual cohomological grading, the second is by scaling dimension. Let
$$H^{i,j}(\mathscr{O}_{loc}(\mathscr{Y} \otimes \mathfrak{g}[1])^{\mathbb{R}^4}, Q + \{I^{(0)}, -\})$$
denote the cohomology in cohomological degree i and scaling dimension j.

Let $\widehat{\mathscr{Y}}$ denote the formal completion of the dga $\mathscr{Y}$ at 0. Thus,
$$\widehat{\mathscr{Y}} = Y[[x_1, x_2, x_3, x_4]]$$
with the natural differential Q.

Observe that $\widehat{\mathscr{Y}}$ is acted on by the algebra $\mathbb{R}[\partial_1, \ldots, \partial_4]$. The generators ∂_i act by derivations. In a similar way, any tensor power of $\widehat{\mathscr{Y}}$ is acted on by $\mathbb{R}[\partial_1, \ldots, \partial_4]$.

The following lemma is a special case of Lemma 6.7.1 in Chapter 5, Section 6.

LEMMA 7.0.1. *There is an isomorphism of complexes*
$$\mathscr{O}_{loc}(\mathscr{Y} \otimes \mathfrak{g}[1])^{\mathbb{R}^4} \cong C^*_{red}(\widehat{\mathscr{Y}} \otimes \mathfrak{g}) \otimes^{\mathbb{L}}_{\mathbb{R}[\partial_1, \ldots, \partial_4]} \mathbb{R}.$$
Here C^*_{red} denotes the reduced Gel'fand-Fuchs cohomology. The action of $\mathbb{R}[\partial_1, \ldots, \partial_4]$ on $\widehat{\mathscr{Y}} \otimes \mathfrak{g}$ is the obvious one; the action on $\mathbb{R}$ is the one where each ∂_i acts trivially.

Further, the subcomplex of $\mathscr{O}_{loc}(\mathscr{Y} \otimes \mathfrak{g}[1])^{\mathbb{R}^4}$ of scaling dimension k corresponds to the subcomplex of $C^*_{red}(\widehat{\mathscr{Y}} \otimes \mathfrak{g})_{\mathbb{R}^4}$ of scaling dimension $k - 4$.

We will apply this to prove Theorem 5.0.1, which is restated here.

THEOREM. *If the Lie algebra $\mathfrak{g}$ is semi-simple, then*
$$H^{i,0}(\mathscr{O}_{loc}(\mathscr{Y} \otimes \mathfrak{g}[1])^{\mathbb{R}^4}, Q + \{I^{(0)}, -\})^{SO(4)} = \begin{cases} 0 & \text{if } i < 0 \\ H^0(\mathfrak{g}, \text{Sym}^2 \mathfrak{g}) & \text{if } i = 0 \\ H^5(\mathfrak{g}) & \text{if } i = 1 \end{cases}$$

Further,
$$H^{i,j}(\mathscr{O}_{loc}(\mathscr{Y} \otimes \mathfrak{g}[1])^{\mathbb{R}^4}, Q + \{I^{(0)}, -\})^{SO(4)} = 0 \text{ if } j > 0 \text{ and } i \leq 2.$$

All these isomorphisms are compatible with the action of the group of outer automorphisms of $\mathfrak{g}$ which preserve the chosen invariant pairing on $\mathfrak{g}$.

This is the key result that allows us to prove renormalizability of Yang-Mills. Again, this theorem holds only for fixed non-zero values of the coupling constant c.

PROOF. There is a spectral sequence converging to the cohomology of $\mathbb{R} \otimes^{\mathbb{L}}_{\mathbb{R}[\partial_1,\ldots,\partial_4]} C^*_{red}(\widehat{\mathscr{Y}} \otimes \mathfrak{g})$ whose first term is $\mathbb{R} \otimes^{\mathbb{L}}_{\mathbb{R}[\partial_1,\ldots,\partial_4]} H^*_{red}(\widehat{\mathscr{Y}} \otimes \mathfrak{g})$. Thus, the next step is to compute the reduced Lie algebra cohomology $H^*_{red}(\widehat{\mathscr{Y}} \otimes \mathfrak{g})$. We are interested, ultimately, only in the cohomology of $\mathscr{O}_{loc}(\mathscr{Y} \otimes \mathfrak{g}[1])^{\mathbb{R}^4}$ in non-negative scaling dimension. This implies that we only need to compute the cohomology groups $H^*_{red}(\widehat{\mathscr{Y}} \otimes \mathfrak{g})$ in scaling dimension ≥ -4.

Thus, let $H^{i,j}_{red}(\widehat{\mathscr{Y}} \otimes \mathfrak{g})$ denote the ith reduced Lie algebra cohomology group in scaling dimension j.

LEMMA 7.0.2.
$$H^{i,0}_{red}(\widehat{\mathscr{Y}} \otimes \mathfrak{g}) = H^i_{red}(\mathfrak{g})$$
$$H^{i,-1}_{red}(\widehat{\mathscr{Y}} \otimes \mathfrak{g}) = 0$$
$$H^{i,-2}_{red}(\widehat{\mathscr{Y}} \otimes \mathfrak{g}) = 0$$
$$H^{i,-3}_{red}(\widehat{\mathscr{Y}} \otimes \mathfrak{g}) = 0$$
$$H^{i,-4}_{red}(\widehat{\mathscr{Y}} \otimes \mathfrak{g}) = H^i\left(\mathfrak{g}, \text{Sym}^2\left(\mathfrak{g}^\vee \otimes \wedge^2 \mathbb{R}^4\right)\right)$$

All isomorphisms are $\text{Aut}(\mathfrak{g}) \times SO(4)$ equivariant.

PROOF. Let
$$\widehat{\mathscr{Y}}(k) \subset \widehat{\mathscr{Y}}$$
be the finite-dimensional subspace consisting of elements of scaling dimension k. Then, $\mathscr{Y}$ is a direct product
$$\widehat{\mathscr{Y}} = \prod_{k \geq 0} \widehat{\mathscr{Y}}(k).$$
Also,
$$\widehat{\mathscr{Y}}(0) = \mathbb{R}.$$

7. COHOMOLOGY CALCULATIONS

A simple computation shows that
$$H^*(\widehat{\mathscr{Y}}(1)) = 0$$
$$H^*(\widehat{\mathscr{Y}}(2)) = \wedge^2 \mathbb{R}^4[-1].$$

(This second equation only holds when the coupling constant c is non-zero).

We will first compute the ordinary (non-reduced) Lie algebra cohomology $H^*(\widehat{\mathscr{Y}} \otimes \mathfrak{g})$. The reduced Lie algebra cohomology is obtained by modifying this in a simple way.

Let
$$C^*(\widehat{\mathscr{Y}} \otimes \mathfrak{g})(k) \subset C^*(\widehat{\mathscr{Y}} \otimes \mathfrak{g})$$
be the subcomplex of scaling dimension k. There is a direct product decomposition
$$C^*(\widehat{\mathscr{Y}} \otimes \mathfrak{g}) = \prod_{k \le 0} C^*(\widehat{\mathscr{Y}} \otimes \mathfrak{g})(k)$$
of complexes. Each subcomplex $C^*(\widehat{\mathscr{Y}} \otimes \mathfrak{g})(k)$ is finite dimensional.

Recall that $\widehat{\mathscr{Y}}(0) \otimes \mathfrak{g} = \mathfrak{g}$, and $\widehat{\mathscr{Y}}(k) = 0$ if $k < 0$. Let
$$\mathscr{Y}_+ \subset \widehat{\mathscr{Y}}$$
be the subspace consisting of elements of scaling dimension > 0.

Observe that
$$C^*(\widehat{\mathscr{Y}} \otimes \mathfrak{g})(k) = \operatorname{Sym}^*(\mathfrak{g}^*[-1]) \otimes C^*(\widehat{\mathscr{Y}}_+ \otimes \mathfrak{g})(k)$$
$$= \operatorname{Sym}^*(\mathfrak{g}^*[-1]) \otimes \operatorname{Sym}^*(\widehat{\mathscr{Y}}_+ \otimes \mathfrak{g}[1])^\vee(k).$$

Filter the complex $C^*(\widehat{\mathscr{Y}} \otimes \mathfrak{g})(k)$ by saying
$$F^r C^*(\widehat{\mathscr{Y}} \otimes \mathfrak{g})(k) = \operatorname{Sym}^*(\mathfrak{g}^*[-1]) \otimes \operatorname{Sym}^{\ge r}(\widehat{\mathscr{Y}}_+ \otimes \mathfrak{g}[1])^\vee(k).$$

The filtration is finite:
$$C^*(\widehat{\mathscr{Y}} \otimes \mathfrak{g})(k) = F^0 C^*(\widehat{\mathscr{Y}} \otimes \mathfrak{g})(k) \supset \cdots \supset F^{k+1} C^*(\widehat{\mathscr{Y}} \otimes \mathfrak{g})(k) = 0.$$

Thus, we find a spectral sequence converging to the cohomology of the complex $C^*(\widehat{\mathscr{Y}} \otimes \mathfrak{g})(k)$ whose first term is
$$H^*(\mathfrak{g}, \operatorname{Sym}^*(H^*(\widehat{\mathscr{Y}}_+) \otimes \mathfrak{g}[1])^\vee(k)).$$

The space $\operatorname{Sym}^*\left(\prod_{k>0} H^*(\widehat{\mathscr{Y}}(k)) \otimes \mathfrak{g}[1]\right)^\vee$ is viewed simply as a module for the Lie algebra $\mathfrak{g}$.

Putting these spectral sequences together for varying k, we find a spectral sequence converging to $H^*(\widehat{\mathscr{Y}} \otimes \mathfrak{g})$ whose first term is
$$H^*\left(\mathfrak{g}, \operatorname{Sym}^*\left(\prod_{k>0} H^*(\widehat{\mathscr{Y}}(k)) \otimes \mathfrak{g}[1]\right)^\vee\right).$$

We can do a little better, because
$$H^*(\widehat{\mathscr{Y}}(1)) = 0.$$

Also, because $\mathfrak{g}$ is semi-simple,
$$H^*(\mathfrak{g}, \mathfrak{g}) = H^*(\mathfrak{g}, \mathfrak{g}^\vee) = 0$$
so that
$$H^*\left(\mathfrak{g}, \left(\prod_{k>0} H^*(\widehat{\mathscr{Y}}(k)) \otimes \mathfrak{g}[1]\right)^\vee\right) = H^*(\mathfrak{g}, \mathfrak{g}^\vee) \otimes \prod_{k>0} H^*(\widehat{\mathscr{Y}}(k))[-1] = 0.$$

Thus, the first term of the spectral sequence can be rewritten as
$$H^*\left(\mathfrak{g}, \mathbb{R} \oplus \mathrm{Sym}^{\geq 2}\left(\prod_{k\geq 2} H^*(\widehat{\mathscr{Y}}(k)) \otimes \mathfrak{g}[1]\right)^\vee\right).$$

Now
$$\mathrm{Sym}^{\geq 2}\left(\prod_{k\geq 2} H^*(\widehat{\mathscr{Y}}(k)) \otimes \mathfrak{g}[1]\right)^\vee$$
consists entirely of terms of scaling dimension ≤ -4, and the scaling dimension -4 part is $\mathrm{Sym}^2(H^*(\widehat{\mathscr{Y}}(2)) \otimes \mathfrak{g}[1])^\vee$.

As before, let $H^{i,j}(\widehat{\mathscr{Y}} \otimes \mathfrak{g})$ denote the ith Lie algebra cohomology in scaling dimension j. We see that
$$H^{i,j}(\widehat{\mathscr{Y}} \otimes \mathfrak{g}) = \begin{cases} H^i(\mathfrak{g}) & \text{if } j = 0 \\ 0 & \text{if } j = -1, -2, -3 \\ H^i(\mathfrak{g}, \mathrm{Sym}^2(H^*(\widehat{\mathscr{Y}}(2)) \otimes \mathfrak{g}[1])^\vee) & \text{if } j = -4. \end{cases}$$

If we take reduced Lie algebra cohomology, we find the same expression except that $H^*(\mathfrak{g})$ is replaced by $H^*_{red}(\mathfrak{g})$.

The final thing we need is that
$$H^*(\widehat{\mathscr{Y}}(2)) = \wedge^2 \mathbb{R}^4[-1].$$

This equation, which only holds when the coupling constant c appearing in the differential on $\mathscr{Y}$ is non-zero, is straightforward to prove. $\square$

Next, we should look again at the complex $\mathbb{R} \otimes^{\mathrm{L}}_{\mathbb{R}[\partial_1,\ldots,\partial_4]} C^*_{red}(\widehat{\mathscr{Y}} \otimes \mathfrak{g})$. This complex can be identified explicitly as
$$\oplus \wedge^i \mathbb{R}^4 \otimes C^*_{red}(\widehat{\mathscr{Y}} \otimes \mathfrak{g})[i]$$
with a differential which is a sum of the differential on $C^*_{red}(\widehat{\mathscr{Y}} \otimes \mathfrak{g})$ and maps
$$\wedge^i \mathbb{R}^4 \otimes C^*_{red}(\widehat{\mathscr{Y}} \otimes \mathfrak{g}) \to \wedge^{i-1} \mathbb{R}^4 \otimes C^*_{red}(\widehat{\mathscr{Y}} \otimes \mathfrak{g})$$
arising from the action of the Lie algebra $\mathbb{R}^4$ on $C^*_{red}(\widehat{\mathscr{Y}} \otimes \mathfrak{g})$, via the operators ∂_i.

7. COHOMOLOGY CALCULATIONS

We will use the spectral sequence for the cohomology of this double complex, whose first term is

$$\oplus \wedge^i \mathbb{R}^4 \otimes H^*_{red}(\widehat{\mathscr{Y}} \otimes \mathfrak{g})[i].$$

We are only interested in the part of scaling dimension ≥ -4. Since $\wedge^i \mathbb{R}^4$ comes with scaling dimension $-i$, we are interested in the subcomplex given by

$$\oplus_{j-i \geq -4} \wedge^i \mathbb{R}^4 \otimes H^{*,j}_{red}(\widehat{\mathscr{Y}} \otimes \mathfrak{g})[i].$$

Lemma 7.0.2 implies that this can be rewritten as

$$\left(\wedge^0 \mathbb{R}^4 \otimes H^*\left(\mathfrak{g}, \operatorname{Sym}^2\left(\mathfrak{g}^\vee \otimes \wedge^2 \mathbb{R}^4\right)\right)\right) \oplus \bigoplus_{i \geq 0} \left(\wedge^i \mathbb{R}^4 \otimes H^*_{red}(\mathfrak{g})[i]\right).$$

The first summand is in scaling dimension -4, the remaining summands are in scaling dimension between 0 and -4.

We would like to compute the differentials of the spectral sequence. All differentials preserve scaling dimension; and, if we assume that all previous differentials are zero, the differential on the kth page maps

$$\wedge^i \mathbb{R}^4 \otimes H^*_{red}(\widehat{\mathscr{Y}} \otimes \mathfrak{g})[i] \to \wedge^{i-k} \mathbb{R}^4 \otimes H^*_{red}(\widehat{\mathscr{Y}} \otimes \mathfrak{g})[i-k].$$

From this, we see that the only possibly non-zero differential, in scaling dimension ≥ -4, is on the fourth page of the scaling dimension -4 part. This differential is a map

$$d_4 : \wedge^4 \mathbb{R}^4 \otimes H^*_{red}(\mathfrak{g})[4] \to H^*\left(\mathfrak{g}, \operatorname{Sym}^2\left(\mathfrak{g}^\vee \otimes \wedge^2 \mathbb{R}^4\right)\right).$$

of cohomological degree one. After this, all spectral sequence differentials must vanish.

We are ultimately interested in the $SO(4)$-invariant part of the cohomology. Thus, we will calculate the cohomology of d_4 after taking $SO(4)$-invariants.

The following lemma completes the proof of the theorem.

LEMMA 7.0.3. *The differential* d_4 *on the fourth page of the spectral sequence*

$$d_4 : \wedge^4 \mathbb{R}^4 \otimes H^3_{red}(\mathfrak{g}) \to H^0\left(\mathfrak{g}, \operatorname{Sym}^2\left(\mathfrak{g}^\vee \otimes \wedge^2 \mathbb{R}^4\right)\right)^{SO(4)}$$

is injective. Further, the cokernel of this map is naturally isomorphic to

$$H^0(\mathfrak{g}, \operatorname{Sym}^2 \mathfrak{g}^\vee).$$

PROOF. Let us write

$$\mathfrak{g} = \mathfrak{g}_1 \oplus \cdots \oplus \mathfrak{g}_k$$

as a direct sum of simple Lie algebras. Note that

$$H^3(\mathfrak{g}) = \oplus H^3(\mathfrak{g}_i)$$

and

$$H^0(\mathfrak{g}, \mathfrak{g} \otimes \mathfrak{g}) = \oplus H^0(\mathfrak{g}_i, \mathfrak{g}_i^\vee \otimes \mathfrak{g}_i^\vee).$$

Since all constructions we are doing are functorial in the Lie algebra $\mathfrak{g}$, the differential in the spectral sequence maps

$$\wedge^4 \mathbb{R}^4 \otimes H^3(\mathfrak{g}_a) \to H^0\left(\mathfrak{g}_a, \operatorname{Sym}^2\left(\mathfrak{g}_a \otimes \wedge^2 \mathbb{R}^4\right)\right).$$

for each simple direct summand $\mathfrak{g}_a$ of $\mathfrak{g}$.

Thus, it suffices to prove the statement with $\mathfrak{g}$ replaced by one of its constituent simple Lie algebras $\mathfrak{g}_a$. First, we show that the map is injective, or equivalently non-zero (as $H^3(\mathfrak{g}_a) = \mathbb{R}$).

This can be seen by an explicit computation. We will write the generator of $H^3(\mathfrak{g}_a)$ as $\langle [x,y], z \rangle$ where $x, y, z \in \mathfrak{g}_a$. Let $e_1, \ldots, e_4$ be a basis for $\mathbb{R}^4$. Then, the generator of $\wedge^4 \mathbb{R}^4 \otimes H^3(\mathfrak{g}_a)$ can be written as

$$\sum \varepsilon_{ijkl} e_i e_j e_k e_l \otimes \langle [x,y], z \rangle$$

where ε_{ijkl} is the alternating symbol on the indices shown.

To compute how this element is mapped by the differentials of the spectral sequence, first we map it to $\wedge^3 \mathbb{R}^4 \otimes C^3_{red}(\widehat{\mathscr{Y}} \otimes \mathfrak{g}_a)$ under the natural map. The resulting element will be exact; we pick a bounding cochain, which is then mapped to $\wedge^2 \mathbb{R}^4 \otimes C^2_{red}(\widehat{\mathscr{Y}} \otimes \mathfrak{g}_a)$. Again, this element is exact, so we pick a bounding cochain, map it to $\wedge^1 \mathbb{R}^4 \otimes C^1_{red}(\widehat{\mathscr{Y}} \otimes \mathfrak{g}_a)$, and so forth.

The element of $\wedge^3 \mathbb{R}^4 \otimes C^3_{red}(\widehat{\mathscr{Y}} \otimes \mathfrak{g}_a)$ is given by the formula

$$\sum \varepsilon_{ijkl} e_i e_j e_k \otimes \langle [x,y], z \partial_l \rangle$$

where, as before, ∂_i is short for $\frac{\partial}{\partial x_i}$, which we think of as an element of the dual of $\mathbb{R}[[x_1, \ldots, x_4]]$.

This is exact, and a bounding cochain is given by the expression

$$\sum \varepsilon_{ijkl} e_i e_j e_k \otimes \langle x, y \partial_l \rangle$$

(up to a non-zero constant). This expression is mapped to

$$\sum \varepsilon_{ijkl} e_i e_j \otimes \langle x \partial_k, y \partial_l \rangle \in \wedge^2 \mathbb{R}^4 \otimes C^2_{red}(\widehat{\mathscr{Y}} \otimes \mathfrak{g}_a).$$

Again, this is exact; a bounding cochain is given (up to sign) by

$$\sum \varepsilon_{ijkl} e_i e_j \otimes \langle x \partial_k, y \mathrm{d} x_l^\vee \rangle$$

where $\mathrm{d} x_l^\vee$ is an element of degree -1 in $\widehat{\mathscr{Y}}^\vee$, dual to the element $\mathrm{d} x_l \in \Omega^1 \subset \widehat{\mathscr{Y}}^1$.

This expression is now mapped to

$$\sum \varepsilon_{ijkl} e_i \otimes \langle x \partial_k, y \mathrm{d} x_l^\vee \partial_j \rangle \in \wedge^1 \mathbb{R}^4 \otimes C^1_{red}(\widehat{\mathscr{Y}} \otimes \mathfrak{g}_a).$$

A bounding cochain for this is

$$\sum \varepsilon_{ijkl} e_i \otimes \langle x \mathrm{d} x_k^\vee, y \mathrm{d} x_l^\vee \partial_j \rangle.$$

Finally this element is mapped to

$$\sum \varepsilon_{ijkl} \otimes \langle x \mathrm{d} x_k^\vee \partial_i, y \mathrm{d} x_l^\vee \partial_j \rangle \in \wedge^0 \mathbb{R}^4 \otimes C^0_{red}(\widehat{\mathscr{Y}} \otimes \mathfrak{g}_a).$$

7. COHOMOLOGY CALCULATIONS

This is non-zero in the cohomology
$$H^{0,-4}_{red}(\widehat{\mathcal{Y}} \otimes \mathfrak{g}_a) = H^0\left(\mathfrak{g}_a, \mathrm{Sym}^2\left(\mathfrak{g}_a^\vee \otimes \wedge^2 \mathbb{R}^4\right)\right)$$

We have seen that the map
$$H^3(\mathfrak{g}_a) \to H^0\left(\mathfrak{g}_a, \mathrm{Sym}^2\left(\mathfrak{g}_a^\vee \otimes \wedge^2 \mathbb{R}^4\right)\right)^{SO(4)}$$
is injective. It remains to show that the cokernel of this map is naturally isomorphic to $H^0\left(\mathfrak{g}_a, \mathrm{Sym}^2\left(\mathfrak{g}_a^\vee\right)\right)$.

It is straightforward to calculate that the map
$$\left(\mathrm{Sym}^2\left(\wedge^2 \mathbb{R}^4\right)\right)^{SO(4)} \to \left(\wedge^2 \mathbb{R}^4 \otimes \wedge^2 \mathbb{R}^4\right)^{SO(4)}$$
is an isomorphism. Thus,

$$H^0\left(\mathfrak{g}_a, \mathrm{Sym}^2\left(\mathfrak{g}_a^\vee \otimes \wedge^2 \mathbb{R}^4\right)\right)^{SO(4)}$$
$$= H^0\left(\mathfrak{g}_a, \mathrm{Sym}^2\left(\mathfrak{g}_a^\vee\right)\right) \otimes \left(\mathrm{Sym}^2\left(\wedge^2 \mathbb{R}^4\right)\right)^{SO(4)}.$$

The map
$$H^3(\mathfrak{g}_a) \to H^0\left(\mathfrak{g}_a, \mathrm{Sym}^2\left(\mathfrak{g}_a^\vee\right)\right) \otimes \left(\mathrm{Sym}^2\left(\wedge^2 \mathbb{R}^4\right)\right)^{SO(4)}$$
arises from the tensor product of a natural isomorphism
$$H^3(\mathfrak{g}_a) \cong H^0(\mathfrak{g}_a, \mathrm{Sym}^2 \mathfrak{g}_a^\vee)$$
with a canonical invariant element of $\mathrm{Sym}^2\left(\wedge^2 \mathbb{R}^4\right)$. To complete the proof, it thus suffices to check that the vector space
$$\left(\mathrm{Sym}^2\left(\wedge^2 \mathbb{R}^4\right)\right)^{SO(4)}$$
is two-dimensional; this is straightforward.

□

□

Appendix 1: Asymptotics of graph integrals

In this Appendix, we will show that certain integrals attached to graphs admit asymptotic expansions of a certain form. This is a key result in our construction of counterterms, and so underlies all the results of this book.

Everything in this chapter will depend on some auxiliary manifold with corners X, equipped with a sheaf A of commutative superalgebras over the sheaf of algebras C_X^∞, with the following properties:

(1) A is locally free of finite rank as a C_X^∞-module. In other words, A is the sheaf of sections of some super vector bundle on X.
(2) A is equipped with an ideal I such that $A/I = C_X^\infty$, and $I^k = 0$ for some $k > 0$. The ideal I, its powers I^l, and the quotient sheaves A/I^l, are all required to be locally free sheaves of C_X^∞-modules.

We will fix the data (X, A) throughout. The algebra $\Gamma(X, A)$ of C^∞ global sections of A will be denoted by $\mathscr{A}$.

In addition, we will fix a compact manifold M with a super vector bundle E on M, whose space of global sections will be denoted by $\mathscr{E}$.

1. Generalized Laplacians

DEFINITION 1.0.1. *A differential operator*

$$\mathrm{D} : \Gamma(M, E^\vee \otimes \mathrm{Dens}(M)) \otimes \mathscr{A} \to \Gamma(M, E) \otimes \mathscr{A}$$

is called a generalized Laplacian *if it has the following properties:*

(1) D *is $\mathscr{A}$-linear, and is of order 2 as a differential operator.*
(2) *There exists a smooth family g of Riemannian metrics on M, parameterized by X, and an isomorphism*

$$\alpha : E^\vee \otimes \mathrm{Dens}(M) \to E$$

of vector bundles on M, such that the symbol $\sigma(\mathrm{D})$ of D is

$$\sigma(\mathrm{D}) = g \otimes \alpha \in \Gamma(\mathrm{Sym}^2 T^*M, \mathrm{Hom}(E^\vee \otimes \mathrm{Dens}(M), E)) \otimes \mathscr{A}.$$

A generalized Laplacian on the vector bundle E on M (whose global sections is $\mathscr{E}$) is specified by a metric on M, a connection on E and a potential $F \in \Gamma(M, \mathrm{End}(E))$. A smooth family of generalized Laplacians is a family where this data varies smoothly; this is equivalent to saying that the operator varies smoothly. We are dealing with a smooth family depending on the manifold X with the bundle of algebras A. The metric on M has coefficients only in $C^\infty(X)$ – and thus has no dependence on A – but the parameters for the connection and the potential will have coefficients in $\mathscr{A} = \Gamma(X, A)$.

The results of (BGV92) show that there is a heat kernel

$$K_t \in \mathscr{E} \otimes \mathscr{E} \otimes \mathscr{A}$$

for the generalized Laplacian D. Further, this heat kernel has a well-behaved small t asymptotic expansion, as we will explain.

Let $y \in M$. Let $U \subset M \times X$ denote the open subset of points (z, x) where $d_\sigma(z, y) < \varepsilon$, where ε is less than the injectivity radius of M for any of the metrics arising from points in X. Normal coordinates on U gives an isomorphism

$$U \cong B_\varepsilon^n \times X$$

where B_ε^n is the ball of radius ε in $\mathbb{R}^n$, and $n = \dim M$.

Thus, we get an isomorphism

$$C^\infty(U) = C^\infty(B_\varepsilon^n) \otimes C^\infty(X).$$

of $C^\infty(X)$-algebras.

The vector bundle E becomes a vector bundle (still called E) on B_ε^n, and we find

$$\Gamma(U, E) = \Gamma(B_\varepsilon^n, E) \otimes C^\infty(X).$$

The following is a variant of a result proved in (BGV92), following (MP49; BGM71; MS67).

THEOREM 1.0.2. *There exists a small t asymptotic expansion of K_t which, in these coordinates, is of the form*

$$K_t \simeq t^{-\dim M/2} e^{-\|x-y\|^2/t} \sum_{i \geq 0} t^i \phi_i$$

where x, y denote coordinates on the two copies of B_ε^n, and

$$\phi_i \in \Gamma(B_\varepsilon^n, E) \otimes \Gamma(B_\varepsilon^n, E) \otimes \mathscr{A}.$$

If we denote

$$K_t^N = t^{-\dim M/2} e^{-\|x-y\|^2/t} \sum_{i=0}^N t^i \phi_i$$

then for all $l \in \mathbb{Z}_{\geq 0}$,

$$\left\| K_t - K_t^N \right\|_l = O(t^{N-(\dim M)/2 - l}).$$

Here $\|\cdot\|_l$ denotes the C^l norm on the space $C^\infty(B_\varepsilon^n) \otimes C^\infty(B_\varepsilon^n) \otimes \mathscr{A}$.

This precise statement is not proved in (BGV92), as they do not use the auxiliary nilpotent ring A. But, as Ezra Getzler explained to me, the proof in (BGV92) goes through *mutatis mutandis*.

2. Polydifferential operators

DEFINITION 2.0.1. *Let* $\mathrm{Diff}(E, E')$ *denote the infinite rank vector bundle on* M *of differential operators between two vector bundles* E *and* E' *on* M. *Let*

$$\mathrm{PolyDiff}(\Gamma(E)^{\otimes n}, \Gamma(E')) = \Gamma(M, \mathrm{Diff}(E, \mathbb{C})^{\otimes n} \otimes E') \subset \mathrm{Hom}(\Gamma(E)^{\otimes n}, \Gamma(E'))$$

where $\mathbb{C}$ *denotes the trivial vector bundle of rank 1. All tensor products in this expression are fibrewise tensor products of vector bundles on* M.

In a similar way, let

$$\mathrm{PolyDiff}^{\leq k}(\Gamma(E)^{\otimes n}, \Gamma(E')) = \Gamma(M, \mathrm{Diff}^{\leq k}(E, \mathbb{C})^{\otimes n} \otimes E')$$

denote the space of polydifferential operators of order $\leq k$.

It is clear that $\mathrm{PolyDiff}(\Gamma(E)^{\otimes n}, \Gamma(E'))$ is the space of sections of an infinite rank vector bundle on M. Further, this infinite rank vector bundle is a direct limit of the finite rank vector bundles of polydifferential operators of order $\leq k$. Thus, $\mathrm{PolyDiff}(\Gamma(E)^{\otimes n}, \Gamma(E'))$ is a nuclear Fréchet space.

If F is a super vector bundle on another manifold N, let

$$\mathrm{PolyDiff}(\Gamma(E)^{\otimes n}, \Gamma(E')) \otimes \Gamma(F)$$

denote the completed projective tensor product, as usual, so that

$$\mathrm{PolyDiff}(\Gamma(E)^{\otimes n}, \Gamma(E')) \otimes \Gamma(F) = \Gamma(M \times N, (\mathrm{Diff}(E, \mathbb{C})^{\otimes n} \otimes E') \boxtimes F).$$

If X is a manifold, we can think of $\mathrm{PolyDiff}(\Gamma(E)^{\otimes n}, \Gamma(E')) \otimes C^\infty(X)$ as the space of smooth families of polydifferential operators parameterized by X.

One can give an equivalent definition of polydifferential operators in terms of local trivialisations $\{e_i\}, \{e'_j\}$ of E and E', and local coordinates $y_1, \ldots, y_l$ on M. A map $\Gamma(E)^{\otimes n} \to \Gamma(E')$ is a polydifferential operator if, locally, it is a finite sum of operators of the form

$$f_1 e_{i_1} \otimes \cdots \otimes f_n e_{i_n} \mapsto \sum_j e'_j \Phi^j_{i_1 \ldots i_n}(y_1, \ldots, y_l)(D_{I_1} f_1) \cdots (D_{I_n} f_n)$$

where $\Phi^j_{i_1 \ldots i_n}(y_1, \ldots, y_l)$ are smooth functions of the y_i, the I_k are multi-indices, and the operators D_{I_k} are the corresponding partial derivatives with respect to the y_i.

3. Periods

Let us recall the definition of the subalgebra

$$\mathscr{P}((0,1)) \subset C^\infty((0,1))$$

of periods.

Suppose we have the following data.
  (1) an algebraic variety X over $\mathbb{Q}$;
  (2) a normal crossings divisor $D \subset X$;
  (3) a Zariski open subset $U \subset \mathbb{A}^1_\mathbb{Q}$, defined over $\mathbb{Q}$, such that $U(\mathbb{R})$ containts $(0,1)$.

(4) a smooth map $X \to U$, of relative dimension d, also defined over $\mathbb{Q}$, whose restriction to D is flat.
(5) a relative d-form $\omega \in \Omega^d(X/U)$, defined over $\mathbb{Q}$, and vanishing along D.
(6) a homology class $\gamma \in H_d((X_{1/2}(\mathbb{C}), D_{1/2}(\mathbb{C})), \mathbb{Q})$, where $X_{1/2}$ and $D_{1/2}$ are the fibres of X and D over $1/2 \in U(\mathbb{R})$. We assume that γ is invariant under the complex conjugation map on the pair $(X_{1/2}(\mathbb{C}), D_{1/2}(\mathbb{C}))$.

Let us assume that the maps
$$X(\mathbb{C}) \to U(\mathbb{C})$$
$$D(\mathbb{C}) \to U(\mathbb{C})$$
are locally trivial fibrations. For $t \in (0,1) \subset U(\mathbb{R})$, we will let $X_t(\mathbb{C})$ and $D_t(\mathbb{C})$ denote the fibre over $s(t) \in U(\mathbb{R})$.

We can transfer the homology class $\gamma \in H_*(X_{1/2}(\mathbb{C}), D_{1/2}(\mathbb{C}))$ to any fibre $(X_t(\mathbb{C}), D_t(\mathbb{C}))$ for $t \in (0,1)$. This allows us to define a function f on $(0,1)$ by
$$f(t) = \int_{\gamma_t} \omega_t.$$
The function f is real analytic, and real valued.

DEFINITION 3.0.1. *Let $\mathscr{P}_\mathbb{Q}((0,1)) \subset C^\infty((0,1))$ be the subalgebra of functions of this form. Elements of this subalgebra will be called rational periods.*

Let
$$\mathscr{P}((0,1)) = \mathscr{P}_\mathbb{Q}((0,1)) \otimes \mathbb{R} \subset C^\infty((0,1))$$
be the real vector space spanned by the space of rational periods. Elements of $\mathscr{P}((0,1))$ will be called periods.

4. Integrals attached to graphs

DEFINITION 4.0.1. *A labelled graph is a connected, oriented graph γ, with some number of tails (or external edges).*

For each vertex v of γ, let $H(v)$ denote the set of half edges (or germs of edges) emanating from v; a tail attached at v counts as a half edge.

Also, γ has an ordering on the sets of vertices, edges, tails and on the set of half edges attached to each vertex.

Each vertex v of γ is labelled by a polydifferential operator
$$I_v \in \operatorname{PolyDiff}(\mathscr{E}^{\otimes H(v)}, \operatorname{Dens}(M)) \otimes \mathscr{A}.$$

Let $O(v)$ denote the order of I_v.

Let $T(\gamma)$ denote the set of tails of γ. Fix $\alpha \in \mathscr{E}^{\otimes T(\gamma)}$, and fix $t_e \in (0, \infty)$ for each $e \in E(\gamma)$. Define a function $f_\gamma(t_e, \alpha)$ as follows.

Let $H(\gamma)$ denote the set of half edges of γ, so $H(\gamma) = \cup_{v \in V(\gamma)} H(v)$. By putting K_{t_e} at each edge of γ, and α at the tails, we get an element
$$\alpha \otimes_{e \in E(\gamma)} K_{t_e} \in \mathscr{E}^{\otimes H(\gamma)}.$$
On the other hand, the linear maps I_v at the vertices define a map
$$\int_{M^{V(\gamma)}} \otimes I_v : \mathscr{E}^{\otimes H(\gamma)} \to \text{Dens}(M)^{\otimes V(\gamma)} \otimes \mathscr{A} \xrightarrow{\int_{M^{V(\gamma)}}} \mathscr{A}.$$
Let
$$f_\gamma(t_e, \alpha) = \int_{M^{V(\gamma)}} \otimes I_v \left(\alpha \otimes_{e \in E(\gamma)} K_{t_e} \right) \in \mathscr{A}.$$

THEOREM 4.0.2. *The integral*
$$F_\gamma(\varepsilon, T, \alpha) = \int_{t_e \in (\varepsilon, T)^{E(\gamma)}} f_\gamma(t_e, \alpha) \prod dt_e$$
has an asymptotic expansion as $\varepsilon \to 0$ of the form
$$F_\gamma(\varepsilon, T, \alpha) \simeq \sum f_r(\varepsilon) \Psi_r(T, \alpha)$$
where $f_r(\varepsilon) \in \mathscr{P}((0,1)_\varepsilon)$, and the Ψ_r are continuous linear maps
$$\mathscr{E}^{\otimes T(\gamma)} \to C^\infty((0, \infty)_T) \otimes \mathscr{A}$$
where T is the coordinate on $(0, \infty)$.

Further, each $\Psi_r(T, \alpha)$ has a small T asymptotic expansion
$$\Psi_r(T, \alpha) \simeq \sum g_r(T) \int_M \Phi_{r,k}(\alpha)$$
where
$$\Phi_{r,k} \in \text{PolyDiff}(\mathscr{E}^{\otimes T(\gamma)}, \text{Densities}(M)) \otimes \mathscr{A}.$$
and g_r are smooth functions of $T \in (0, \infty)$.

I should make a few comments clarifying the statement of this theorem.

(1) The functions $f_r(\varepsilon)$ appearing in this small ε asymptotic are sums of integrals of the form
$$\int_{t_1, \ldots, t_k \in U \subset (\varepsilon, 1)^k} \frac{F_1(t_1, \ldots, t_k)^{1/2}}{F_2(t_1, \ldots, t_k)^{1/2}}$$
where $F_i, G_i \in \mathbb{Z}_{\geq 0}[t_1, \ldots, t_k]$ are polynomials with positive integer coefficients, and $U \subset (\varepsilon, 1)^k$ is an open subset cut out by inequalities $f_1(t_i) > 0, \ldots, f_n(t_i) > 0$ where each f_m is of the form $t_i^r - t_j$, for some $1 \leq i, j \leq k$ and some $r \in \mathbb{Z}$.

Functions of this form are always periods, in the sense we use. Indeed, we can define an algebraic variety
$$X = \{\varepsilon, t_1, \ldots, t_k, y, z \mid F_2(t_i) \neq 0, \; y^2 = F_1(t_i), \; z^2 = F_2(t_i)\},$$

on which the function
$$\frac{F_1^{1/2}}{F_2^{1/2}} = \frac{y}{z}$$
is defined.

We define a divisor $D \subset X$, as the zero locus of the function
$$\left(\prod_{i=1}^{n} f_i\right)\left(\prod_{j=1}^{k}(t_j - \varepsilon)(t_j - 1)\right).$$

The cycle
$$\gamma \in H_k(X_\varepsilon, D_\varepsilon)$$
is the locus where $f_i \geq 0$, $\varepsilon \leq t_j 1$. The form $\omega_\varepsilon \in \Omega^k(X_\varepsilon, D_\varepsilon)$ is
$$\omega_\varepsilon = \frac{y}{z} dt_1 \ldots dt_k.$$

(2) Asymptotic expansion means the following.

For every compact subset $K \subset X$, let $\|\cdot\|_{K,l}$ denote the C^l norm on the space $\mathscr{A} = \Gamma(X, A)$, defined by taking the supremum over K of the sum of all order $\leq l$ derivatives.

Let us consider $F_\gamma(\varepsilon, T, \alpha)$ as a linear map
$$\mathscr{E}^{\otimes T(\gamma)} \to \mathscr{A} \otimes C^\infty(\{0 < \varepsilon < T\})$$
$$\alpha \mapsto F_\gamma(\varepsilon, T, \alpha).$$

The precise statement is that for all $R, l \in \mathbb{Z}_{\geq 0}$, all compact subsets $K \subset X$ and $L \subset (0, \infty)$, there exists $m \in \mathbb{Z}_{\geq 0}$ such that
$$\sup_{T \in L} \left\| F_\gamma(\varepsilon, T, \alpha) - \sum_{r=0}^{R} f_r(\varepsilon) \Psi_r(T, \alpha) \right\|_{K,l} < \varepsilon^{R+1} \|\alpha\|_m$$
for all T sufficiently small. Here $\|\alpha\|_m$ denotes the C^m norm on the space $\mathscr{E}^{\otimes T(\gamma)}$.

(3) The small T asymptotic expansion in part (2) has a similar definition.

4.1. There is a variant of this result which holds when M is $\mathbb{R}^n$. In that case, E is a trivial super vector bundle on $\mathbb{R}^n$; all of the polydifferential operators associated to vertices of γ are translation invariant; and the kernel K_t is an element of $C^\infty(\mathbb{R}^n \times \mathbb{R}^n) \otimes E^{\otimes 2}$ of the form
$$K_t = \sum e_i \otimes e_j f(\|x-y\|^2) P_{i,j}(x-y, t^{1/2}, t^{-1/2}) e^{-\|x-y\|^2/t}$$
where e_i is a basis of the vector space E, $P_{i,j}$ are polynomials in the variables $x_k - y_k$ and $t^{\pm 1/2}$, and f is a smooth function on $[0, \infty)$ of compact support which takes value 1 in a neighbourhood of zero.

This case is in fact much easier than the general case. The general case, of a compact manifold, is reduced to the case of $\mathbb{R}^n$ using the small t

asymptotic expansion of the heat kernel. Thus, the proof given on a compact manifold applies for $\mathbb{R}^n$ also.

4.2. A second variant of this result, which we will occasionally need, incorporates a propagator of the form $P(\varepsilon, T) + \delta K_\varepsilon$, where δ is an odd parameter of square zero. In this case, the graphs we use may have one special edge, on which we put K_ε instead of K_{t_e}. Thus, instead of integrating over the parameter t_e for this edge, we specialise to ε.

5. Proof of Theorem 4.0.2

The function $F_\gamma(\varepsilon, T, \alpha)$, whose small ε asymptotic behaviour we want to understand, is an integral

$$F_\gamma(\varepsilon, T, \alpha) = \int_{t_e \in (\varepsilon, T)^{E(\gamma)}} f_\gamma(t_e, \alpha) \prod \mathrm{d}t_e.$$

Theorem 4.0.2 will be an immediate corollary of the existence of a small asymptotic expansion for the integrand, $f_\gamma(t_e, \alpha)$. What we will show is that we can write the region of integration $(0, T)^{E(\gamma)}$ as a union of a finite number of closed sets, and that on each such closed set the function f_γ has an asymptotic expansion in terms of rational functions.

More precisely, we will show the following.

PROPOSITION 5.0.1. *Let us choose a total ordering of the set of edges $E(\gamma)$ of γ. This allows us to enumerate this set of edges as $e_1, \ldots, e_k$, and let t_i denote t_{e_i}.*

Then there exist

(1) a finite number $C_1, \ldots, C_r$ of closed sets which cover $(0, T)^{E(\gamma)}$,
(2) $F_i, G_i \in \mathbb{Z}_{\geq 0}[t_1, \ldots, t_k] \setminus \{0\}$ for $i \in \mathbb{Z}_{\geq 0}$,
(3) polydifferential operators

$$\Psi_i \in \mathrm{PolyDiff}(\mathscr{E}^{\otimes T(\gamma)}, \mathrm{Dens}(M)) \otimes C^\infty(X)$$

for $i \in \mathbb{Z}_{\geq 0}$,
(4) $d_r, m_r \in \mathbb{Z}_{\geq 0}$ where $d_r \to \infty$,

such that

$$\left| f_\gamma(t_1, \ldots, t_k) - \sum_{i=1}^{r} \frac{F_i(t_1, \ldots, t_k)^{1/2}}{G_i(t_1, \ldots, t_k)^{1/2}} \int_M \Psi_i(\alpha) \right| \leq \|\alpha\|_{m_r} (\max(t_i))^{d_r}$$

for all

$$\{t_1, \ldots, t_k\} \in C_j \subset (0, T)^k$$

with $\max(t_i)$ sufficiently small. Here, $\max(t_i)$ denotes the largest of the $t_1, \ldots, t_k$;

The proof of this proposition will occupy the remainder of Appendix 1. Theorem 4.0.2 is an immediate corollary of this proposition, and requires no further proof.

The strategy of the proof of proposition 5.0.1 is to first show that the integrand $f_\gamma(t_e, \alpha)$ has a good asymptotic expansion when all of the t_e are roughly of the same size. "Roughly of the same size" means that there is some R such that for all e, e', $t_e^R \le t_e$. This leaves us to understand the regions where some of the t_e are much smaller than others. These regions occur when the lengths of the edges of some subgraph $\gamma' \subset \gamma$ collapses. These regions are dealt with by an inductive argument.

5.1. Without loss of generality, we may assume that X is a compact manifold with boundary. Further, the sheaf of algebras A on X plays only a combinatorial role, and has no effect on the analysis. Thus, we may assume, without loss of generality, that A is $\mathbb{C}$.

Further, we will often avoid mention of the parameter space X. Thus, if f is some expression which has depends on X, I will often abuse notation and write $|f|$ for the C^l norm of f as a function on X.

For a function $I : E(\gamma) = \{1, \ldots, k\} \to \mathbb{Z}_{\ge 0}$, let $|I| = \sum I(i)$. Let $t^I = \prod t_i^{I(i)}$. Similarly, if $n \in \mathbb{Z}$, let $t^n = \prod t_i^n$.

Let
$$O(\gamma) = \sum_{v \in V(\gamma)} O(v)$$
where $O(v)$ is the order of the polydifferential operator I_v we place at the vertex v.

For $R > 1$ let
$$A_{R,T} \subset (0, T)^{E(\gamma)}$$
be the region where $t_i^R \le t_j$ for all i, j. This means that the t_i are all of a similar size.

The first lemma is that the function f_γ has a good asymptotic expansion on the region $A_{R,T}$.

PROPOSITION 5.1.1. *Fix any $R > 1$.*
There exists
(1) $F_i, G_i \in \mathbb{Z}_{\ge 0}[t_1, \ldots, t_k] \setminus \{0\}$ *for* $i \in \mathbb{Z}_{\ge 0}$,
(2) *polydifferential operators*
$$\Psi_i \in \mathrm{PolyDiff}(\mathscr{E}^{\otimes T(\gamma)}, \mathrm{Dens}(M)) \otimes C^\infty(X)$$
for $i \in \mathbb{Z}_{\ge 0}$,
(3) $d_r \in \mathbb{Z}_{\ge 0}$ *tending to infinity,*
(4) *A constant $C(\gamma)$, depending only on the combinatorial structure of the graph and the dimension of M (but independent of the local action functionals placed at the vertices of the graph),*

such that
$$\left| f_\gamma(t_1, \ldots, t_k) - \sum_{i=1}^r \frac{F_i(t_1, \ldots, t_k)^{1/2}}{G_i(t_1, \ldots, t_k)^{1/2}} \int_M \Psi_i(\alpha) \right| \le \|\alpha\|_{d_r + RC(\gamma) + 5RO(\gamma)} \, t_1^{d_r}$$

for all $\{t_1, \ldots, t_k\}$ *with* $t_1 \ge t_2 \ge \cdots \ge t_k$, $t_1^R \ge t_k$, *and t_1 sufficiently small.*

5. PROOF OF THEOREM 4.0.2

PROOF. As above, let

$$K_t^N(x,y) = t^{-\dim M/2}\Psi(x,y)e^{-\|x-y\|^2/t}\sum_{i=0}^{N} t^i \phi_i(x,y)$$

be the approximation to the heat kernel to order t^N. This expression is written in normal coordinates near the diagonal in M^2. The $\phi_i(x,y)$ are sections of $E \boxtimes E$ defined near the diagonal on M^2. $\Psi(x,y)$ is a cut-off function, which is 1 when $\|x-y\| < \varepsilon$ and 0 when $\|x-y\| > 2\varepsilon$.

We have the bound

$$\left\|K_t(x,y) - K_t^N(x,y)\right\|_l = O(t^{N-\dim M/2 - l}).$$

The first step is to replace each K_{t_i} by $K_{t_i}^N$ on each edge of the graph. Thus, let $f_\gamma^N(t_i, \alpha)$ be the function constructed like $f_\gamma(t_i, \alpha)$ except using $K_{t_i}^N$ in place of K_{t_i}.

Each time we replace K_{t_i} by the approximation $K_{t_i}^N$, we get a contribution of $t_i^{N-O(\gamma)-(\dim M)/2}$ from the edge e_i, times the $O(\gamma)$ norm of the contribution of the remaining edges, times $\|\alpha\|_{O(\gamma)}$.

The contribution of the remaining edges to the $\|-\|_{O(\gamma)}$-norm can be calculated to be $\prod_{j \neq i} t_j^{-O(\gamma)-(\dim M)/2}$. We are thus left with the bound

$$\left|f_\gamma(t_i,\alpha) - f_\gamma^N(t_i,\alpha)\right| < C t^{-O(\gamma)-(\dim M)/2} t_1^N \|\alpha\|_{O(\gamma)}$$

where C is a constant. (Recall our notation : t^n denotes $\prod t_i^n$).

In particular, if the t_i are in $A_{R,T}$, so that $t_i > t_1^R$, and if N is chosen to be sufficiently large, we find that

$$\left|f_\gamma(t_i,\alpha) - f_\gamma^N(t_i,\alpha)\right| < \|\alpha\|_{O(\gamma)} t_1^{N-|E(\gamma)|(O(\gamma)+(\dim M)/2)R+1}$$

for t_1 sufficiently small.

Next, we construct a small t_i asymptotic expansion of $f_\gamma^N(t_i, \alpha)$. Recall that $f_\gamma^N(t_i, \alpha)$ is defined as an integral over a small neighbourhood of the small diagonal in $M^{V(\gamma)}$. Let

$$n = \dim M.$$

By using a partition of unity we can consider $f_\gamma^N(t_i, \alpha)$ as an integral over a small neighbourhood of zero in $\mathbb{R}^{nV(\gamma)}$. This allows us to express $f_\gamma^N(t_i, \alpha)$ as a finite sum of integrals over $\mathbb{R}^{nV(\gamma)}$, of the following form.

For each vertex v of γ, we have a coordinate map $x_v : \mathbb{R}^{nV(\gamma)} \to \mathbb{R}^n$. Fix any $\varepsilon > 0$, and let $\chi : [0,\infty) \to [0,1]$ be a smooth function with $\chi(x) = 1$ if $x < \varepsilon$, and $\chi(x) = 0$ if $x > 2\varepsilon$. Let us define a cut-off function ψ on $\mathbb{R}^{nV(\gamma)}$ by the formula

$$\psi = \chi(\|\sum x_v\|^2)\chi\left(\sum_{v' \in V(\gamma)} \left\|x_{v'} - |V(\gamma)|^{-1}\sum_{v \in V(\gamma)} x_v\right\|^2\right).$$

Thus, the function ψ is zero unless all points x_v are near their centre of mass $|V(\gamma)|^{-1} \sum x_v$, and ψ is zero when this centre of mass is too far from the origin.

For each $1 \le i \le k$, let Q_i be the quadratic for on $\mathbb{R}^{n|V(\gamma)|}$ defined by

$Q_i(x) = 0$ if the edge e_i is a loop, i.e. is attached to only one vertex

$Q_i(x) = \|x_{v_1} - x_{v_2}\|^2$ if v_1, v_2 are the vertices attached to the edge e_i

We can write f_γ^N as a finite sum of integrals of the form

$$\int_{\mathbb{R}^{nV(\gamma)}} \psi e^{-\sum Q_i/t_i} \sum_{I,K} t^{I - (\dim M)/2 - O(\gamma)} \Phi_{I,K} \partial_{K,x} \alpha.$$

In this expression,

- The sum is over $I : E(\gamma) \to \mathbb{Z}_{\ge 0}$, with all $I(e) \le N + O(\gamma) + 1$, and multi-indices $K : V(\gamma) \times \{1, \ldots, n\} \to \mathbb{Z}_{\ge 0}$, with $\sum K(v, i) \le O(\gamma)$. The notation $\partial_{K,x}$ denotes

$$\partial_{K,x} = \prod_{v \in V(\gamma), 1 \le i \le n} \frac{\partial}{\partial x_{v,i}^{K(v,i)}}.$$

  In this notation, we are pretending (by trivialising the vector bundle E on some small open sets in M) that α is a function on $\mathbb{R}^{nV(\gamma)}$.

- The $\Phi_{I,K}$ are smooth functions on $\mathbb{R}^{nV(\gamma)}$.

Next, we will use Wick's lemma to compute the asymptotics of this integral. Let $c = (1/|V(\gamma)|) \sum x_v$ be the centre of mass function $\mathbb{R}^{nV(\gamma)} \to \mathbb{R}^n$. We can perform the integral in two steps, first integrating over the coordinates $y_v = x_v - c$, and secondly by integrating over the variable c. (Of course, there are $|V(\gamma)| - 1$ independent y_v coordinates). The quadratic form $\sum Q_i/t_i$ on $\mathbb{R}^{nV(\gamma)}$ is non-degenerate on the subspace of $\mathbb{R}^{nV(\gamma)}$ of vectors with a fixed centre of mass, for all $t_i \in (0, \infty)$. Thus, the integral over the variables y_v can be approximated with the help of Wick's lemma.

Let us order the set $V(\gamma)$ of vertices as $v_1, v_2, \ldots, v_m$. We will use the coordinates $y_1, \ldots, y_{m-1}$, and c on $\mathbb{R}^{nV(\gamma)}$. Then f_γ^N is a finite sum of integrals of the form

$$\int_{w \in \mathbb{R}^n} \chi(|w|^2) \int_{y_1, \ldots, y_{m-1} \in \mathbb{R}^n} \left\{ \chi(\sum \|y_v\|^2) \right.$$
$$\left. e^{-\sum Q_i(y)/t_i} \sum t^{I - O(\gamma) - (\dim M)/2} \Phi_{I,K} \partial_{K,w,y_i} \alpha \right\}.$$

Here we are using the same notation as before, in these new coordinates.

To get an approximation to the inner integral, we take a Taylor expansion of the functions α and $\Phi_{I,K}$ around the point $\{y_i = 0, w\}$, only expanding in the variables y_i. We take the expansion to order N'. We find, as an approximation to the inner integral, an expression of the form

$$\int_{y_i \in \mathbb{R}^n} e^{-\sum Q_i(y)/t_i} \sum t^{I-(\dim M)/2 - O(\gamma)} y^K c_{K,I,L} \left(\prod \frac{\partial^{L_i}}{\partial^{L_i} y_i} \frac{\partial^{L_w}}{\partial^{L_w} w} \alpha \right)_{y_i = 0}$$

where the sum is over a finite number of multi-indices I, K, L, and $c_{K,I,L}$ are constants.

We can calculate each such integral by Wick's lemma. The application of Wick's lemma involves inverting the quadratic form $\sum Q_i(y)/t_i$. Let $A = A(t_i)$ denote the matrix of the quadratic form $\sum Q_i(y)/t_i$; this is a square matrix of size $(\dim M)|V(\gamma)|$, whose entries are sums of t_i^{-1}. Note that $\left(\prod_{i=1}^k t_i \right) A$ has polynomial entries.

Let

$$P_\gamma(t_i) = \det \left(\left(\prod_{i=1}^k t_i \right) A \right).$$

This is the graph polynomial associated to γ (see (BEK06)). One important property of P_γ is that it is a sum of monomials, each with a non-negative integer coefficient.

We can write

$$A^{-1} = P_\gamma^{-1} B$$

where the entries of B are polynomial in the t_i. Note also that

$$\det A = P_\gamma t^{-(\dim M)(|V(\gamma)|-1)/2}.$$

(The expansion from Wick's lemma gives an overall factor of $(\det A)^{-1/2}$).

Thus, we find, using Wick's lemma, an approximation of the form

$$f_\gamma^N(t_i) \simeq P_\gamma^{-1/2} \sum_{l \geq 0, I : E(\gamma) \to \frac{1}{2}\mathbb{Z}_{\geq 0}} P_\gamma^{-l} t^{I-(\dim M)/2 - O(\gamma)} \int_M \Psi_{l,I}(\alpha)$$

where the $\Psi_{l,I}$ are polydifferential operators

$$\Psi_{l,I} : \mathscr{E}^{\otimes T(\gamma)} \to \mathrm{Densities}(M)$$

and the sum is finite (i.e. all but finitely many of the $\Psi_{l,I}$ are zero).

This expansion is of the desired form; it remains to bound the error term.

LEMMA 5.1.2. *The error term in this expansion is bounded by*

$$\|\alpha\|_{N'+1+O(\gamma)} t_1^{(N'+1)/2 - RC(\gamma) - 2RO(\gamma)}$$

for N' sufficiently large and t_1 sufficiently small. Here $C(\gamma)$ is a constant depending only on the dimension of M and the combinatorial structure of the graph.

PROOF. The error in this expansion arises from the error in the Taylor expansion of the functions $\alpha, \Phi_{I,K}$ around 0. Recall that everything is Taylor expanded to order N'. Thus, if N' is sufficiently large, the magnitude of the error in the expansion can be bounded by an expression of the form

$$t^{-(\dim M)/2 - O(\gamma)} \|\alpha\|_{N'+1+O(\gamma)} \int_{w \in \mathbb{R}^n} \int_{y_1, \ldots, y_{m-1} \in \mathbb{R}^n} \prod_{i=1}^k e^{-\sum Q_i(y)/t_i} \sum_K |P(y)|$$

where $P(y)$ is a polynomial of order $N' + 1 - O(\gamma)$.

The Gaussian factor

$$e^{-\sum Q_i(y)/t_i}$$

increases if we increase each t_i. Since $t_i \leq t_1$, we find we can bound the integral by the corresponding integral where we use the Gaussian using the quadratic form $\sum Q_i(y)/t_1$.

Also, since each $t_i > t_1^R$, we have the bound

$$t^{-(\dim M)/2 - O(\gamma)} = \prod_{i=1}^k t_i^{-(\dim M)/2 - O(\gamma)} \leq t_1^{-R|E(\gamma)|((\dim M)/2 + O(\gamma))}.$$

Putting these two inequalities together, we find that our error is bounded by

$$t_1^{-R|E(\gamma)|((\dim M)/2 + O(\gamma))} \|\alpha\|_{N'+1+O(\gamma)} \int_{w \in \mathbb{R}^n} \int_{y_i \in \mathbb{R}^n} \prod_{i=1}^k e^{-\sum Q_i(y)/t_1} |P(y)|.$$

Since $P(y)$ is a polynomial of order $N' + 1 - O(\gamma)$, we see that

$$\int_{y_1, \ldots, y_{m-1} \in \mathbb{R}^n} \prod_{i=1}^k e^{-\sum Q_i(y)/t_1} |P(y)| = C t_1^{\frac{1}{2}(\dim M |V(\gamma)| + N' + 1 - O(\gamma))}$$

for some constant C.

Thus, we find our error is bounded by

$$\|\alpha\|_{N'+1+O(\gamma)} t_1^{(N'+1)/2 - RC(\gamma) - 2RO(\gamma)}$$

for some $C(\gamma)$ which only depends on the combinatorial structure of the graph and on $\dim M$. □

By choosing N and N' appropriately, we can use this bound to complete the proof of the proposition. □

5.2. Next, we will discuss the asymptotic expansion of the function $f_\gamma(t_e)$ when some of the t_e are much smaller than others.

A *subgraph* γ' of a graph γ is given by the set of edges $E(\gamma') \subset E(\gamma)$. The vertices of the subgraph γ' are the ones that adjoin edges in $E(\gamma')$. The tails of γ' are the half-edges of γ which adjoin vertices of γ', but which are not part of any edge of γ'.

Let us fix a proper subgraph γ' of γ. We will consider the function $f_\gamma(t_1, \ldots, t_k)$ on the region where the t_i corresponding to edges in γ' are much smaller than the remaining t_i.

5. PROOF OF THEOREM 4.0.2

Let us enumerate the edges of γ as $e_1, \ldots, e_k$, where $e_1, \ldots, e_l \in E(\gamma) \setminus E(\gamma')$ and $e_{l+1}, \ldots, e_k \in E(\gamma')$. Let $t_i = t_{e_i}$.

Let $A_{R,T}^{\gamma'} \subset (0,T)^{E(\gamma)}$ be the subset where

$$t_e^R \geq t_{e'} \text{ if } e \in E(\gamma) \setminus E(\gamma') \text{ and } e' \in E(\gamma')$$
$$t_e^R \leq t_{e'} \text{ if } e, e' \in E(\gamma').$$

This means that the lengths of the edges of the subgraph γ' are all around the same size, and are all much smaller than the lengths of the other edges.

LEMMA 5.2.1. *Let γ, γ' be as above. Assume that γ' contains at least one edge. Let $R > l$.*

Then, there exists a finite number of closed subsets $C_1, \ldots, C_r$ of $A_{R,T}^{\gamma'}$, which cover $A_{R,T}^{\gamma'}$, such that for each C_j, there exists

(1) $F_i, G_i \in \mathbb{Z}_{\geq 0}[t_1, \ldots, t_k] \setminus \{0\}$ for $i \in \mathbb{Z}_{\geq 0}$,
(2) polydifferential operators

$$\Psi_i \in \text{PolyDiff}(\mathcal{E}^{\otimes T(\gamma)}, \text{Dens}(M)) \otimes C^\infty(X)$$

for $i \in \mathbb{Z}_{\geq 0}$,
(3) $d_r \in \mathbb{Z}_{\geq 0}$ tending to infinity,

such that

$$\left| f_\gamma(t_1, \ldots, t_k) - \sum_{i=1}^r \frac{F_i(t_1, \ldots, t_k)^{1/2}}{G_i(t_1, \ldots, t_k)^{1/2}} \int_M \Psi_i(\alpha) \right|$$
$$\leq \|\alpha\|_{d_r + RC(\gamma) + 5RO(\gamma)} \max(t_i)^{d_r}$$

for all $\{t_1, \ldots, t_k\} \in C_j$ with $\max(t_i)$ sufficiently small. Here, $\max(t_i)$ refers to the largest of the t_i.

PROOF. Without loss of generality let us assume that $t_1 \geq t_2 \geq \cdots \geq t_k$. We can write

$$f_\gamma(t_1, \ldots, t_k, \alpha) = f_{\gamma'}(t_{l+1}, \ldots, t_k, \alpha \otimes K_{t_1} \otimes \cdots \otimes K_{t_l}).$$

The right hand side of this equation denotes the graph integral for γ' with inputs being tensor products of the heat kernels K_{t_i}, for $1 \leq i \leq l$, and α.

The starting point of the proof is Proposition 5.1.1 applied to each connected component of the graph γ'. This proposition implies that we can approximate the contribution of each such connected component by a local action functional applied to the tails of that connected component. If we approximate the contribution of each connected component in this way, we are left with a graph integral $\Psi_{\gamma/\gamma'}(t_1, \ldots, t_l, \alpha)$ for the quotient graph γ/γ', with certain local action functionals at the vertices of γ/γ'. (The quotient graph is defined to be the quotient of γ by the equivalence relation saying that two vertices are equivalent if they are in the same connected component of the subgraph γ').

More formally, Proposition 5.1.1, applied to the connected components of the subgraph γ', implies that there exist

(1) local action functionals $I_{v,i}$ at each vertex v of γ/γ', for each $i \in \mathbb{Z}_{\geq 0}$
(2) there exists $F_i, G_i \in \mathbb{Z}_{\geq 0}[t_{l+1}, \ldots, t_k] \setminus \{0\}$, for each $i \in \mathbb{Z}_{\geq 0}$

with the following properties. Let $\Psi^i_{\gamma/\gamma'}(t_1, \ldots, t_l, \alpha)$ denote the graph integral obtained by putting the local action functional $I_{v,i}$ at the vertex v of γ/γ', the heat kernel K_{t_i} at the edges of γ/γ', and α at the tails of γ/γ' (which are the same as the tails of γ).

Then, there exists a sequence $d_r \in \mathbb{Z}_{\geq 0}$, tending to infinity, such that

$$(\dagger) \quad \left| f_\gamma(t_1, \ldots, t_k, \alpha) - \sum_{i=1}^r \frac{F_i(t_{l+1}, \ldots, t_k)^{1/2}}{G_i(t_{l+1}, \ldots, t_k)^{1/2}} \Psi^i_{\gamma/\gamma'}(t_1, \ldots, t_l, \alpha) \right|$$
$$< t_{l+1}^{d_r} \| \alpha \otimes K_{t_1} \otimes \cdots \otimes K_{t_l} \|_{d_r + RC(\gamma') + 5RO(\gamma')}$$

for all $t_1, \ldots, t_k$ with $t_{l+1} \geq \cdots \geq t_k$, $t_{l+1}^R \geq t_k$, and t_{l+1} sufficiently small. Here $C(\gamma')$ is some combinatorial factor only depending on the combinatorics of the graph γ', and $O(\gamma')$ is the sum of the orders of the local action functionals on the vertices of γ'.

We are only interested in the region where $t_i^R \geq t_{l+1}$ if $1 \leq i \leq l$. Note that
$$\|K_t\|_n = O(t^{-(\dim M)/2 - n}).$$

Thus,
$$\|\alpha \otimes K_{t_1} \otimes \cdots \otimes K_{t_l}\|_{d_r + RC(\gamma') + 5RO(\gamma')}$$
$$\leq \|\alpha\|_{d_r + RC(\gamma') + 5RO(\gamma')} t_l^{-ld_r - RC(\gamma, \gamma') - 5lRO(\gamma')}$$

if t_1 is sufficiently small. Here $C(\gamma, \gamma')$ is some constant depending only on the combinatorics of the graph γ, its subgraph γ', and on $\dim M$. The right hand side of this expression only has t_l, because we are assuming that $t_1 \geq \cdots \geq t_l$.

Putting this into equation ($\dagger$), we find that the right hand side of ($\dagger$) can be bounded by
$$t_{l+1}^{d_r} t_l^{-ld_r - RC(\gamma, \gamma') - 5lRO(\gamma')} \|\alpha\|_{d_r + RC(\gamma') + 5RO(\gamma')}.$$

Recall that $t_l^R \geq t_{l+1}$. Thus, $t_{l+1}^{d_R} \leq t_l^{Rd_R}$, so we find that the right hand side of equation ($\dagger$) can be bounded by
$$t_l^{Rd_r - ld_r - RC_1(\gamma, \gamma') - RC_2(\gamma, \gamma')O(\gamma)} \|\alpha\|_{d_r + RC(\gamma') + 5RO(\gamma')}.$$

where $C_i(\gamma, \gamma')$, as before, only depend on the combinatorics of the graphs.

If we take R to be greater than l, the exponent of t_l in this expression tends to infinity as $r \to \infty$. In addition, since $t_l < t_1$, for r sufficiently large, we have
$$t_l^{Rd_r - ld_r - RC_1(\gamma, \gamma') - RC_2(\gamma, \gamma')O(\gamma)} \leq t_1^{Rd_r - ld_r - RC_1(\gamma, \gamma') - RC_2(\gamma, \gamma')O(\gamma)}.$$

5. PROOF OF THEOREM 4.0.2

Thus, we have shown that for all $R > l$, we can find some F_i, G_i and Ψ_i as above, together with a sequence d_r, m_r tending to ∞, such that
(‡)
$$\left| f_\gamma(t_1, \ldots, t_k, \alpha) - \sum_{i=1}^{r} \frac{F_i(t_{l+1}, \ldots, t_k)^{1/2}}{G_i(t_{l+1}, \ldots, t_k)^{1/2}} \Psi^i_{\gamma/\gamma'}(t_1, \ldots, t_l, \alpha) \right| < t_1^{d_r} \|\alpha\|_{m_r}.$$

We can assume, by induction on the number of vertices of a graph, that Proposition 5.0.1 holds for the graph γ/γ'. This implies that we can cover the set $\{t_1 \geq t_2 \cdots \geq t_k\}$ by a finite number of closed sets, $C_1, \ldots, C_r$; and that for each such closed set C_m, we can find functions $H_{ij}, L_{ij} \in \mathbb{Z}[t_1, \ldots, t_l]$, local action functionals Φ_{ij} of α, such that we can approximate $\Psi^i_{\gamma/\gamma'}$ on the set C_m by
$$\sum_{j \geq 0} \frac{H_{ij}(t_1, \ldots, t_l)^{1/2}}{L_{ij}(t_1, \ldots, t_l)^{1/2}} \Phi_{ij}(\alpha).$$

More precisely, there exists some sequences $n_s, e_s \in \mathbb{Z}_{\geq 0}$ where e_s tends to ∞ such that
$$\left| \Psi^i_{\gamma/\gamma'}(t_1, \ldots, t_l) - \sum_{j=0}^{s} \frac{H_{ij}(t_1, \ldots, t_l)^{1/2}}{L_{ij}(t_1, \ldots, t_l)^{1/2}} \Phi_{ij}(\alpha) \right| \leq t_1^{e_s} \|\alpha\|_{n_s}$$

for t_1 sufficiently small, and $(t_1, \ldots, t_l) \in C_m$.

Plugging this estimate for $\Psi^i_{\gamma/\gamma'}$ into equation (‡) yields the desired result. □

We have shown that Proposition 5.0.1 holds on the following subsets of $(0, T)^{E(\gamma)}$:
(1) The regions $A_{R,T}$, where $t_1 \geq \cdots \geq t_k$ and $t_1^R \leq t_k$, for any $R > 1$.
(2) The regions $A^{\gamma'}_{R,T}$, where $t_1 \geq \cdots \geq t_k$, $t_l^R \geq t_{l+1}$, and $t_{l+1}^R \leq t_k$, for any subgraph $\gamma' \subset \gamma$ which contains at least one edge, and any $R > k$. Here $t_{l+1}, \ldots, t_k$ correspond to the edges of γ in the subgraph γ'.

It remains to show that we can cover $(0, T)^{E(\gamma)}$ with a finite number of regions of this form.

LEMMA 5.2.2. *Fix any $R > k = \#E(\gamma)$, and any $0 < T < \infty$. Then the regions $A^{\gamma'}_{R^{2m}, T}$, where $0 \leq m \leq |E(\gamma)|$ and $\gamma' \subset \gamma$ is non-empty, cover $(0, T)^{E(\gamma)}$. (The regions $A_{R^{2k}, T}$ appear as $A^\gamma_{R^{2k}, T}$, where γ is considered as a subgraph of itself).*

PROOF. Let $\{t_e\} \in (0, T)^{E(\gamma)}$. As before, label the elements of $E(\gamma)$ by $\{1, 2, \ldots, k\}$, in such a way that $t_1 \geq t_2 \geq \cdots \geq t_k$.

Either $t_j^R \geq t_k$ for all $j < k$, or there is a smallest $i_1 < k$ such that $t_{i_1}^R \leq t_k$. In the first case, we are done, as then $\{t_e\} \in A^{\gamma'}_{R,T}$ where γ' is the subgraph with the single edge corresponding to t_k.

Suppose the second possibility holds. Then either for all $j < i_1$, $t_j^R \geq t_{i_1}$, or there exists a smallest $i_2 < i_1$ with $t_{i_2}^R \leq t_{i_1}$. In the first case, we're done, as we are in $\mathscr{A}_{R,T}^{\gamma'}$ where γ' is the subgraph whose edges correspond to $t_{i_1}, t_{i_1+1}, \ldots, t_k$.

Again, let's suppose the second possibility holds. Then $t_{i_2}^{R^2} \leq t_{i_1}^R \leq t_k$. Either, for all $j < i_2$, $t_j^{R^2} \geq t_{i_2}$, and then we are in $A_{R^2,T}^{\gamma'}$ where γ' is the subgraph whose edges correspond to $t_{i_2}, t_{i_2+1}, \ldots, t_k$.

Otherwise, there is some smallest $i_3 < i_2$ with $t_{i_3}^{R^2} \leq t_{i_2}$. Then $t_{i_3}^{R^4} \leq t_k$. And so forth.

We eventually end up either finding ourselves in one of the regions $A_{R^{2j},T}^{\gamma'}$, for some non-empty proper subgraph $\gamma' \subset \gamma$, or we find that some $i_{k+1} = 1$, so we are in $A_{R^{2k},T}^{\gamma} = A_{R^{2k},T}$. □

This completes the proof of Proposition 5.0.1. Theorem 4.0.2 is an immediate corollary.

Appendix 2 : Nuclear spaces

Throughout this book, we freely use linear algebra in the symmetric monoidal category of nuclear spaces. This appendix will recall some basic facts about the theory of nuclear spaces, and in particular will explain how the topological vector spaces we use throughout (such as spaces of smooth functions and distributions) form nuclear spaces.

1. Basic definitions

Recall that a *topological vector space* (sometimes abbreviated to TVS) over $\mathbb{R}$ or $\mathbb{C}$ is a vector space equipped with a topology that makes scalar multiplication and addition continuous. All of the topological vector spaces we consider are Hausdorff. The most important class of TVS (containing both Banach and nuclear spaces) are the *locally convex spaces*. Such a space has a basis (for the topology) given by convex sets. Finally, a Fréchet space is a complete, metrizable locally convex spaces.

The subtleties in functional analysis often arise from the freedom to choose different topologies on dual spaces or tensor products of spaces. For V a TVS, let V^* denote the space of *all* linear maps from V to the base field, and let $V^\vee \subset V^*$ denote the subspace of *continuous* linear functionals. There are several natural topologies on the $V^\vee$, of which two are the following. The *weak* topology on $V^\vee$ is the topology of pointwise convergence: a sequence of functionals $\{\lambda_n\}$ converges to zero if the sequence $\{\lambda_n(x)$ converges to zero for any $x \in V$. The *strong* topology on $V^\vee$ is the topology of bounded convergence: a filter of the origin converges in the strong topology if it converges uniformly on every bounded subset in V. We will use the strong topology here.

There are also several natural topologies one can put on the space $\mathrm{Hom}(E, F)$ of continuous linear maps from $E \to F$. The topology of interest here is the topology of bounded convergence, described as follows. Let $U \subset F$ be an open subset of F, and $B \subset E$ a bounded subset of E. A basis of neighbourhoods of 0 in $\mathrm{Hom}(E, F)$ consists of those subsets

$$\mathcal{U}(B, U) = \{f : E \to F \mid f(B) \subset U\} \subset \mathrm{Hom}(E, F).$$

With this topology, as long as both E and F are locally convex and Hausdorff, so is $\mathrm{Hom}(E, F)$.

We define the projective topological tensor product of two locally convex spaces V and W as follows. Let $V \otimes_{alg} W$ denote the algebraic tensor product

of V and W. The *projective* topology on $V \otimes_{alg} W$ is the finest locally convex topology so that the canonical map
$$V \times W \to V \otimes_{alg} W$$
is continuous.

We will let $V \otimes_\pi W$ (or just $V \otimes W$ when there is no ambiguity) denote the completion of $V \otimes_{alg} W$ with respect to this topology.

There is another natural topology on $V \otimes_{alg} W$, called the injective topology, which is coarser than the projective topology; every reasonable topology on $V \otimes_{alg} W$ lies between these two. (We will not recall the precise definition of the injective topology, as it is a little technical: see (Trè67) or (Gro52) for details).

We will let $V \otimes_i W$ denote the completion of $V \otimes_{alg} W$ with respect to the injective topology.

We now give the most elegant definition of a nuclear space:

DEFINITION 1.0.1. *A locally convex Hausdorff space V is* nuclear *if the canonical map $V \hat{\otimes}_p W \to V \hat{\otimes}_i W$ is an isomorphism for any locally convex Hausdorff space W.*

There are several equivalent definitions, although these equivalences are highly nontrivial. For more details, see chapter 50 of (Trè67) or (Gro52).

DEFINITION 1.0.2. *Let* Nuc *denote the category, enriched in ordinary vector spaces, whose objects are complete nuclear spaces and whose morphisms are linear continuous maps.*

REMARK. Although the space $\mathrm{Hom}(E, F)$ of continuous linear maps between nuclear spaces does have a natural topology, as described above, Nuc does *not* form a symmetric monoidal category enriched in topological vector spaces.

If E, F are nuclear spaces, then so is their completed tensor product $E \otimes F$. Thus, we have

PROPOSITION 1.0.3. Nuc *is a symmetric monoidal category with respect to the completed tensor product.*

2. Examples

Many function spaces in geometry are nuclear:
- smooth functions on an open set in $\mathbb{R}^n$,
- smooth functions with compact support on an open set, or on a compact set, in $\mathbb{R}^n$,
- distributions (the continuous dual space to compactly-supported smooth functions) on an open set in $\mathbb{R}^n$,
- distributions with compact support on an open set in $\mathbb{R}^n$,
- Schwartz functions on $\mathbb{R}^n$,
- Schwartz distributions on $\mathbb{R}^n$,

- holomorphic functions on an open set in $\mathbb{C}^n$,
- formal power series in n variables (with the inverse limit topology),
- polynomials in n variables.

In consequence, we see that, for example, smooth sections of a vector bundle on a compact manifold form a nuclear Fréchet space.

Nuclear spaces enjoy the following useful properties:

- A closed linear subspace of a nuclear space is nuclear.
- The quotient of a nuclear space by a closed linear subspace is nuclear.
- A countable direct sum of nuclear spaces (equipped with finest locally convex topology) is nuclear.
- Likewise, a countable colimit of nuclear spaces is nuclear, as long as the colimit is taken in the category of Hausdorff locally convex topological vector spaces.
- A direct product of nuclear spaces (equipped with the product topology) is nuclear.
- Likewise, a limit of nuclear spaces is nuclear.
- The completed tensor product is nuclear.
- The tensor product of nuclear spaces commutes with all limits. That is, if $E = \lim E_i$, $F = \lim F_j$ are nuclear spaces written as limits of nuclear spaces E_i, F_j, then

$$E \otimes F = \lim_{i,j}(E_i \otimes F_j).$$

3. Subcategories

Let us list three important classes of nuclear spaces:

(1) $(\mathcal{NF})$ spaces, that is, nuclear Fréchet spaces.
(2) $(\mathcal{NLF})$ spaces, which are countable colimits of $\mathcal{NF}$-spaces.
(3) $(\mathcal{NDF})$ spaces, that is, nuclear $(\mathcal{DF})$ spaces. A space is $(\mathcal{NDF})$ if and only if its strong dual is of type $(\mathcal{NF})$.

Nuclear spaces which are $(\mathcal{NF})$ or $(\mathcal{NDF})$ are particularly well behaved.

- If E is a nuclear Fréchet space (that is, of type $(\mathcal{NF})$), then the strong dual $E^\vee$ is a space of type $(\mathcal{NDF})$.
- If E is of type $(\mathcal{NDF})$, then the strong dual $E^\vee$ is of type $(\mathcal{NF})$.
- Countable products of spaces of type $(\mathcal{NF})$ are again of type $(\mathcal{NF})$.
- Closed subspaces and Hausdorff quotients of spaces of type $(\mathcal{NF})$ are again of type $(\mathcal{NF})$ (thus, spaces of type $(\mathcal{NF})$ are closed under countable limits and finite colimits).
- Similarly, spaces of type $(\mathcal{NDF})$ are closed under finite limits and countable colimits.
- If E, F are both of type $(\mathcal{NF})$, or both of type $(\mathcal{NDF})$, then

$$(E \otimes F)^\vee = E^\vee \otimes F^\vee.$$

- If E, F are both of type $(\mathcal{NF})$, or both of type $(\mathcal{NDF})$, then so is $E \otimes F$.
- If $E = \operatorname{colim} E_i$ is a countable colimit of nuclear spaces, and F is of type $(\mathcal{NDF})$, then
$$E \otimes F = \operatorname{colim}(E_i \otimes F).$$

These properties can be rewritten in categorical terms, as follows.

PROPOSITION 3.0.1. *Let $\mathcal{NF}$ (respectively, $\mathcal{NDF}$) denote the full subcategories of Nuc consisting of spaces of type $(\mathcal{NF})$ (or type $(\mathcal{NDF})$, respectively). Then the categories $\mathcal{NF}$ and $\mathcal{NDF}$ are symmetric monoidal subcategories of Nuc. $\mathcal{NF}$ is closed under countable limits and finite colimits; $\mathcal{NDF}$ is closed under countable colimits and finite limits. Sending a nuclear space to its strong dual induces equivalences of symmetric monoidal categories*
$$^\vee : \mathcal{NF}^{op} \cong \mathcal{NDF}$$
$$^\vee : \mathcal{NDF}^{op} \cong \mathcal{NF}.$$

3.1. Let us now list some further useful properties of nuclear spaces.
- Any nuclear space is semi-reflexive: this means that the natural map $E \to (E^\vee)^\vee$ is a linear isomorphism (although not necessarily a homeomorphism).
- Any nuclear space E of type $(\mathcal{NF})$ is reflexive: this means that the natural map $E \to (E^\vee)^\vee$ is a linear homeomorphism, where duals are given the strong topology. This follows from the facts that $(\mathcal{NF})$ spaces are barrelled ((Trè67), page 347); barrelled nuclear spaces are Montel ((Trè67), page 520); and Montel spaces are reflexive ((Trè67), page 376).
- If E is a nuclear space of type $(\mathcal{NF})$, then the strong and weak topologies on bounded subsets of $E^\vee$ coincide ((Trè67), page 357).
- Let E, F be nuclear Fréchet spaces. In the following equalities, Hom-spaces between nuclear spaces will be equipped with the topology of uniform convergence on bounded subsets, as above. Then, as we see from (Trè67), page 525:
$$E \otimes F = \operatorname{Hom}(E^\vee, F)$$
$$E^\vee \otimes F = \operatorname{Hom}(E, F)$$
$$E^\vee \otimes F^\vee = (E \otimes F)^\vee$$
$$E^\vee \otimes F^\vee = \operatorname{Hom}(E, F^\vee)$$
- If E is of type $(\mathcal{NLF})$ and F is any nuclear space, then
$$E^\vee \otimes F = \operatorname{Hom}(E, F).$$
- Finally, for any nuclear spaces E and F, we have the following isomorphism ((Trè67), page 522, proposition 50.4) *only of vector spaces without topology*
$$E^\vee \otimes F = \operatorname{Hom}(E, F).$$

This last isomorphism is, in fact, a topological isomorphism if we equip $\mathrm{Hom}(E,F)$ with a certain topology which is in general different from the topology of uniform convergence on bounded subsets.

Caution: the evaluation map
$$E^\vee \times E \to \mathbb{R}$$
is not, in general, continuous for the product topology on $E^\vee \times E$; it is only separately continuous. One can see this easily when $E = C^\infty(M)$, where M is a compact manifold. Then, $E^\vee = \mathcal{D}(M)$ is the space of supported distributions on M. If the evaluation map
$$C^\infty(M) \times \mathcal{D}(M) \to \mathbb{R}$$
were continuous, it would extend to a continuous linear map
$$C^\infty(M) \otimes \mathcal{D}(M) = \mathrm{Hom}(C^\infty(M), C^\infty(M)) \to \mathbb{R}.$$
On the subspace
$$C^\infty(M) \otimes \Omega^n(M),$$
the putative evaluation map must be given by restriction to the diagonal $M \subset M \times M$ and integrating. However, $C^\infty(M) \otimes \mathcal{D}(M)$ contains the δ-distribution on the diagonal; and we cannot restrict this to the diagonal.

This means that, although the spaces of linear maps between nuclear spaces are topological vector spaces in a natural way, Nuc *does not* form a symmetric monoidal category enriched in topological vector spaces with the projective tensor product.

4. Tensor products of nuclear spaces from geometry

- If M is a manifold, and F is any locally convex Hausdorff topological vector space, then $C^\infty(M) \otimes F$ is the space of smooth maps $M \to F$.
- If M and N are manifolds, then
$$C^\infty(M) \otimes C^\infty(N) = C^\infty(M \times N).$$
- Let $\mathcal{D}(M)$ be the dual of $C^\infty(M)$, that is, the space of compactly supported distributions on M. Then,
$$\mathcal{D}(M) \otimes \mathcal{D}(N) = \mathcal{D}(N \times M).$$

Similar identities hold when we consider sections of vector bundles on M and N instead of just smooth functions.

5. Algebras of formal power series on nuclear Fréchet spaces

Since the category Nuc is a symmetric monoidal category, we can perform most of the familiar constructions of linear algebra in this category. In particular, if E is a nuclear space, we will let
$$\mathrm{Sym}^n E = \left(E^{\otimes n}\right)_{S_n}$$
be the coinvariants for the natural symmetric group action on $E^{\otimes n}$.

Suppose that E is a nuclear Fréchet space. Then, $E^\vee$ is a space of type $(\mathcal{NDF})$, and in particular a nuclear space. We will let
$$\mathscr{O}(E) = \prod_{n \geq 0} \operatorname{Sym}^n E^\vee.$$
We will view $\mathscr{O}(E)$ as the algebra of formal power series on E. Note that, since the category of nuclear spaces is closed under limits, $\mathscr{O}(E)$ is a nuclear space, and indeed a commutative algebra in the symmetric monoidal category of nuclear spaces.

If $A \in \operatorname{Nuc}$ is a commutative algebra in the category of nuclear spaces, then we will view
$$\mathscr{O}(E) \otimes A$$
as the algebra of A-valued formal power series on E. Thus, we will often write $\mathscr{O}(E, A)$ instead of $\mathscr{O}(E) \otimes A$. Note that, because the tensor product commutes with limits,
$$\mathscr{O}(E) \otimes A = \prod_n (\operatorname{Sym}^n E^\vee, A)$$
$$= \prod_n \operatorname{Hom}(E^{\otimes n}, A)_{S_n}$$
where the space $\operatorname{Hom}(E^{\otimes n}, A)$ is equipped with the strong topology (that of uniform convergence on bounded subsets). The last equality holds because, for all nuclear Fréchet spaces E and all nuclear spaces F, $\operatorname{Hom}(E, F) = E^\vee \otimes F$ where $\operatorname{Hom}(E, F)$ is equipped with the strong topology.

As an example, if $E = C^\infty(M)$, then
$$\mathscr{O}(C^\infty(M)) = \prod_{n \geq 0} \mathcal{D}(M^n)_{S_n}$$
where $\mathcal{D}(M^n)$ refers to the space of distributions on M^n.

Bibliography

[BD04] Alexander Beilinson and Vladimir Drinfeld, *Chiral algebras*, American Mathematical Society Colloquium Publications, vol. 51, American Mathematical Society, Providence, RI, 2004.

[BEK06] S. Bloch, H. Esnault, and D. Kreimer, *On motives associated to graph polynomials*, Comm. Math. Phys. **267** (2006), no. 1, 181–225.

[BF00] Romeo Brunetti and Klaus Fredenhagen, *Microlocal analysis and interacting quantum field theories: renormalization on physical backgrounds*, Comm. Math. Phys. **208** (2000), no. 3, 623–661.

[BF09] _____, *Quantum field theory on curved backgrounds*, Quantum field theory on curved spacetimes, Lecture Notes in Phys., vol. 786, Springer, Berlin, 2009, pp. 129–155.

[BGM71] M. Berger, P. Gauduchon, and E. Mazet, *Le spectre d'une variété riemannienne*, Lecture Notes in Mathematics, no. 194, Springer-Verlag, 1971.

[BGV92] Nicole Berline, Ezra Getzler, and Michèle Vergne, *Heat kernels and Dirac operators*, Grundlehren der Mathematischen Wissenschaften [Fundamental Principles of Mathematical Sciences], vol. 298, Springer-Verlag, Berlin, 1992.

[BP57] N.N. Bogoliubov and O.S. Parasiuk, *On the multiplication of the causal function in the quantum theory of fields*, Acta Math. **97** (1957), 227–266.

[CCRF+98] A.S. Cattaneo, P. Cotta-Ramusino, F. Fucito, M. Martellini, M. Rinaldi, A. Tanzini, and M. Zeni, *Four-dimensional Yang-Mills theory as a deformation of topological BF theory*, Communications in Mathematical Physics **197** (1998), no. 3, 571–621.

[CG10] Kevin Costello and Owen Gwilliam, *Factorization algebras in perturbative quantum field theory*, Available at http://math.northwestern.edu/~costello/ (2010).

[CK98] Alain Connes and Dirk Kreimer, *Hopf algebras, renormalization and noncommutative geometry*, Comm. Math. Phys. **199** (1998), no. 1, 203–242.

[CK99] _____, *Renormalization in quantum field theory and the Riemann-Hilbert problem*, J. High Energy Phys. (1999), no. 9, Paper 24, 8 pp. (electronic).

[CM04] Alain Connes and Matilde Marcolli, *Renormalization and motivic Galois theory*, Int. Math. Res. Not. (2004), no. 76, 4073–4091.

[CM08a] A.S. Cattaneo and P. Mnëv, *Remarks on Chern-Simons invariants*, Communications in Mathematical Physics **293** (2008), 803–836.

[CM08b] Alain Connes and Matilde Marcolli, *Noncommutative geometry, quantum fields and motives*, American Mathematical Society Colloquium Publications, vol. 55, American Mathematical Society, Providence, RI, 2008.

[Cos07] Kevin Costello, *Renormalisation and the Batalin-Vilkovisky formalism*, arXiv:0706.1533 (2007).

[DF01] M. Dütsch and K. Fredenhagen, *Algebraic quantum field theory, perturbation theory, and the loop expansion*, Comm. Math. Phys. **219** (2001), no. 1, 5–30.

[DGMS75] Pierre Deligne, Phillip Griffiths, John Morgan, and Dennis Sullivan, *Real homotopy theory of Kähler manifolds*, Invent. Math. **29** (1975), no. 3, 245–274.

[EG73] H. Epstein and V. Glaser, *The role of locality in perturbation theory*, Ann. Inst. H. Poincaré Sect. A (N.S.) **19** (1973), 211–295 (1974).

[Fey50] R. P. Feynman, *Mathematical formulation of the quantum theory of electromagnetic interaction*, Physcal review **80** (1950), no. 3, 440–457.

[Gro52] A. Grothendieck, *Résumé des résultats essentiels dans la théorie des produits tensoriels topologiques et des espaces nucléaires*, Ann. Inst. Fourier Grenoble **4** (1952), 73–112 (1954).

[GSW88] Michael B. Green, John H. Schwarz, and Edward Witten, *Superstring theory. Vol. 1*, second ed., Cambridge Monographs on Mathematical Physics, Cambridge University Press, Cambridge, 1988, Introduction.

[GW73] D. Gross and F. Wilczek, *Ultraviolet behiavour of non-Abelian gauge theories*, Physical review letters **30** (1973), 1343–1346.

[Haa92] Rudolph Haag, *Local quantum physics*, Springer, 1992.

[Hep66] K. Hepp, *Proof of the Bogolyubov-Parasiuk theorem on renormalization*, Comm. Math. Phys. **2** (1966), no. 4, 301–326.

[HO09] Stefan Hollands and Heiner Olbermann, *Perturbative quantum field theory via vertex algebras*, J. Math. Phys. **50** (2009), no. 11, 112304, 42.

[Hol09] Stefan Hollands, *Axiomatic quantum field theory in terms of operator product expansions: general framework, and perturbation theory via Hochschild cohomology*, SIGMA Symmetry Integrability Geom. Methods Appl. **5** (2009), Paper 090, 45.

[HW10] Stefan Hollands and Robert M. Wald, *Axiomatic quantum field theory in curved spacetime*, Comm. Math. Phys. **293** (2010), no. 1, 85–125.

[Iac08] V. Iacovino, *Master equation and perturbative Chern-Simons theory*, arXiv:0811.2181 (2008).

[Kad66] Leo P. Kadanoff, *Scaling laws for Ising models near t_c*, Physics 2 (1966), 263–272.

[KL09] D. Krotov and A. Losev, *Quantum field theory as effective BV theory from Chern-Simons*, Nuclear Phys. B **806** (2009), no. 3, 529–566.

[KS] Maxim Kontsevich and Yan Soibelman, *Deformation theory, volume I*, Available at http://www.math.ksu.edu/~soibel/

[KZ01] M. Kontsevich and D. Zagier, *Periods*, Mathematics unlimited—2001 and beyond, Springer, 2001, pp. 771–808.

[Man99] Yuri I. Manin, *Frobenius manifolds, quantum cohomology, and moduli spaces*, American Mathematical Society Colloquium Publications, vol. 47, American Mathematical Society, Providence, RI, 1999.

[Mnë09] P. Mnëv, *Notes on simplicial bf theory*, Moscow Mathematical Journal **9** (2009), no. 2, 371–410.

[MP49] I. Minakshisundaram and A. Pleijel, *Some properties of the eigenfunctions of the Laplace-operator on Riemannian manifolds*, Canadian J. Math. **1** (1949), 242–256.

[MS67] H.P. McKean and I. M. Singer, *Curvature and the eigenvalues of the Laplacian*, J. Differential Geometry **1** (1967), no. 1, 43–69.

[Pol73] H. D. Politzer, *Reliable perturbative results from strong interactions?*, Physical review letters **30** (1973), 1346–1349.

[Pol84] J. Polchinski, *Renormalization and effective Lagrangians*, Nuclear Phys. B (1984), no. 231-269.

[Sch93] A. Schwarz, *Geometry of Batalin-Vilkovisky quantization*, Comm. Math. Phys. **155** (1993), no. 2, 249–260.

[Seg99] Graeme Segal, *Notes on quantum field theory.*

[Trè67] François Trèves, *Topological vector spaces, distributions and kernels*, Academic Press, New York, 1967.

[vS07] Walter D. van Suijlekom, *Renormalization of gauge fields: a Hopf algebra approach*, Comm. Math. Phys. **276** (2007), no. 3, 773–798.

[Tam03] D. Tamarkin, *A formalism for the renormalization procedure*, arXiv:math/0304211(2003).

[Wil71] K.G. Wilson, *Renormalization group and critical phenomena, I.*, Physical review B **4** (1971), no. 9, 3174–3183.

[Wil72] _____, *Renormalizaton of a scalar field theory in strong coupling*, Physical review D **6** (1972), no. 2, 419–426.

Titles in This Series

171 **Leonid Pastur and Mariya Shcherbina,** Eigenvalue distribution of large random matrices, 2011

170 **Kevin Costello,** Renormalization and effective field theory, 2011

169 **Robert R. Bruner and J. P. C. Greenlees,** Connective real K-theory of finite groups, 2010

168 **Michiel Hazewinkel, Nadiya Gubareni, and V. V. Kirichenko,** Algebras, rings and modules: Lie algebras and Hopf algebras, 2010

167 **Michael Gekhtman, Michael Shapiro, and Alek Vainshtein,** Cluster algebra and Poisson geometry, 2010

166 **Kyung Bai Lee and Frank Raymond,** Seifert fiberings, 2010

165 **Fuensanta Andreu-Vaillo, José M. Mazón, Julio D. Rossi, and J. Julián Toledo-Melero,** Nonlocal diffusion problems, 2010

164 **Vladimir I. Bogachev,** Differentiable measures and the Malliavin calculus, 2010

163 **Bennett Chow, Sun-Chin Chu, David Glickenstein, Christine Guenther, James Isenberg, Tom Ivey, Dan Knopf, Peng Lu, Feng Luo, and Lei Ni,** The Ricci flow: Techniques and applications, Part III: Geometric-analytic aspects, 2010

162 **Vladimir Maz'ya and Jürgen Rossmann,** Elliptic equations in polyhedral domains, 2010

161 **Kanishka Perera, Ravi P. Agarwal, and Donal O'Regan,** Morse theoretic aspects of p-Laplacian type operators, 2010

160 **Alexander S. Kechris,** Global aspects of ergodic group actions, 2010

159 **Matthew Baker and Robert Rumely,** Potential theory and dynamics on the Berkovich projective line, 2010

158 **D. R. Yafaev,** Mathematical scattering theory: Analytic theory, 2010

157 **Xia Chen,** Random walk intersections: Large deviations and related topics, 2010

156 **Jaime Angulo Pava,** Nonlinear dispersive equations: Existence and stability of solitary and periodic travelling wave solutions, 2009

155 **Yiannis N. Moschovakis,** Descriptive set theory, 2009

154 **Andreas Čap and Jan Slovák,** Parabolic geometries I: Background and general theory, 2009

153 **Habib Ammari, Hyeonbae Kang, and Hyundae Lee,** Layer potential techniques in spectral analysis, 2009

152 **János Pach and Micha Sharir,** Combinatorial geometry and its algorithmic applications: The Alcála lectures, 2009

151 **Ernst Binz and Sonja Pods,** The geometry of Heisenberg groups: With applications in signal theory, optics, quantization, and field quantization, 2008

150 **Bangming Deng, Jie Du, Brian Parshall, and Jianpan Wang,** Finite dimensional algebras and quantum groups, 2008

149 **Gerald B. Folland,** Quantum field theory: A tourist guide for mathematicians, 2008

148 **Patrick Dehornoy with Ivan Dynnikov, Dale Rolfsen, and Bert Wiest,** Ordering braids, 2008

147 **David J. Benson and Stephen D. Smith,** Classifying spaces of sporadic groups, 2008

146 **Murray Marshall,** Positive polynomials and sums of squares, 2008

145 **Tuna Altinel, Alexandre V. Borovik, and Gregory Cherlin,** Simple groups of finite Morley rank, 2008

144 **Bennett Chow, Sun-Chin Chu, David Glickenstein, Christine Guenther, James Isenberg, Tom Ivey, Dan Knopf, Peng Lu, Feng Luo, and Lei Ni,** The Ricci flow: Techniques and applications, Part II: Analytic aspects, 2008

143 **Alexander Molev,** Yangians and classical Lie algebras, 2007

142 **Joseph A. Wolf,** Harmonic analysis on commutative spaces, 2007

141 **Vladimir Maz'ya and Gunther Schmidt,** Approximate approximations, 2007

TITLES IN THIS SERIES

140 **Elisabetta Barletta, Sorin Dragomir, and Krishan L. Duggal,** Foliations in Cauchy-Riemann geometry, 2007

139 **Michael Tsfasman, Serge Vlăduţ, and Dmitry Nogin,** Algebraic geometric codes: Basic notions, 2007

138 **Kehe Zhu,** Operator theory in function spaces, 2007

137 **Mikhail G. Katz,** Systolic geometry and topology, 2007

136 **Jean-Michel Coron,** Control and nonlinearity, 2007

135 **Bennett Chow, Sun-Chin Chu, David Glickenstein, Christine Guenther, James Isenberg, Tom Ivey, Dan Knopf, Peng Lu, Feng Luo, and Lei Ni,** The Ricci flow: Techniques and applications, Part I: Geometric aspects, 2007

134 **Dana P. Williams,** Crossed products of C^*-algebras, 2007

133 **Andrew Knightly and Charles Li,** Traces of Hecke operators, 2006

132 **J. P. May and J. Sigurdsson,** Parametrized homotopy theory, 2006

131 **Jin Feng and Thomas G. Kurtz,** Large deviations for stochastic processes, 2006

130 **Qing Han and Jia-Xing Hong,** Isometric embedding of Riemannian manifolds in Euclidean spaces, 2006

129 **William M. Singer,** Steenrod squares in spectral sequences, 2006

128 **Athanassios S. Fokas, Alexander R. Its, Andrei A. Kapaev, and Victor Yu. Novokshenov,** Painlevé transcendents, 2006

127 **Nikolai Chernov and Roberto Markarian,** Chaotic billiards, 2006

126 **Sen-Zhong Huang,** Gradient inequalities, 2006

125 **Joseph A. Cima, Alec L. Matheson, and William T. Ross,** The Cauchy Transform, 2006

124 **Ido Efrat, Editor,** Valuations, orderings, and Milnor K-Theory, 2006

123 **Barbara Fantechi, Lothar Göttsche, Luc Illusie, Steven L. Kleiman, Nitin Nitsure, and Angelo Vistoli,** Fundamental algebraic geometry: Grothendieck's FGA explained, 2005

122 **Antonio Giambruno and Mikhail Zaicev, Editors,** Polynomial identities and asymptotic methods, 2005

121 **Anton Zettl,** Sturm-Liouville theory, 2005

120 **Barry Simon,** Trace ideals and their applications, 2005

119 **Tian Ma and Shouhong Wang,** Geometric theory of incompressible flows with applications to fluid dynamics, 2005

118 **Alexandru Buium,** Arithmetic differential equations, 2005

117 **Volodymyr Nekrashevych,** Self-similar groups, 2005

116 **Alexander Koldobsky,** Fourier analysis in convex geometry, 2005

115 **Carlos Julio Moreno,** Advanced analytic number theory: L-functions, 2005

114 **Gregory F. Lawler,** Conformally invariant processes in the plane, 2005

113 **William G. Dwyer, Philip S. Hirschhorn, Daniel M. Kan, and Jeffrey H. Smith,** Homotopy limit functors on model categories and homotopical categories, 2004

112 **Michael Aschbacher and Stephen D. Smith,** The classification of quasithin groups II. Main theorems: The classification of simple QTKE-groups, 2004

111 **Michael Aschbacher and Stephen D. Smith,** The classification of quasithin groups I. Structure of strongly quasithin K-groups, 2004

110 **Bennett Chow and Dan Knopf,** The Ricci flow: An introduction, 2004

109 **Goro Shimura,** Arithmetic and analytic theories of quadratic forms and Clifford groups, 2004

For a complete list of titles in this series, visit the
AMS Bookstore at **www.ams.org/bookstore/**.